Generic Systems Engineering

Nadine Schlüter

Generic Systems Engineering

A Methodical Approach to Complexity Management

Springer

Nadine Schlüter
School of Mechanical Engineering
and Safety Engineering
Bergische Universität Wuppertal
Wuppertal, Northrhine Westphalia, Germany

ISBN 978-3-662-67993-7 ISBN 978-3-662-67994-4 (eBook)
https://doi.org/10.1007/978-3-662-67994-4

This book is a translation of the original German edition "Generic Systems Engineering" by Schlüter, Nadine, published by Springer-Verlag GmbH, DE in 2023. The translation was done with the help of an artificial intelligence machine translation tool. A subsequent human revision was done primarily in terms of content, so that the book will read stylistically differently from a conventional translation. Springer Nature works continuously to further the development of tools for the production of books and on the related technologies to support the authors.

This Springer imprint is published by the registered company Springer-Verlag GmbH, DE, part of Springer Nature.
The registered company address is: Heidelberger Platz 3, 14197 Berlin, Germany

Paper in this product is recyclable.

Preface to the 3rd edition

The essence of time consists in the change of things.—Helmar Nahr

Time not only changes, it also creates many new insights. Since the 2nd edition of Generic Systems Engineering (GSE), another six years have passed, during which a large number of dissertations as well as national and international research projects have used and further developed the GSE. The number of citations continues to rise, scientists and practitioners from different disciplines pick up the ideas and philosophies or even the entire GSE for their work. IT tools are also increasingly adapting to the needs of a universal, cross-disciplinary but also standardized Systems Engineering.

Reason enough to give an update on the GSE, which also includes the current results of the project team around the Department of Product Safety and Quality (PSQ) at the Bergische Universität Wuppertal.

Accordingly, my thanks go to my department head, Prof. Dr.-Ing. Manuel Löwer, and my colleagues and team members of the last six years. To the doctors, whose dissertations have contributed significantly to the further development of the GSE, also many thanks—not only for their research but also for the co-authorships in the practical chapter.

Special thanks also to Prof. Dr.-Ing. habil. Petra Winzer. She created the GSE book in 2013 and further maintained it in the second edition. Thank you for trusting me to hand over the third edition of the GSE. The productive and motivating discussions as well as the constructive cooperation with Prof. Winzer and Dr. Marian Mistler made this third edition possible.

The following book is written in the masculine form—exclusively for easy readability. When, for example, the project leader is written about, the female project leader is of course also meant.

And thus I would like to agree with the opinion of my predecessor with this book:

"When ideas live on in young scientists—what better can happen to an author."—Prof. Dr.-Ing. habil. Petra Winzer

Wuppertal
in November 2022

Nadine Schlüter

Preface to the 2nd edition

No one can change, everyone can improve. (Ernst Freiherr von Fechtersleben)
In this sense, the revision of the book on Generic Systems Engineering, published by Springer Verlag in 2013, was carried out.

In the context of research projects, dissertations, numerous national and international conferences and discussion forums, my research group and I were able to present, test and refine the developed Generic Systems Engineering approach. These insights have flowed into this book. Two new examples show how and where the Generic Systems Engineering can be used with what results.

I am pleased that the project team has successfully used the Generic Systems Engineering to develop a solution approach for customer integration into the process of virtual product development and to test it with companies.

The second example shows how the field data feedback into the design and development process can be made more systematic using the Generic Systems Engineering.

My team at the Department of Product Safety and Quality Engineering at the Bergische Universität Wuppertal and my dear husband, Andreas Peschke, encouraged me to this second edition of the book. They all, as well as Springer Verlag, supported me vigorously in the present revision. For this I am very grateful. My special thanks go to my friend, Mrs. Gabriele Seider, who always held all the threads in her hand. Dr. Nadine Schlüter and Dr. Michel Mamrot were always constructively argumentative discussion partners for me, who have proven the practicability of the Generic Systems Engineering with their committed research activity and, fortunately, want to continue working in this research field.

When ideas live on in young scientists—what better can happen to an author.

Wuppertal
April 2016

Petra Winzer

Preface to the 1st edition

The fools make the same mistakes every day, the smart ones make new mistakes every day.

Mastering these is the concern of Generic Systems Engineering.

The present book on Generic Systems Engineering represents a first summary by my research group and me. It is an attempt to systematize and present over twenty years of collected experiences and research results. During my research activity, I experienced time and again that it is difficult to communicate together in the problem-solving process in teams consisting of specialists from various scientific disciplines. The Systems Engineering approach, i.e. thinking in systems, has always been particularly helpful to me in solving problems in a structured and systematic way.

Since Systems Engineering developed in various directions, it was a personal concern of mine to contribute to restoring the universal character of Systems Engineering. In this, I was greatly supported by my former and current employees at the Department of Product Safety and Quality Engineering at the Bergische Universität Wuppertal.

But this book would not have been written if two people in particular had not motivated me to do so. These are my colleague and friend Gabriele Seider and my husband Andreas Peschke. They in particular, as well as my family, gave me courage and strength, so that I can finally publish this book after years of work together with Springer Verlag. I am deeply grateful to them all.

Wuppertal
February 2013

Petra Winzer

Contents

List of Figures

List of Tables

Introduction 1

Systems Engineering (SE) was and is a structured approach to reduce complexity. Its declared goal is to break down complex issues into individual aspects, to network them and to develop detailed solutions without losing sight of the whole. However, numerous new specialized SE approaches have emerged in the course of the development of SE, so that despite intensive efforts by the German Society for Systems Engineering (GfSE 2022) and INCOSE (2022), there is currently no unified direction. In addition, recall actions in the automotive industry, delays in the commissioning of new ICE trains, delays in deliveries in the aircraft industry or shifts in planned shuttle launches in aerospace repeatedly show that failurescontinue to occur in development. Solving these and many other problems requires the collaboration of various specialists. They need a systematic approach that enables them to jointly identify, understand and systematically eliminate problems. This is the reason for the further development of SE to Generic Systems Engineering approach (GSE approach).

Why this is necessary, how it was derived based on the original foundations of SE, what standardized modules it consists of, how these interact and how it can be applied, is presented in this book, as shown in overview Fig. 1.1.

The strengths of SE become clearly visible in dealing with the new dimensions of complexity problems. These have their origin in various developments of today and are characterized by:

- the individualization of products (increase in components and functions, shortening of innovation and product life cycles, dynamization of systems),
- the conservation of resources (miniaturization of products, merging of system boundaries) and
- globalization (increase in international division of labor and specialization, increase in the number of stakeholders and their requirements as well as their networking).

N. Schlüter, *Generic Systems Engineering*, https://doi.org/10.1007/978-3-662-67994-4_1

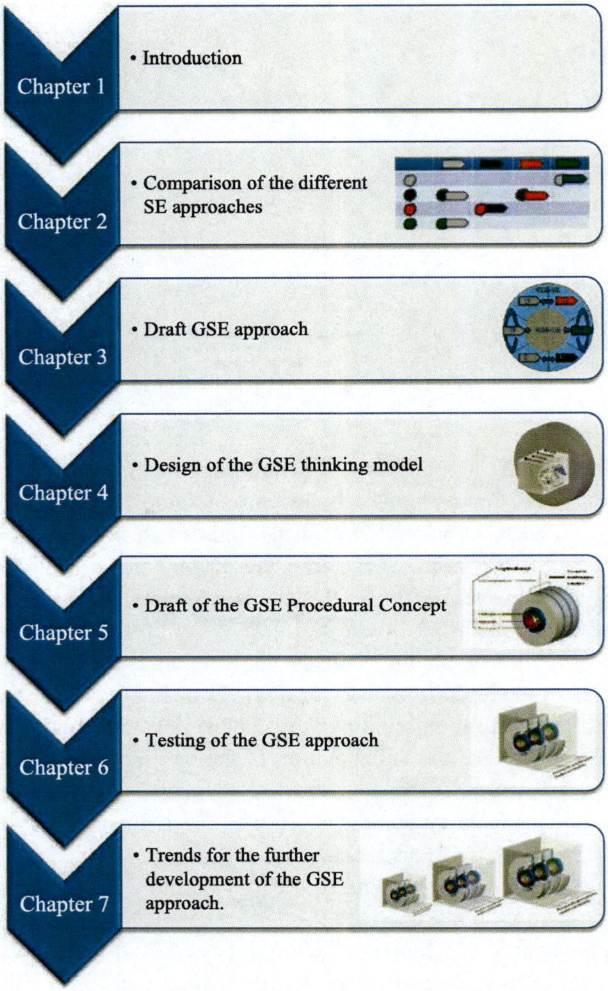

Fig. 1.1 The structure of the book

However, this requires reforming the existing SE. This proves to be difficult because the original universal concept of SE mutated into a multitude of thinking models and procedural concepts, as outlined in Chap. 2. Unifying these again to develop a general SE approach, namely the GSE approach, is the goal of this book.

A prerequisite for a structured problem-solving process is to structure reality (system thinking). On this basis, it is possible to develop an image of reality, i.e., a system model. This is referred to as a thinking model in GSE. The temporally logical linking of the problem-solving steps to solve a complex task is the subject of the procedural concept.

This book makes it clear that the first draft of the GSE approach is a universal solution approach that allows problems of various kinds to be solved and the solution space to be systematically scanned in order to develop the most efficient solution for the respective problem (see Chap. 3).

Following this, numerous possibilities for creating models of reality are compared with each other and a GSE thinking model is developed from this (see Chap. 4). This uses standardized views on the one hand to create an model of reality (component, function, requirement and process view) and attributed relations in the views and between the views via formalized templates on the other hand.

The benefit of developing the GSE thinking model for the problem-solving process is illustrated using four examples:

- the drive of a logistical system,
- the development of a mechatronic system,
- the KitVes system for using wind energy and
- the further development of a production company with regard to new norm requirements.

Based on the GSE model tested on examples, the GSE procedural concept further could now be elaborated. By comparing the different procedural concepts of SE, which were modified in the individual disciplines (such as software development, product development, manufacturing engineering and safety engineering) over the past years, a general procedural concept (GSE procedural concept) could be developed (Chap. 5). It consists of standardized modules (the GSE analysis, the GSE goal formation the GSE design and the GSE project management module). The GSE modules use various methods and procedures in parallel, problem-specifically, whose temporal logical coupling is supported by the GSE project management module with its phases, i.e., the planning, implementation and realization phase. In the context of the problem-solving process, it is essential that the GSE thinking model and the GSE procedural concept form a synergistic unit. At the beginning of problem solving, the problem must be assigned to a system. For this, the problem-related facts in reality must be meaningfully delimited as a system. Then a system model can be created, which can be represented in different granularity helpful for problem solving. This in turn is input for the corresponding GSE procedural concept, which must be adapted to the problem specification. The results of the individual partial steps that arise during the realization of the GSE procedural concept must necessarily contribute to the precision of the GSE thinking model. The principles of systematic thinking and action must be used in a case-specific manner. In order to be able to practically show the synergistic effects of the GSE thinking model and the GSE procedural concept, the examples of Chap. 4 are picked up again in Chap. 5. This can show at the GSE analysis, GSE goal formation or GSE design modules or when using the GSE project management module, how methods and procedures can be modified in an application-oriented manner and integrated into the GSE procedural concept. This is played through in a focused and exemplary manner, as a continuous illustration would go beyond the scope of this book. The interlocking of the GSE analysis module with the GSE thinking model is exemplified for the reliability analysis of the kite's rope. How the GSE design module is to be coupled with the GSE thinking model is shown by the simulation of the efficient driving mode of a drive of a logistical system. The functionality of the GSE project management module is illustrated for the classic project management by controlling the solution process for ensuring the reliability of the pantograph. The use of agile project management is explained using the sociotechnical example of organizational development.

Chapter 6 outlines eight case studies on how the GSE thinking model with the GSE procedural concept contributes to continuous problem solving. These are:

- the requirement-oriented product development (first example, Sect. 6.1),
- the development of mechatronic systems (second example, Sect. 6.2),
- the reliable design of products over the product life cycle (third example, Sect. 6.3),
- the identification of failurs in potentially critical usage processes (fourth example, Sect. 6.4),
- the field data feedback into the design and development process (fifth example, Sect. 6.5),
- the failure cause search and solution algorithm (sixth example, Sect. 6.6),
- the customer-integrated product development (seventh example, Sect. 6.7) and
- the requirement-oriented organizational development using the REMOt procedural concept (eighth example, Sect. 5.8).

Chapter 7 summarizes the essential findings and outlines possibilities for transferring the GSE approach from technical systems to sociotechnical systems, i.e., specifically to companies and company networks, as well as emerging questions to be solved by research. The GSE approach developed in this book is to be seen as a contribution to restoring the universal character of SE in order to better cope with the new dimensions of complexity in the present and future. Transdisciplinary teams are currently working together internationally to solve problems together. In order to be able to master the problem-solving process efficiently in a team, they urgently need a common thinking model and procedural concept, which the GSE approach offers.

References

GfSE (2022): Homepage der Gesellschaft für Systems Engineering e. V. Gesellschaft für Systems Engineering e. V. Online verfügbar unter www.gfse.de, zuletzt geprüft am 07.11.2022.

INCOSE (2022): Systems Engineering Vision 2035. Hg. v. INCOSE. USA. Online verfügbar unter https://www.incose.org/about-systems-engineering/se-vision-2035, zuletzt geprüft am 07.11.2022.

Systems Engineering (SE)—Old Thinking in a New Guise

2

Sincethere have been humansthey have tried to implement ideas or solve problems. This includes the desire to do this quickly and flawlessly. For example, the people of the Stone Age had the problem of satisfying their hunger. The organization of communal hunting was supposed to solve it. Their need to protect themselves from extreme weather conditions led to the construction of communal settlements. Today, one desire is to cover large distances in the shortest possible time. The construction of ever faster airplanes is one possible solution—another is the development of high-speed trains. The exploration of space is another challenge, the joint construction of the ISS space station is a related solution. It is also a current goal to develop new products and services faster and faster and to make them available to the end customer efficiently. The Internet of Things and Industry 4.0 are mentioned here as representatives (Bauernhansl 2014). Artificial intelligence, digitization, smart products that are also sustainable, or green technologies are further aspects from which new challenges arise (Dumitrescu et al. 2021).

What is the commonality of all these examples? They are very complex tasks that are solved by a group of people from different disciplines. The simpler and more goal-oriented these complex tasks are solved together, the greater the success. One means of achieving this success is system thinking. Thinking is a function of the human brain, i.e. the conceptual reflection of the general essential laws in the objects and processes of objective reality. Thinking is the highest form of human activity (Encyclopedia of Economics 1982; JuraForum.de 2021). Structuring complex facts using systems is the essence of system thinking, which is the basis of system theory. System theory represents "the theory between the elements of a system, the relationships between sub- and overall system" (Ackermann 2007, p. 19). In this way, the detail can be solved first without losing the context for the whole. This makes a complex task more manageable and therefore simpler—and also divisible—to solve. In relation to the settlement construction

N. Schlüter, *Generic Systems Engineering*, https://doi.org/10.1007/978-3-662-67994-4_2

of our ancestors, this meant that one group took care of the building material in a division of labor, while others dealt with the preparation of the building ground.

The system-theoretical approach can be seen as a basis for modeling, analysis and synthesis of complex structures (Bruijn and Herder 2009; Weilkiens 2019; Haberfellner et al. 2019; Gausemeier et al. 2013a). When this is applied to (socio-) technical systems and linked with a design process, be it the design and construction of an airplane, a high-speed train or the ISS space station, then we speak of Systems Engineering (SE). Thus, SE is part of the general system theory (Ropohl 2012). SE was and is a structured approach to reduce complexity in the design and realization of products, services or even product-service systems. Consequently, it contributes to the control of complexity. SE is a cross-disciplinary approach to the development and design of multidisciplinary systems (Gausemeier et al. 2013a), such as mechatronic systems, bionic systems or socio-technical systems, as they represent companies (Haberfellner et al. 2019; Mistler et al. 2021).

The basic philosophy of Systems Engineering (SE), i.e. system thinking, is based on the belief that humans and everything that surrounds them have system character. Each of these systems can be described. Basically, SE enables the development of concepts and the provision of methods for the analysis and design of complex situations, processes or structures. This is referred to as the SE procedure concept. Accordingly, SE combines thinking in systems, which is part of system theory, with an procedure concept that serves systematic problem solving.

But what exactly does Systems Engineering (SE) encompass?
SE helps to manage complexity or to provide transparency to complex problems for all involved persons, making the problems more manageable. This in turn allows to recognize sub-problems and their interactions with each other including their relationship with the system environment. Only such a basis enables a goal-oriented search for solution approaches for complex problems. However, in this context, the explanation of the system itself remains open. This question will be examined in detail in Sect. 2.1. After the conceptual clarification, the importance of SE in the past will be considered in Sect. 2.2. Do today's dimensions of complexity differ from those of the past? If so, does this result in new requirements for SE? Such questions are answered in Sect. 2.3. When, where and why the SE approach is used today is illuminated in Sect. 2.4. This includes the analysis of whether the SE approaches of today meet the current and future requirements that arise from the new dimensions of complexity. Finally, Sect. 2.5 will examine whether the SE approaches of the past and present contain ideas that could meet the requirements of the future. At the same time, requirements for SE in the "new guise" are formulated in summary, in order to contribute to mastering complexity in the new dimensions.

2.1 SE as a Scientific Discipline

SE is part of system theory, i.e. thinking in systems. It covers the design of technical systems over their entire product life cycle, i.e. starting from the idea to the recycling of the respective technical system. Increasingly, socio-technical systems are also considered in the engineering disciplines using SE (Dumitrescu et al. 2021; Schlueter et al. 2019; Nicklas 2016; Schlueter 2016; Mistler et al. 2021; Züst 2004; Dumitrescu et al. 2014; Maurer and Schulze 2014; Beyerer and Winzer 2018). In extension of this, system theory "deals in an abstract way with the basic and formalizable properties of systems, with the philosophical and mathematical aspects in the foreground" (Ludwig 2001, p. 30).

SE can also be defined by its own combination of terms, i.e. by the term "system" and the term "engineering". But what is a system? Can an airplane, a car, a robot, a human, a process, a service, a company each be called a system?

A **"system"**is an artifact, an image of reality in a very abstract form. It cannot be easily recognized as a "system" because it is a mental construct of the observer who uses systemic thinking (Heinrich 2015). The system is something composed or related, which is determined by its function, its behavior, its structure or its state (Schnieder and Schnieder 2013). In the case of very complex systems, these can in turn be broken down into subsystems. The smallest components of the system are the elements and their interrelationship. Every system has a system boundary and a system environment. It can be separated from its environment as a black box system. This system definition is very comprehensive, universal and general. It is initially considered sufficient to explain the term SE, well aware that there are numerous system definitions in the literature (Luhmann 1980; Ludwig 2001; Hanenkamp 2004; Haberfellner et al. 2019; Schnieder and Schnieder 2013; Gausemeier et al. 2013a; Ehrlenspiel and Meerkamm 2017; Ebert 2019; Rupp 2021).

In order to be able to represent systems, models are needed. They are the simplified representation of a planned or real existing object. The purpose of models is to abstractly reproduce a complex fact to its essence (Schnieder and Schnieder 2013; Mamrot 2014; Nicklas 2016; Schlueter et al. 2018; Bielefeld 2020). Their development takes place in SE in a problem-solving oriented manner. This means that when a problem is identified, the system that underlies the problem is defined first and foremost, and this system is to be modeled. Since systems are increasingly analyzed and designed by interdisciplinary teams, the demand for a transdisciplinary meta-model that can be used as a common basis for the involved disciplines is growing (Heinrichsmeyer et al. 2020; Mistler et al. 2021; Schlueter et al. 2019; Dumitrescu et al. 2021; Gausemeier et al. 2013a; Huber 2014; Albers et al. 2014). But what should this look like? Who develops it and how? How can such a model be transparently represented and updated? This book aims to find an answer to these questions.

"Engineering" refers to a discipline that uses theories or structured tools to develop or change products or services. In engineering, the sub-disciplines are differentiated according to the object of consideration. Expressions of this are Mechanical Engineering, Electrical Engineering, Software Engineering, Manufacturing Engineering, Safety Engineering or Quality Engineering, to name just a few. In the sense of system theory, these objects are certain types of systems, such as the software system in Software Engineering, the drive in Electrical Engineering or the factory in Manufacturing Engineering. On the other hand, safety in Safety Engineering or quality in Quality Engineering correspond to specific aspects under which the various objects or system types, i.e. the drive, the software or the factory, can be considered.

When the two terms "system" and "engineering" are put back together, a discipline is created that uses methods and structured tools to design complex systems. According to the definition of the International Council of Systems Engineering (Incose) (INCOSE 2015), SE is an interdisciplinary approach to enable the realization of successful systems. It focuses on defining customer needs and required functionality early in the development cycle, documenting requirements, and then proceeding with design synthesis and system validation, considering the entire problem: operation, cost and schedule, performance, training and support, testing, manufacturing and disposal. SE takes into account both the business and technical requirements of all customers with the goal of delivering a quality product that meets user needs (INCOSE 2015). INCOSE defines five steps for such a problem-solving process: Preparing a business or mission analysis (step 1). Defining a problem or opportunity space (step 2). Characterizing a solution space (step 3). Evaluating alternative solution classes (step 4) and Managing the business or mission analysis (step 5) (INCOSE 2015).

SE can connect the most diverse scientific disciplines. This is particularly promoted by a modification of SE, the **Model-based SystemsEngineering (MBSE)** (Weilkiens 2019; Gausemeier et al. 2013a). System thinking itself forms the basis for interdisciplinary product development. If, for example, a logistic system for an airport is to be developed, logisticians, software engineers, mechanical engineers, electrical engineers, industrial engineers, etc. are to be integrated into the development team. Each of these team members speaks their own language, i.e. they are shaped by the thinking and action of their respective discipline, i.e. mechanical engineering, electrical engineering, logistics, business administration or computer science. However, the design of the logistic system requires an interdisciplinary thinking and procedure approach. This is offered by MBSE through its generalistic thinking model, i.e. thinking in systems (Haberfellner et al. 2019). The team, consisting of specialists from different disciplines, thus has the opportunity to describe the overall system of the logistic system together using system theory as a model. The result is the unified system understanding of the logistic system, which in turn forms the basis for the product design, which can certainly be done in a division of labor. If new controls for the drives of the logistic system are created on this basis or if the roles are dimensioned anew, the respective specific subsystem is optimized with a view to the model of the entire logistic system and reinserted into the overall model.

SE as well as MBSE therefore differ from the traditional, specific engineering disciplines in that the complex system is first considered as a whole, also in interaction with its environment, e.g. also the user of the system, but also in the interaction with its subsystems or elements, in a cross-disciplinary manner. Consequently, the main purpose of SE is to coordinate the activities of those involved in the problem-solving process and thus to build a bridge between the disciplines (Mistler et al. 2021; Winzer 2015; Nicklas 2016; Ott 2009).

However, SE is also seen as a generalized problem-solving approach (Bahill and Gissing 1998). BAHILL and GISSING describe SE as a way for interdisciplinary teams to collectively identify problems, assign them to a system, and then solve them accordingly. Such a problem could be, for example, the poor efficiency of a drive in a logistics system. The discipline-specific solution approach of the electrical engineer is to optimize the running behavior of the drive through simulation. In contrast, the interdisciplinary team using the SE approach recognizes that the efficiency of the drive can also be improved in relation to the rollers, the belt, and the cargo, to name just a few adjacent subsystems of the logistics system. Thus, when applying SE, this team, for example, examines and correlates the interplay between the efficiency of the drive and the running behavior of the rollers, the friction of the conveyor belt, the weight of the cargo, or the starting and stopping of the roller conveyor. Through this approach of SE, the efficiency of the subsystem "drive" can be optimized in relation to the overall system "logistics system". The holistic solution to the problem just described requires that all team members have the same system understanding of the logistics system and, derived from this, a uniform system model. Through system thinking, which is the basis of SE, it becomes possible for the interdisciplinary team to create a complex image of the logistics system. This image of the logistics system, also called a model, forms the basis for the multidisciplinary team to search for solutions to efficiently design the efficiency of the drive.

For the concept of problem-solving, SE recommends the use of methods, procedures, and structured tools. However, when these are bundled into a basic, universal solution concept, opinions diverge significantly (Lindemann 2016; Winzer 2015; Nicklas 2016; Haberfellner et al. 2019; Bahill and Gissing 1998; Ott 2009; Bender and Gericke 2021; Haberfellner and Daenzer 1999).

In the case of more complex problems to be solved, SE recommends the application of different basic principles of systemic thinking and action. While HABERFELLNER (Haberfellner et al. 2019) fundamentally demands thinking from the general to the detail when solving problems, other authors leave this to the user (Bender and Gericke 2021; Lindemann 2016; Ott 2009).

The SE approach proves to be universal, i.e., transferable to any problem. Since everything that surrounds humans can be described as a system, it is possible to assign a system to every problem exactly. Now the system, its system structure, the system elements and their relationships, as well as the interrelationships between the system and its system environment can be analyzed in relation to the problem. This step is equivalent to a problem analysis, which is highly systematic and comprehensible using the

SE approach. Causes and effects can be made visible in the system model and thus recognized more quickly. The step-by-step scanning of the solution space focuses on the goal of developing solution variants, comparing them, and selecting the best solution for the specific problem. Consequently, the SE approach is considered a global or universal solution approach.

However, critics fear that due to the universality and abstraction of the SE approach, it does not lead quickly and efficiently to practical solutions. This particularly concerns the need to first precisely describe or define the system with its system boundaries, subsystems, elements, and their interrelationships. During this time, according to the critics, intuitive solutions could already be found.

But is a spontaneous, quick solution always really the best?
The attempt to answer this question is comparable, for example, to the search for an answer to the question of the meaningfulness of a document management system in a company. These critics can be countered by the fact that various scientific disciplines have used and continue to use the SE approach to systematically solve complex problems. The results they came to in the past or are coming to in the present will be examined more closely in the following chapters.

SE can be understood as a scientific discipline that primarily uses a system-theoretical approach and a procedural concept in conjunction with basic principles to solve complex problems and tasks in a transdisciplinary manner (Dumitrescu et al. 2021; Weilkiens 2019). The subject of SE is the system and the procedural approach is the problem-solving cycle for solving problems in the system. Various scientific disciplines can use this universal subject and the universally valid approach to problem solving. Consequently, SE can also be seen as a kind of patronage science, as Fig. 2.1 illustrates.

An initial summary shows that SE fundamentally uses system thinking. HABERFELLNER describes it as a way to structure facts and situations and to present them in

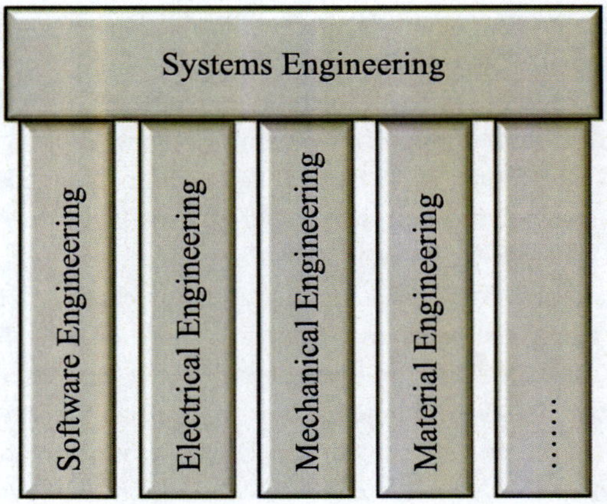

Fig. 2.1 SE disciplines (Weilkiens 2007, p. 15)

their contexts in order to better understand them (Haberfellner et al. 2019). Through system thinking, an image, i.e., a model of the system, of objective reality, can be created—which can then be designed in a goal-oriented and requirement-appropriate manner. The approach to designing the system can be efficiently supported by methods, procedures, or structured tools. The temporal logical coupling of the methods, procedures, and structured tools is referred to by LINDEMANN as the procedural model of SE (Lindemann 2016). HABERFELLNER refers to the entire problem-solving process as a procedural model. He divides this into a substantive part, the system design, and an organizational part, project management (Haberfellner et al. 2019). Consequently, SE uses a conceptual framework, i.e., the SE philosophy, to derive a universal problem-solving approach based on this using an SE thinking and SE procedural model (Haberfellner et al. 2019). It is used to make complex facts transparent, to simplify them, and thus to make them manageable (Winzer 2015; Dalhöfer and Rall 2009; Sitte and Winzer 2011).

Complex facts that need a solution exist today and will exist in the future. But they also existed in the past. What significance did system thinking have in earlier times? This question will be pursued in the following chapter.

2.2 Systems Thinking as an Opportunity for Complexity Management in the Past

Neither the search for fundamental solution approaches nor the search for a universal approach to solving complex problems is something new.

Can't the seven wonders of the world also be understood as "complex systems" within the value creation of that time? How complex were the plans of the ancient Egyptians when they "organized" thousands of people to realize a construction like the Cheops Pyramid?

What difficulties did the Incas face in the 15th century when they created such a fundamental structure as Machu Picchu, which still attracts many tourists from all over the world today? Did the builders of the time also excuse mistakes with the complexity of the construction project?

But what is complexity anyway?

In its form as an adjective, the word "complex" usually characterizes terms such as problem, structure, context, etc., or categories such as system or process. The word "complexity" is derived from its Latin origin "complecti" and means as much as "embrace", "encompass". But the use of the terms "complex" or "complexity" is associated with more. It is primarily related to the description of the different and multi-layered worlds of life surrounding humans.

In the traditional understanding, complexity stands for a property of a system or object that makes it difficult to control its overall behavior, even if complete information about individual components and their interactions is available (Wildemann 2004; Domenico and Sayama 2019; Flückiger and Rauterberg 1995). Against this background, dealing with the "complexity" of these systems means designing them as desired, despite the problematic analysis of their overall behavior. Following these considerations, the Cheops Pyramid and the Inca buildings are also complex systems. This is also supported by LUHMANN. For him, complexity is a system property with two dimensions. These can be described by the variety of elements (variety), i.e., the types and number of elements, and the variety of relationships (connectivity), i.e., the types and number of relationships (Luhmann 1980). Accordingly, systems—since there have been humans—are to be classified as complex. Recognizing, understanding, and designing them was the goal of many scientists in the past and still is.

First thoughts about complex relationships or their first attempts at explanation can be traced back to the philosophy of ancient Greece. Already Aristotle formulated in Book 7, Chap. 17 of his Metaphysics (Detel et al. 2009): "That which is composed of parts in such a way that it forms a unified whole, not in the manner of a heap, but like a syllable, is obviously more than just the sum of its parts." This quote is better known in a shorter, more concise version: "The whole is more than the sum of its parts", i.e., that the elements of such a whole—or a system—are in interactions and thus these interactions also characterize the whole as well as the individual elements themselves (Detel et al. 2009).

Without even remotely claiming a complete outline of system-theoretical treatises, some significant milestones in the development of human thinking should prove that systems thinking in the attempt to explain the complex world surrounding humans has a very long tradition. Already the 7000-year-old Chinese Book of Changes, handed down by Confucius (551 BC–479 BC), contains hexagrams based on the principle of the evolution of system structures in heaven, in man, and in nature. In ancient Greece, especially treatises with the aim of presenting the totality of the real in a coherent form were among the merits of Aristotle (384–322 BC) (Detel 2005).

Centuries later, in the philosophy of modern times, scholars such as the Dutchman Spinoza (1632–1677), the Frenchman Descartes (1596–1650), and the German Leibniz (1646–1716)—to name just a few—shaped a systemic understanding of different complexities. However, they reduced the complexity of living systems by transferring them to mechanical systems or understanding them as machines. LUHMANN assigns the "equilibrium metaphor" to the 17th century. These theories assume that a system can be disturbed, but the system's sensitivity to disturbance is determinable and influenceable (Luhmann 1980). In the theory of evolution since Darwin, i.e., the Darwinian distinction of variants and structural changes, the theory of open systems is explained. Open systems are described by the input-output model. Relationships between the systems themselves, but also between the system and its environment can be described (Luhmann 1980).

Despite the historical dimensions of systemic thinking, considerations of (socio-)technical systems only came into focus with the industrialization of economies. However, initially, the (socio-)technical systems were still manageable, or their subsystems still had a high degree of autonomy.

Only with their growing complexity did new thinking models and methods become necessary to better control (socio-) technical systems, their development, their production, and their use.

According to JACKSON (Jackson 2000), the term "SE" was first used in 1940 in the Bell Telephone Laboratories. The management of the Bell Telephone Laboratories specifically developed SE into a method in the 1950s to control the interface problem. As a result, the "Society for General Systems Research" (today: "International Society for the Systems Sciences") was founded in the USA in 1954. The first fundamental works on general system theory appeared in its yearbook "General Systems" (Luhmann 1980). Especially in American space travel, the "SE" was used and further developed by NASA within the framework of the Apollo Program. Arthur David Hall considered the system as real and divisible (Hall 1965). The method he developed for system recognition and design contributed to the optimization of a system while maintaining objectivity.

The transfer of system theory into cybernetics is essentially based on the theory of WIENER (Wiener 1994). It describes the interactions between the systems or between the system and its environment as an input and output model, which can be influenced by control variables (Luhmann 1980). Consequently, cybernetics assumes the targeted control of systems. This naturally did not remain without influence on the SE. For example, the SE according to JENKINS (Jenkins and Youle 1971) focused on the design of hardware systems on the one hand and on the design of company parts up to entire companies on the other hand. The SE here received the role of a tool or a supporting instrument for the optimal use of various resources such as money, people, machines, and material. The 1980s were characterized by a critical approach to the SE, while in the 1990s the focus increasingly shifted to the consideration of dynamic systems. The SE served, for example, as a basis for the design of learning organizations or knowledge management (Foerster et al. 1993; Senge 1999).

With the concept of SE, an engineering-specific system approach was created, which on the one hand is based on the universally applicable method of thinking in systems and on the other hand supports,

- addressing real problems,
- developing practically relevant design solutions for systems of any kind and/or
- optimizing the necessary resource use within the framework of value creation.

In summary, it can be stated that systemic thinking fundamentally served to control complexity.

Complex systems have always existed. In antiquity, they were characterized, among other things, by a high degree of location-bound division of labor and specialization. The number of stakeholders making demands on these complex systems grew with industrialization. While in antiquity individual products were essentially created for a regional market, this changed with the industrial revolution towards global markets. Consequently, it was not the complexity itself that changed over the course of history, but the dimensions of the complexity of systems changed. To recognize this and develop corresponding solution ideas, thinking in systems was also used, which itself evolved. The equilibrium theory, the input-output theory, and the theory of system control should be mentioned here because they clarify basic tendencies of the design possibilities of systems regardless of their complexity (Luhmann 2000). They all have in common that they make complex relationships transparent and unravel them. Using simple rules, complex systems could be designed. Thus, SE emerged as a separate scientific discipline. Its goal was to develop a general interdisciplinary problem-solving approach through the systemic thinking approach (Luhmann 2000).

Currently, there is often talk of the increase or the new, difficult-to-control dimensions of the complexity of systems.

But do the approaches of WIENER, HALL, JACKSON, to name just a few, still apply today? Is it the same complexity as in the past? What demands are made on the SE today? These questions will be answered in the following chapter in order to decide on this basis whether the SE approach can also be used today to solve current problems, or whether it needs to be reformed.

2.3 The New Dimensions of Complexity and Their Requirements for SE

It is clear that systems thinking and later SE were simple means of dealing with complexity in the past. The problem of increasing complexity is also often mentioned in current literature. Consequently, many authors deal with it (Schuh and Riesener 2018; Vester 2003; Wildemann 2004; Dalhöfer and Rall 2009; Nicklas 2016; Mistler 2021; Heinrichsmeyer 2020; Schlueter et al. 2019; Lanza et al. 2018; Bielefeld 2020; Haberfellner et al. 2019).

While complexity in the past was defined by characteristics of the system structure, variety (diversity of elements), and connectivity (diversity of relationships) (Luhmann 1980), today dynamics or the uncertainty of future system states are considered as further dimensions of complexity, as illustrated in Fig. 2.2 (Westphal and Kummer 2001; Wildemann 2004; Schnieder and Schnieder 2013; Bauernhansl 2014; Gausemeier et al. 2013b).

But does complexity really have new dimensions today and how can these be characterized?

The answer to this question helps to decide whether the SE approach of the past is usable in the present and in the future.

The current character of value creation processes can be described using a variety of new developments and trends that now influence all areas of life. An expression of globalization (Focus Online 2011) is the following example: "After the recall of more than seven million cars due to floor mat and gas pedal problems, Toyota also has to repair almost half a million Prius hybrid vehicles due to unsafe brakes." Toyota, as a globally

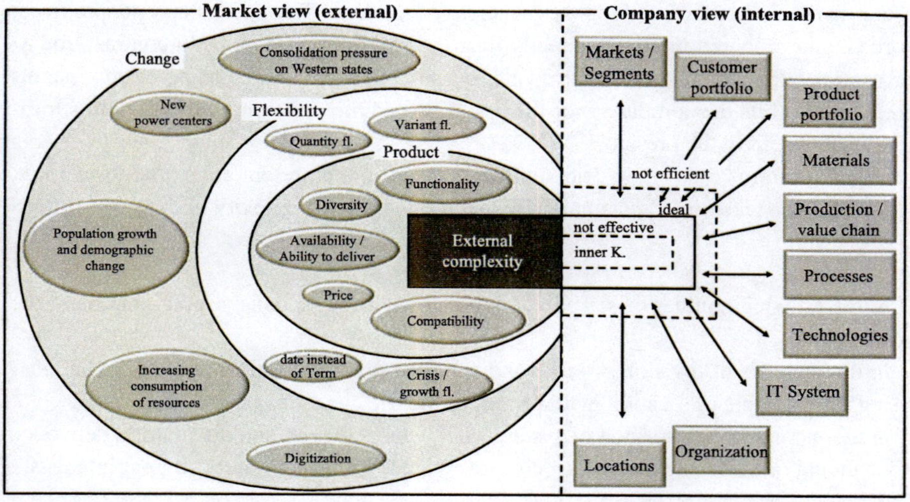

Fig. 2.2 Comparison of external—internal complexity. (© Fraunhofer IPA after Bauernhansl 2014, p. 14)

operating company, faces special challenges here. Due to the international division of labor, this recall action affects not only Toyota models but many other car brands. A large number of stakeholders, i.e., not only the end users but all those who were harmed in this network by Toyota, including the dealers, the workshops, the insurance companies, etc., will now make demands on Toyota.

What does this example illustrate?

A significant dimension of complexity today is that of globalization. It goes hand in hand with an increasing international division of labor or specialization. For example, a Toyota supplier specializes in brake systems, while others focus on drive systems, locking systems, etc. This specialization, in turn, leads to an increasing global networking in the automotive industry. This increases the number of stakeholders. While in the past, for example during the construction of the Cheops Pyramid, only a few stakeholders, mostly only the Pharaoh or selected members of the Pharaoh's family, made demands on the builder, this is no longer the case today. The increasing division of labor and the globalization of product manufacturing increase the number of stakeholders making demands on the final producer. Not only the different laws that must be observed in the respective country for product approval, but also the country-specific customer interests or distribution systems, the increased number of system, part, and material suppliers are, among other things, an expression of this.

The increasing division of labor and specialization are often seen as a current challenge in controlling complex systems. However, in the past, for example during the construction of the Cheops Pyramid, there was also a high division of labor. Consequently, this is not a new challenge. What is new, however, is that in the past, the implementation of the division of labor followed the principle of location. This means that people lived, worked, and procured materials directly near the pyramid to be built. In contrast, today, for example, Toyota produces worldwide, but also delivers components, or subsystems worldwide. While the complexity of the past followed the principle of location, it is location-independent in the present.

Globalization is thus a new and significant trend in the present compared to the past, which must be taken into account when dealing with the complexity of systems. It finds its expression:

- in the growing number of stakeholders with their simultaneously increasing demands on systems,
- in the increase of division of labor and specialization while simultaneously abandoning the principle of location in all phases of the life cycle of systems, and,
- in the increasing networking of systems, i.e., value creation and distribution networks, financing or ownership networks, etc. are created, with sometimes different interests.

A second essential trend of the complexity of systems is illustrated in Fig. 2.3.

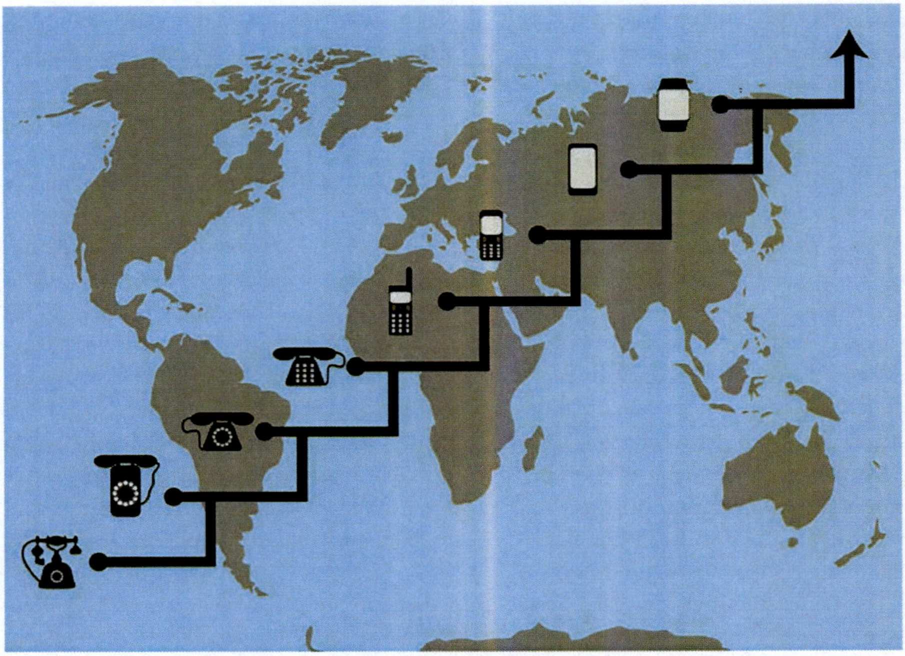

Fig. 2.3 Development of the mobile phone. (After colourbox 2022)

Characteristic of this trend is the increasing individualization of products. The reason for this is the customer's desire to purchase a unique product each time. This trend towards individualization has two main consequences for the complexity of systems. On the one hand, innovation cycles and thus the product life cycle of a product are shortened, and on the other hand, the number of functionsand components per product increases. While the mobile phone of the past only had the "phone function", the current mobile phone has additional functions that enable surfing the internet or taking photos, are equipped with a location and navigation function, and enable the consumption of various services via downloadable apps. The variety of functions logically affects the variety of components. This, in turn, must be mastered over the entire product life cycle, a task that poses special challenges for companies that manufacture very long-lasting products, such as trams. If there are changes to drives, switches, connectors, etc. in the overall "tram" system, this must be traceable on the one hand and on the other hand, the modified subsystems or components must be kept on hand for repair. This can cover a period of 20–40 years. However, it is also the case that the overall "tram" system has a longer innovation and product life cycle than the "drive" subsystem. Nevertheless, the new or modified "drive" subsystem, which is supposed to be more energy-efficient and quieter according to customer wishes, must be coordinated with the overall "tram" system.

Consequently, customer-specific products were and are an expression of complexity both in the past and in the present. In contrast to the past, however, innovation cycles and thus sometimes product life cycles are shortening today. At the same time, the number of functions and components per product is increasing significantly.

Efficiently mastering the variety of functions and components of customer-specific products in very short innovation cycles over the entire product life cycle—this is a current trend in the complexity of the present.

Nanotechnology, microtechnology, biotechnology, mechatronics, and cyber-physical systems are examples of a third trend in the complexity of current systems, "miniaturization". We encounter it every day: household appliances, tools, machines, and systems are getting smaller, more compact, and can communicate via the internet. The increasing miniaturization of electronic components requires corresponding adaptation of necessary mechanical parts. This is made possible by the increasing use of mechatronic components and subsystems. Mechatronics here stands for an engineering sub-discipline that achieves the functionality of a technical system through a close linkage of mechanical, electronic, and data processing components (Bauer 2003). Mechatronic components combine mechanical, electronic, and informational functions. They are now conquering the technologized living environment and are indispensable in a large number of today's products, such as household appliances, motor vehicles, etc. Ultimately, there is no exact boundary in mechatronic components between mechanics, electronics, and computer science. Accordingly, the designers and producers of these components need knowledge and skills from various disciplines. In addition, a consistent modularization of the products has proven to be a helpful concept for reducing complexity in product development.

Modules in mechatronic systems refer to building blocks that form a logical and functional unit and can be developed, tested, maintained, and replaced as such in a division of labor. Two approaches are used: product platforms and modular systems. They differ in the combinability of the components of the product (Gausemeier 2007). Further developments are embedded systems and cyber-physical systems. The latter, with their sensors, can directly perceive their environment, evaluate it with globally available services, and interact accordingly with the environment (Bauernhansl 2014). Consequently, the miniaturization trend merges with the globalization trend at this point, especially with regard to Industry 4.0.

Another consequence of the miniaturization trend is the use of new, intelligent materials while taking into account their environmental compatibility and resource conservation. This miniaturization trend in the complexity of systems is closely linked with the globalization trend. The manufacturer of miniaturized drives must also master an increasing number of stakeholders, a growing dynamization of requirements, an increase in division of labor, or specialization as well as a global networking of the partners involved in the development and production of the drive.

But also the individualization trend of systems is again connected with the miniaturization and globalization trend. The mobile phone is the best example of this. The design of the mobile phone can be individually designed, i.e., the customer can, for example, load various apps onto his mobile phone or choose between different housing colors. The mobile phone should basically ensure location-independent telephony, therefore worldwide dialing into various networks must be possible (Globalization trend). It has already been established that mobile phones are getting smaller and therefore also contain miniaturized components (Miniaturization trend). If the eruptions of the sun become stronger, this can lead to disturbances in radio traffic and the GPS system of the mobile phone (Globalization trend). This example shows that the trends of complexity will continue to network and influence each other in the present and future.

In addition, the dynamization trend is becoming increasingly important in connection with the globalization trend. As SCHUH et al. (Schuh et al. 2020) show, the dynamization trend results from the dynamics of the market. This poses complex problems for companies. In order to remain competitive in the market in the long term, faster and more goal-orientated decision-making processes are required in companies. According to SCHUH et al., companies create the prerequisites for this, for example, through digitalization and industrial change. The basic fields of action are resource use, the use of information systems, the establishment of the organizational structure, and the lived corporate culture (Schuh et al. 2020). HABERFELLNER et al. note that agility is becoming increasingly important in order to be able to flexibly meet the constantly changing requirements in the dynamic market environment (Haberfellner et al. 2019). This is also reflected in a German study on Systems Engineering by BENNO et al., in which a coupling of Systems Engineering with agile approaches such as Scrum is recommended (Benno et al. 2018). How dynamic the market is and how quickly even large corporations in industries can be displaced is shown, for example, by the production of smartphones.

This industry is so fast-paced and competitive that even companies like Microsoft have largely withdrawn from this business.

Furthermore, sustainability is becoming increasingly important in society and the economy. Because of global climate change, it is both a huge societal and industrial challenge to manage this. The questions arise as to how resources can be used effectively and environmentally friendly and how environmentally harmful emissions can be reduced to a minimum. This requires, among other things, the use of environmentally friendly and reusable materials as well as the expansion of green technologies. The sustainability trend thus implies a huge upheaval that can only be achieved globally and in which every country must see itself in the responsibility to also support this change. However, this upheaval also raises social questions. This means how such a change can be implemented nationally and internationally in a socially compatible way and who bears which responsibility to what extent. Consequently, various authors deal with these challenges (Dumitrescu et al. 2021; acatech 2021; acatech et al. 2021).

In a first interim conclusion, the following trends in the complexity of systems are thus recognizable:

the globalization trend, which finds its expression in:

- the increase in the number of stakeholders,
- the growing diversity of requirements,
- the increasing division of labor and specialization,
- the increased networking and globally distributed value creation networks, and
- the location independence of the individual phases of the product life cycle,

the individualization trend of products and services, which results in:

- shortens their innovation cycles,
- increases customer expectations and needs,
- changes the temporal and content phases of the product life cycle, and
- increases the variety of functions and components,

the miniaturization trend, which includes:

- includes a merging of system boundaries,
- transdisciplinarity in all phases of innovation and product life cycles, and
- resource conservation,

the dynamization trend, which requires:

- increasing agility or the necessity of flexibility,
- the integration of digitalization through all areas of social life, for example in the form of Artificial Intelligence,

- the networking of interacting, intelligent technical systems, and
- the increasing dissolution of traditional corporate structures and industry boundaries,

the sustainability trend, which includes:

- resource conservation,
- the reduction of environmentally harmful emissions,
- the use of environmentally friendly and reusable materials,
- the increasing application of green technologies,
- the striving for social equality, and
- responsible social action on a global level.

Thus, it becomes clear that these trends in complexity have changed compared to those of the past.

Although there were also globalization efforts in the past, the number of stakeholders and the diversity of requirements were much lower. The principle of location prevailed in the division of labor and specialization.

Customer-specific manufacturing, in contrast to today, was characterized by long innovation and product life cycles. The products had low functionality. Miniaturization, dynamization, and sustainability are a trend of the modern age. Nevertheless, it must be stated that in the past, in the present, and also in the future, it will always be about recognizing complexity and mastering it on the basis of simple rules. SE has been able to contribute to this in the past. Thus, this scientific discipline has experience in dealing with complexity and the potential to master future tasks. To do this, SE used systems thinking. It helped in history to handle the complexity of the world.

Thus, systems thinking, i.e., the breaking down of complex facts into meaningful parts (systems) in order to better recognize their interactions with each other, with their elements, and their system environment, will also be helpful for the present and future (Vester 2003; Heinrich 2015). This includes creating transparency of systems of all kinds. Without recognizing the interrelationships in the systems and their environment, a goal-oriented system change is not possible. It must—as in the past—also be traceable in the present and future. This can be facilitated by observing basic principles of systemic thinking and action, which are only partially identified as part of SE (Haberfellner et al. 2019). They should be used in the development and construction of products (Lindemann 2016) and in solving complex problems (Haberfellner et al. 2019; Ott 2009; Haberfellner and Daenzer 2002).

Numerous basic principles of systematic thinking and procedures are described in the literature (Ehrlenspiel and Meerkamm 2017; Bender and Gericke 2021; Baumann and Erlenspiel 1981; Ott 2009; Lindemann 2016). They are briefly summarized below because their integration into the problem-solving approach of SE contributes to managing complexity in new dimensions.

The basic principle of thinking in systems (Haberfellner et al. 2019; Haberfellner and Daenzer 2002)
With the help of thinking in systems, complex issues can be understood, divided into systems, and designed. Thus, the system is a conceptual construct that serves a specific purpose (Heinrich 2015). It enables the orderly handling of complexity, the recognition of relationships between the system and its environment, and the description of system elements and their relationships. Thinking in systems can define the solution space and expand it for efficient solution search. It facilitates the development of a model, i.e., a system image, which can be understood and used by various disciplines for the design process. It is assumed that this model of the system was created interdisciplinarily.

The basic principle from the whole to the detail (Haberfellner et al. 2019; Haberfellner and Daenzer 2002)
This basic principle uses the Black-Box model and derives hierarchies for systems. It is also presented in the literature as a top-down approach. For example, if a logistical system at an airport is to be optimized, this system can first be represented as a black box to consider its interactions with the airport's infrastructure, which can be understood as the overall system. Subsequently, drives or curve elements of the logistical system can be examined and optimized in detail. The application of this basic principle allows for the step-by-step reduction of system complexity.

It can be coupled with the basic principle of recurring reflection and the basic principle of structuring. HABERFELLNER incorporates this basic principle into his SE procedure model (Haberfellner et al. 2019). But it is also fundamentally usable for creating and refining system models (Sitte and Winzer 2011).

The basic principle of recurring reflection (Dörner 2007; Badke-Schaub and Frankenberger 2004; Mistler 2021)
This principle is intended to help manage complex tasks without losing sight of the big picture. For example, if a drive for a logistical system has been selected, it can be optimized using this basic principle. This optimization process includes, among other things, the starting and stopping of the logistical system or the interaction of the conveyed goods with the conveyor belt. Considering these interactions of the drive with the logistical system can lead to a critical reflection of the already achieved optimization results of the drive.

The basic principle of structuring (Zilahi-Szabó 2002)
The reduction of complexity is possible, among other things, through hierarchization, group formation, clustering and/or modularization. Hierarchization is based on a formal ranking, such as the subdivision of mechatronic systems into networked and autonomous mechatronic systems. Module and group formation, on the other hand, is freely selectable, so the logistical system can be divided into drive, control, safety modules, etc.

Which type of system structuring appears to be useful depends on the solution path and the team implementing it.

The basic principle from the abstract to the concrete (Bender and Gericke 2021)
When requirements are transformed into new product ideas, it is useful to employ this basic principle. If a robot is to be developed as a household helper, it is initially sufficient to consider which basic functions this robot should fulfill. These could be orientation, movement, and monitoring functions, for example. Once these have been fixed, it can now be examined in detail which components and or subsystems best fulfill these functions and how they are to be combined in the household helper robot. When solving problems, it is useful to use different levels of concretization and to plan feedback loops to avoid getting lost in details from the start.

The basic principle of minimal models (Böhm and Fuchs 2002)
This basic principle involves searching for solutions based on simplified, generally understandable models. Models are, as already emphasized, images of objective reality. When dealing with complex tasks, focus should be on a few model elements. Transferring this basic principle to the complexity of modern times is a particular challenge because system design can only be transdisciplinary. This was already demonstrated using the example of the mechatronic system. But it can also be applied to other examples like the logistical system. Computer scientists, mechanical engineers, electrical engineers, business economists, logisticians, etc. are involved in the development, realization, and maintenance of the logistical system. If the logistical system as a whole is to be optimized, a model is needed that all disciplines can use. This model must be so minimalist that it can be understood and used by everyone. The basic principle of minimal models is very closely linked with the basic principle of comprehensibility outlined below.

The basic principle of comprehensibility (Böhm and Fuchs 2002)
Its core statement is that concepts and models must be consistent and verifiable at all times. It also demands the completeness and redundancy-free nature of concepts and models. The focus here is on the unambiguity of the model's content. Comprehensibility means that the representations and descriptions are comprehensible regardless of the person, time, context, and tool.

In relation to the logistical system, this means that the concept of the logistical system must be understood not only by the logistician but also by the mechanical engineer or the maintenance engineer.

The basic principle of applying multiple perspectives (Böhm and Fuchs 2002)
The diversity of today's systems no longer allows a system to be viewed from just one perspective. For example, the logistical system can be viewed from the perspective of material flow but also from the perspective of information flow or energy flow. The

application of this basic principle is closely linked with the following basic principle of neutrality.

The basic principle of neutrality (Ott 2009)
This basic principle is based on the strict separation of "what" and "how" (Ott 2009, p. 50). In relation to the logistical system, this means first clearly identifying what requirements the logistical system should fulfill. Only then should it be considered how these requirements can be implemented by the logistical system or parts of the logistical system, i.e., for example, by a roller conveyor or by a shuttle.

The basic principle of reusability (Zilahi-Szabó 2002)
This states that complex systems should be broken down into modules in such a way that these modules can be reused. In the logistical system, such modules are, for example, curves, drives, and straights. The arbitrary coupling of these modules can create very differentiated logistical systems for different purposes. This basic principle is closely tied to the following basic principle of standardization.

The basic principle of standardization (Ott 2009)
It serves the goal of being able to use parts of the complex system multiple times under different conditions. In the case of the logistical system at an airport, for example, a roller belt conveyor may be integrated into this system. This roller belt conveyor consists of standardized rollers that could also be used in a roller conveyor system in a lignite open-cast mine. Standardization does not only have to refer to the components and modules of a technical system; it can also be extended to procedures and documents.

The basic principles of reusability and standardization can be combined with that of information encapsulation.

The basic principle of information encapsulation (Bing 2001)
The basic principle supports the interchangeability of components. The drive of a logistical system, which is controlled both centrally and decentrally, must be interchangeable at all times. If a new drive is installed in the logistical system, it must be immediately recognized by the central control and be able to communicate with it. The same applies to sensors that may transmit system states to the drive to enable optimal operation.

The basic principle of discursive approach (Wulf 2002)
The goal-oriented approach is a main component of this basic principle. The goals are constantly critically questioned. Such a general goal as the optimization of a logistical system must be further specified by asking, for example: "In relation to what is the logistical system to be optimized?" "Is its optimization related to the material flow, the energy flow, or the information flow?" The basic principle can be repeated multiple times when dealing with complex tasks.

The basic principle of thinking in alternatives (Haberfellner et al. 2019; Haberfellner and Daenzer 2002)
There are always a multitude of solution alternatives for solving complex tasks. This basic principle demands the development of not just one, but several solution variants when solving problems. For example, goods at an airport can be transported using a multi-shuttle, a forklift, or a conveyor belt. Discussing the advantages and disadvantages of different solutions often leads to new ideas. Thinking in alternatives supports the creativity process.

The basic principle of modality change (Lindemann 2016)
The basic principle aims not to follow only one action strategy. The development of the logistical system, for example, initially proceeded from the "known"—transporting with roller conveyor—concluding to the "unknown"—transporting with linear drive. A synthesis is developed based on this. This can again be "top-down" or "bottom-up". The above-mentioned basic principle does not recommend which "modalities" are to be coupled and when to switch. What is important in its application is that alternative views are used in problem solving to avoid routines.

The basic principle of problem decomposition (Dörner 2007)
The basic principle is due to the recognition that people are only limited in their ability to understand and design complex systems. For this reason, DÖRNER recommends sensibly breaking down problems into sub-problems in order to enable systematic, goal-oriented problem solving. It is assumed that the solution to the sub-problem contributes to the solution of the overall problem. This connection may or may not exist. That is the difficulty in applying this basic principle. It can be applied in conjunction with the following basic principle of minimizing interfaces.

The basic principle of minimizing interfaces (Haberfellner et al. 2019; Haberfellner and Daenzer 1999)
When breaking down complex facts or problems, they should be split in such a way that as few interfaces as possible are created (Haberfellner et al. 2019; Haberfellner and Daenzer 1999). The basic principle is closely related to the basic principles of structuring, applying multiple perspectives, and information encapsulation.

The basic principle of subtraction of systems or system elements (Boyd and Goldenberg 2019)
In the subtraction of systems or system elements, something is eliminated that is considered important and necessary. It can then be decided whether the "lost" can be taken over by something existing. This basic principle can lead to completely new solutions when combined with the basic principles of minimizing interfaces and or information encapsulation, such as replacing the keyboard of a mobile phone with a touchscreen.

The basic principle of multiplication of systems or system elements (Boyd and Goldenberg 2019)
In multiplication, systems or system elements are duplicated and possibly changed if necessary. This basic principle of multiplication is particularly important when ensuring the safety of systems, in a way that a redundant system is available in case of a system failure. But it can also be used to improve system properties, in a way that a mobile phone uses not one but four or five cameras, for example.

The basic principle of division or decomposition of systems or system elements (Boyd and Goldenberg 2019)
By applying this basic principle, systems can be broken down into meaningful subsystems. The cloud-based applications in industry as well as consumer goods are examples of this, i.e., data storage and security are carried out in a completely different system, although they are directly necessary for ensuring the basic function of the computer or mobile phone.

In summary, it can be stated that only very few of the basic principles of systematic thinking and action are currently explicitly associated with SE (Haberfellner et al. 2019). However, in order to effectively manage the new dimensions of complexity in the present and future, the following basic principles of systemic thinking and action must be linked with SE:

- Basic principle of thinking in systems
- Basic principle from the whole to the detail
- Basic principle of recurring reflection
- Basic principle of structuring
- Basic principle from the abstract to the concrete
- Basic principle of minimal models
- Basic principle of comprehensibility
- Basic principle of applying multiple views
- Basic principle of neutrality
- Basic principle of reusability
- Basic principle of standardization
- Basic principle of information encapsulation
- Basic principle of discursive approach
- Basic principle of thinking in alternatives
- Basic principle of modality change
- Basic principle of problem decomposition
- Basic principle of minimizing interfaces
- Basic principle of subtraction
- Basic principle of multiplication
- Basic principle of division or decomposition

On this basis, SE can contribute to systematically solving complex problems effectively.

The mentioned basic principles of systematic thinking and action are to be integrated into both the thinking model and the procedural concept of SE, especially since the dimensions of complexity have changed compared to the past. Nevertheless, the complexity of systems was, is, and will be manageable through SE. To make it manageable, it is necessary to know complex systems and to be able to interact with them. In the past, SE used systemic thinking, which today has partly been preserved in the form of the basic principles of systemic thinking and action, but was not seamlessly connected with SE. This applies both to the use of SE for scientific problem solving and to its application in the management of complex product, process, and service systems or complex organizational systems. The solution is a cross-disciplinary, transparent, traceable reformed SE concept in a holistic sense, which integrates the basic principles of systematic thinking and action into its thinking model and procedural concept in a problem-related manner and which meets the requirements arising from the new dimensions of complexity management.

Consequently, the thinking model and procedural concept of SE must meet at least the following requirements if they are to contribute to the management of system complexity:

- Transdisciplinarity,
- Traceability,
- Transparency,
- Observance and Implementation of the basic principles of thinking and action.

But before the various approaches of SE regarding the fulfillment of these requirements are examined, the following chapter will consider what is understood by SE today, which SE approaches exist, where they are applied, and whether they fully or partially meet the requirements derived from the new trends of complexity.

2.4 The Evolution of SE

To this day, human thinking deals with systems and their complexity. This is particularly true for "system theory" as a scientific discipline, which creates the theoretical foundations for the structural and functional analysis of various systems in order to make possible predictions about system behavior (Haberfellner et al. 2019; Gausemeier et al. 2013a; Hinrichsen and Pritchard 2005; Schuldt 2003). In contrast, SE specifically deals with knowledge about predominantly technical systems, their subsystems, and their interactions with each other with the aim of developing new technical systems or designing existing ones (Haberfellner et al. 2019; Weilkiens 2019; Lindemann 2016; Haberfellner and Daenzer 1999). HABERFELLNER proves that SE is also suitable for the design of socio-technical systems, i.e., companies (Haberfellner et al. 2019). HEINRICH uses SE to better define and control projects, which he considers as temporary systems (Heinrich 2015). MISTLER uses SE to make socio-technical systems like companies agile. This means building a system structure that can be flexibly changed through modularity (Mistler 2021).

As an independent scientific discipline, like system theory, SE has a longer history. Over the course of this history, various approaches, concepts, methods, and tools have proven themselves in practice for solving more or less complex problems (see also Sect. 2.2).

Against this background, it therefore seems sensible to question the origins of system thinking and its development, as well as the resulting perceptions and basic assumptions of SE, including current trends. In addition to this, it is also about analyzing the methodological foundations of the currently practiced SE with regard to managing complexity in the mentioned new dimensions. Only knowledge of the real possibilities and limits of SE offers the chance to find new approaches or methods for solving future complex problems or to develop a universal problem-solving concept. Complex constellations are generally based on both their components and their relationships with each other. Consequently, dealing with complexity requires considering both aspects in their mutual conditionality (Jackson 2000). Therefore, it is important to make the new dimensions of system complexity transparent in the shortest possible time in a transdisciplinary way, so that their representation becomes possible and their components and their interactions with each other become recognizable. In this context, today's system theory and SE both contain theoretical and practice-oriented approaches, which support the search for efficient solution concepts for complex problems.

However, both scientific disciplines also demonstrate that the systemic view obviously holds the key to mastering complexity. Overall, the SE of the present is characterized by the search for practicable ways to combine the philosophy of system thinking with generally useful methods for the practical implementation of the SE claim.

A current and in a certain sense official definition is provided by INCOSE, the International Society for SE. It reads: Systems Engineering is an interdisciplinary approach with the aim of successfully realizing systems. Systems Engineering focuses on defining and documenting system requirements in the early development phase, developing a system design, and verifying the system for the preservation of the stated requirements, taking into account the overall problem: operation, time, test, creation, cost and planning, training and support, and disposal. Systems Engineering integrates all disciplines and forms a structured development process from concept to production to operational phase. Both technical and economic aspects are considered in order to develop a system that meets user needs (INCOSE 2015, p. 265).

An expression of the increasing importance of SE is also the increase in the number of members of the International Council on Systems Engineering (INCOSE). In the USA alone, it rose from 3600 in 2005 to 4900 in 2007 (Weilkiens 2007). In Germany, the number of members is rather low and shows hardly any development during this period. The particularly high number of members in the USA has two causes. On the one hand, INCOSE was founded in the USA in 1990. On the other hand, NASA, Boeing, and General Motors favored the SE approaches that emerged during this time. However, even INCOSE cannot hide the fact that various SE concepts now exist. In Germany, the Society for Systems Engineering disseminates and further develops the SE. This society was founded in 1997 and is the German section of the International Council on Systems Engineering (INCOSE 2015). The annual Day of Systems Engineering promotes the exchange of experience between users in companies and scientists. The current overview of SE activities in Germany is described in Dumitrescu et al. (2021).

The wide variety of SE concepts ranges from universal problem-solving approaches (Gausemeier et al. 2012; Winzer 2015; Mistler 2021; Schlueter et al. 2019; Dumitrescu et al. 2021; Haberfellner et al. 2019; Ehrlenspiel and Meerkamm 2017; Bahill and Gissing 1998; Sell 2013; Rink 2002; Züst 2004; Sage and Rouse 2009; IEEE 1220 2005; Eigner and Stelzer 2009; Haberfellner and Daenzer 1999; Sell 2013) to specific SE approaches that focus exclusively on:

- product development (VDI 2221 1993; Lindemann 2016; Ponn and Lindemann 2011; Schmitt et al. 2014; Gausemeier et al. 2012; Bender and Gericke 2021),
- software development (Fuchs et al. 2001; Sommerville 2007; Partsch 2010; ISO/IEC/IEEE 29148 2018),
- company design (Schenk et al. 2014; Schuh 2004; Wiendahl et al. 2009; VDI 5200 2011; Grundig 2018) or
- the creation of safety concepts (VDI 2247 1994; IEC 61508 2010)

focus.

Thus, there are currently two main groupings in SE. One group pursues a universal problem-solving approach, while the other favors the use of specific SE approaches for solving discipline-specific questions.

The following analysis of selected SE concepts from both directions is intended to clarify to what extent they combine system thinking and engineering on the one hand, and on the other hand, consider the problem-solving process in a multidisciplinary, transparent, and traceable manner using the basic principles of systematic thinking and its procedures to manage the complexity of systems.

2.4.1 Universal SE Concepts

Bahill and Gissing describe SE as a problem-solving discipline for the modern world (Bahill and Gissing 1998). In this view, SE as a scientific discipline offers a way to manage complexities and achieve excellent results. BAHILL and GISSING were the first to deal with the variety of SE approaches that emerged in the sixties and seventies.

As a result of their comparative consideration, they conclude that traditionally the following steps of SE have proven themselves and should therefore be maintained:

- Find out the real problem.
- Describe the problem exactly in its entirety.
- Define goals for solving the problem.
- Develop supportive variants for problem solving.
- Filter the best results for solving the problem.
- Test the best variants for solving the problem.
- Evaluate the test results and control the fulfillment of the goals.
- Possibly improve the found solutions and develop more and better variants for solving the problem.
- Realize the best variant for solving the problem (Jenkins and Youle 1971).

The general problem-solving approach according to Sell is based on the basic pattern of human decision-making processes (Sell 2013). It illustrates the relationship between thinking and acting. The framework, divided into three orthogonal action parts "Orientation", "Execution" and "Control", is shown as a flowchart in Fig. 2.4.

HABERFELLNER (Haberfellner et al. 2019), on the other hand, focuses on three essential steps of the human problem-solving process, i.e., the problem definition, the methodology for problem solving, and the solution itself. In his universal problem-solving approach, he takes into account that people who have to solve problems together have different expertise, different experiences, and their actions are based on different ethical actions.

In clarification of this, HABERFELLNER now assumes that the problem-solving process includes system design and project management. The latter specifically controls the problem-solving process from the general to the detail. The SE philosophy is the conceptual framework that includes system thinking and the procedure model (Haberfellner et al. 2019). RINK emphasizes the necessity of iteration in the problem-solving process (Rink 2002). Basically, he assumes that each problem-solving cycle includes the steps of goal setting, solution finding, and selection of the best solution. These three steps are repeated in each problem life cycle phase (see Fig. 2.5).

There are always feedback loops between the individual product life cycle phases. However, iterations are also necessary within, i.e., in the respective processes of goal setting, solution finding, and selection of solutions, whether for goal specification, which can result from changed requirements or from the process of developing the solution. RINK neglects these iteration loops as well as the system description. While he provides clear cycles for iterative problem solving, WULF relies on a discursive process (Wulf 2002).

The approach of Fig. 2.6 quite corresponds to reality in the problem-solving process. After goals have been defined and the corresponding solutions have been developed, it

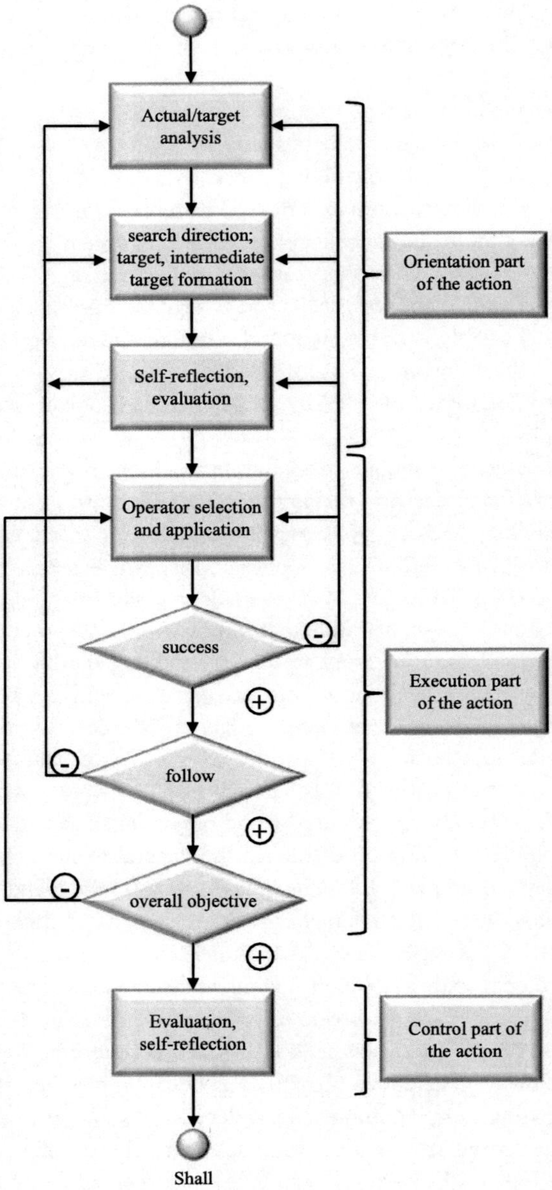

Fig. 2.4 Procedure model for problem solving. (After Sell 2013, p. 70)

must be checked in the testing phase to what extent the set goals could actually be realized. If this is not sufficiently the case, either new solutions must be sought and/or a goal specification must be made. Consequently, a goal specification does not necessarily always have to be made. A clear reference to system thinking is not found in WULF.

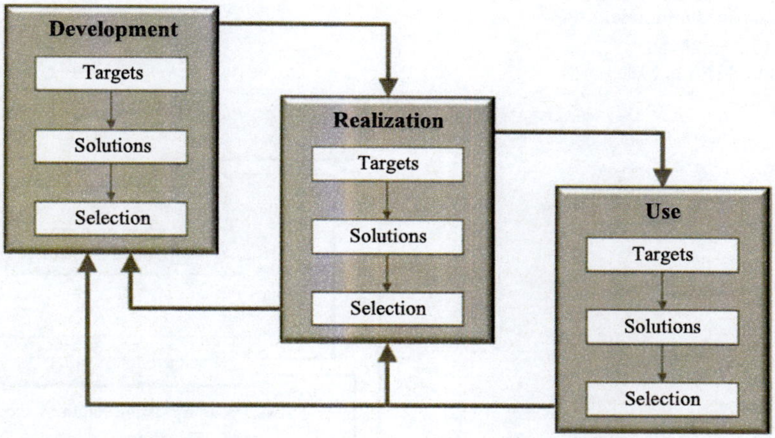

Fig. 2.5 Basic structure for systematic procedure models. (After Rink 2002)

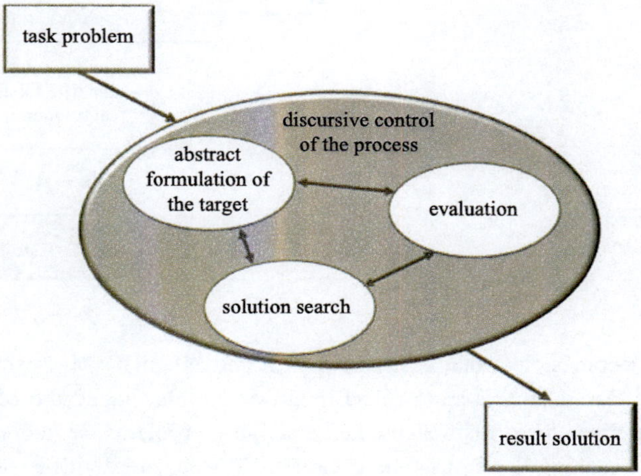

Fig. 2.6 Problem solving as a discursive process. (After Lindemann 2005, p. 36)

The approach of EHRLENSPIEL and MEERKAMM (Ehrlenspiel and Meerkamm 2017) is also divided into three main steps. These include clarifying problems, searching for solutions, and selecting solutions (see Fig. 2.7).

The targeted approach of EHRLENSPIEL and MEERKAMM seems to be a sequential one at first glance, although he recommends its iterative application. To what extent system thinking is integrated into his approach is not directly apparent from the description of the procedure. Compared to the previously discussed problem-solving

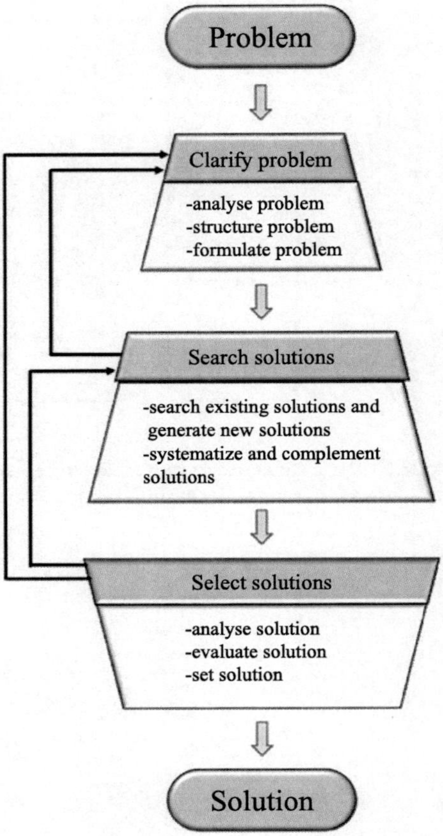

Fig. 2.7 Procedure model. (After Ehrlenspiel and Meerkamm 2017, p. 115)

approaches, it becomes clear that EHRLENSPIEL and MEERKAMM exclude the goal-setting process. Based on this approach, LINDEMANN developed the Munich Model, which also comprises three main steps, i.e., clarifying problems, searching for solution alternatives, and making decisions. These three main steps are divided into further sub-steps, such as planning goals, analyzing goals, structuring, searching for solution alternatives, making decisions, securing goals. They are summarized in Fig. 2.8.

The Munich procedure model is compared to the other universal problem-solving approaches of SE a networked approach. It allows, after a solution has been found, to check the goals again and to search for solutions again. LINDEMANN emphasizes that the "Munich Model" is based on the SE approach, a reference to system thinking is not explicitly stated.

The trend of the inflationary development of universal problem-solving approaches based on SE was tentatively ended by the development of the IEEE Standard 1220 of 2005. The aim of the standard is to provide guidance for the development and design

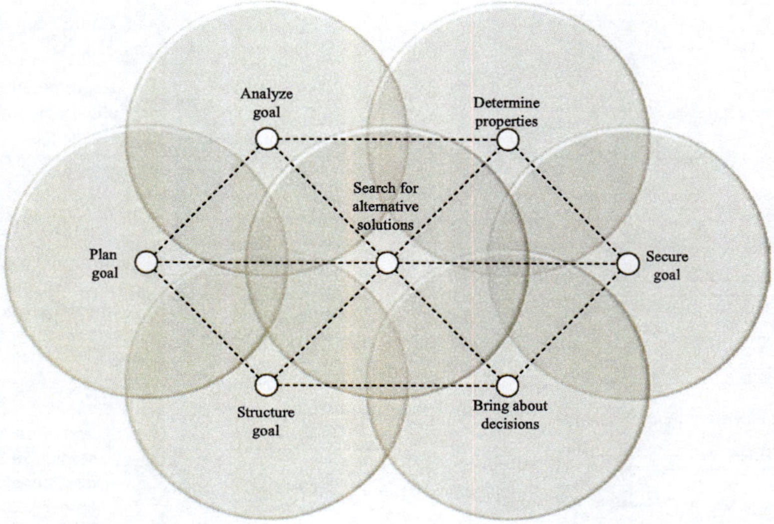

Fig. 2.8 The Munich Model. (After Lindemann 2005, p. 40)

of systems. It describes SE as a cyclical process that can be applied to all phases of the product life cycle. Figure 2.9 illustrates the standardized approach of SE according to IEEE 1220 (2005).

The standardized SE process according to IEEE 1220, 2005 combines the classic system thinking, which finds its expression in system analysis, here in particular with the problem-solving algorithm, which is realized through project management.

In evaluation of the various SE approaches, Sage and Rouse (Sage and Rouse 2009) conclude that SE essentially consists of two main processes, i.e., the technical process and the control process. The technical process of SE includes the definition, analysis, design, testing, and evaluation of solutions. Since these steps can run iteratively, bind different times and resources, this is to be optimized via the SE control process. If SE is applied to the product life cycle in a problem- or product-related manner, various approaches are possible. Sage and Rouse analyzed this and summarized it in the matrix of Table 2.1. It is intended to restore the framework for the universal character of SE.

Sage and Rouse demonstrate that the universal SE approach they developed is suitable not only for product development but also for the design of companies or company networks.

While all previously discussed, universal problem-solving approaches of SE, whether by RINK, Sell, EHRLENSPIEL and MEERKAMM or LINDEMANN, do not explicitly connect thinking in systems with the engineering approach, i.e., with the problem-solving cycle, Haberfellner, Sage and Rouse at least indicate that the definition of the problem is related to its assignment to the system. BAHILL, EHRLENSPIEL and

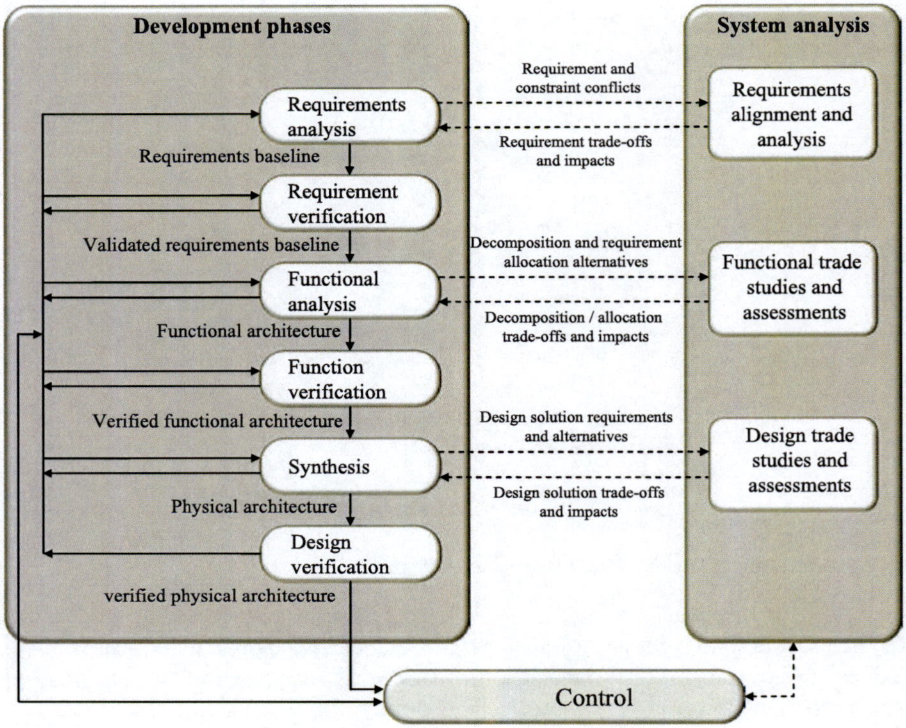

Fig. 2.9 Procedure of SE according to IEEE 1220–2005. (After Ott 2009, p. 71)

Table 2.1 The two-dimensional framework of SE (Sage and Rouse 2009, p. 20)

SE steps	Formulation			Analysis		Interpretation	
Design steps	Problem definition	Value system design	System synthese	System analysis	Alternative refinement	Decision making	Action planning
Requirements and specifications							
System architecture and preliminary conceptual design							
Logical design and functional architecture							
Detailed design, physical architecture and tests							
Operational implementation							
Evaluation and adaptation							
Use and maintenance							

MEERKAMM and WULF demand the description of the problem. But what is the relationship between the problem and the system to be considered? The delimitation of the system significantly influences the more precise characteristics of the problem, as well as the approach in SE. This is proven by the example of optimizing the subsystem "Drive" depending on the overall system "Conveyor system" from Sect. 2.3.

In SE, requirements are initially collected through the requirements analysis (1st step) and a requirements model is developed (Haskin et al. 2011). Various software of Requirements Engineering can be used for this. The best known is DOORS. This creates the so-called problem space. In the following step (2nd step), the functional system is modeled via the function analysis, which must be solution-neutral (Lamm and Weilkiens 2015). The FAS method can be used for this, which represents the functions in functional blocks. But the use of Petri nets and function trees is also common. As a result, according to SE, the solution space is created. Based on this, the system architecture is created in the 3rd step through the selection of components. The implementation of the system architecture can, if it is a mechatronic system, be carried out separately by the computer scientist, electrical engineer or mechanical engineer. In the worst case, three system architecture models are created, which must be merged again. These are referred to as distributed models of SE (Woll et al. 2015). There is hardly any information cross-linking between all these mentioned models. How the system architecture description should be done is not described in the INCOSE Systems Engineering Handbook V3.2.2 (Haskin et al. 2011).

The universal concepts of Systems Engineering represent interdisciplinary problem-solving approaches, which serve the purpose of strengthening the understanding of the system through transparency, in order to be able to design the life cycle of the system efficiently (INCOSE 2007). The transition of Systems Engineering towards Model-Based Systems Engineering (MBSE) makes a significant contribution to this. Model-Based Systems Engineering focuses on the continuous description and analysis of technical systems based on their modeling (Dumitrescu et al. 2014). Model-Based Systems Engineering focuses on an interdisciplinary system model. Various model types such as requirement, function or structure models are connected with each other. However, these distributed models must be consistently linked with each other. This can be created via trace links. But before such trace links are created, companies must be clear about which information from which models should be linked with which systems and how these trace links should be used. Only on this basis can a data integration solution be created. The following model types are linked with each other via Model-Based Systems Engineering:

- Requirement models,
- Behavior models, also referred to as function models,
- Product structures, also known as structure models,
- Workflow models,
- Project models and
- Test models (Woll et al. 2015).

Model-Based Systems Engineering characterizes the change in Systems Engineering from documented and text-based to model-based approaches, see Fig. 2.10.

As this illustration shows, the system model is created in the early stages of system design. While Model-Based Systems Engineering postulates the adaptation and maintenance of the system model throughout the development process, it only describes this process sequentially in the procedural concept. The system model must be adjusted, refined, and maintained throughout its entire lifecycle. This always requires an iteration of the procedural concept with the system model. Model-Based Systems Engineering currently has no widespread application in the German industry. ALBERS et al. see the reason for this in the different levels of abstraction in system modeling (Albers et al. 2014). Different modeling languages and differentiated modeling methods make it difficult for practitioners to create the system model across disciplines. Thus, Model-Based Systems Engineering is a response to current trends in complexity, but it also contains further research tasks to be solved, such as:

- the linking of various system models,
- the interaction between the system model and the procedural concept, and
- the description of how to maintain and document systems engineering throughout the lifecycle of the system.

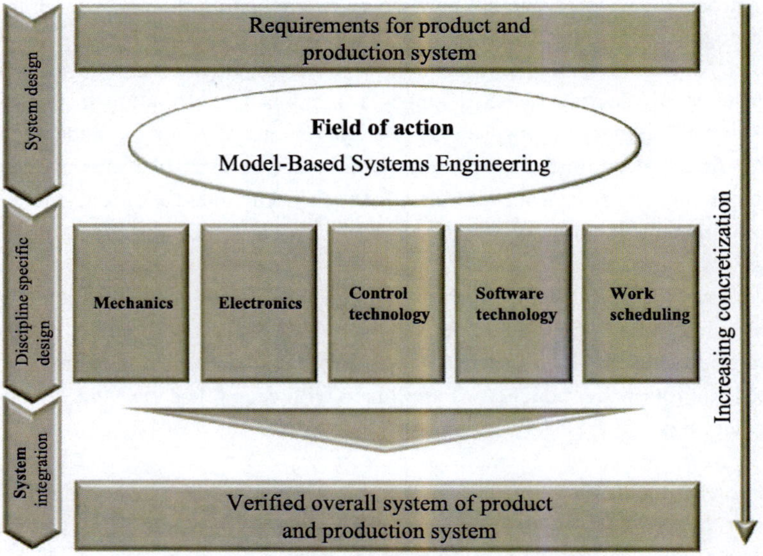

Fig. 2.10 Model-Based Systems Engineering. (After Dumitrescu et al. 2014)

A new branch in the universal approaches to Systems Engineering is Systems of System Engineering (SoSE). It describes the development, design, and transformation of systems that must be integrated into complex systems, thus the focus of Systems of Systems Engineering is the examination of complex systems in their network-like structures, in order to recognize relationships between the systems (Keating et al. 2003; Luzeaux and Ruault 2010).

Systems of Systems is seen by SAHIN and NCUBE as an approach superior to Systems Engineering (Sahin et al. 2009; Ncube et al. 2013).

Through a top-down approach, the Systems of Systems Engineering approach splits complex systems into subsystems and analyzes, designs, and reassembles their interaction into a complete system. However, the focus of Systems of Systems is on the interaction of the systems. The system itself is not the focus. The same applies to the interaction of the complex system with its environment (Jamshidi 2009; Dimario 2010).

Systems of Systems (SoS) is used for both technical systems and complex sociotechnical systems as well as for corporate networks or intelligent systems and production and service systems (DAG 2010; Jing et al. 2013; Weiss 2013; Lim and Ncube 2013; Schlueter et al. 2019).

In summary, it can be stated that the Systems of Systems approach, while using the tools of Systems Engineering, recommends different approaches. There is no uniform or universally valid definition of Systems of Systems Engineering. The approaches are diverse, as Systems of Systems Engineering can consider both technical and complex sociotechnical systems. The interaction of the complex system with its environment and the systems themselves are not primarily considered. The focus is on the interaction of the systems with each other (viewing the system as a network) (Maier 2005; Padilla et al. 2008; Ncube et al. 2013).

Caused by the multitude of universal Systems Engineering approaches, another direction in Systems Engineering developed, i.e., Lean Systems Engineering. The goal of Lean Systems Engineering is to streamline the problem-solving process. Especially in the phase of requirement analysis, according to OPPENHEIM, a lot of time is often wasted (Oppenheim 2011). For this reason, Lean Systems Engineering focuses on the phase of requirement analysis on both the customer and the client defining the requirements together. This way, misunderstandings regarding the requirements are cleared up early on, and necessary coordination can be minimized in the course of system development and design. Lean Systems Engineering describes standardized procedures and recommends tools for the problem-solving process. It combines the wisdom of Lean Thinking with the sequence of steps of Systems Engineering (see Oppenheim 2011). If the Lean Systems Engineering approach could be expanded so that the customer and client define the system model together, i.e., not only the requirement model, but also the function model, structure model, and process model, then a defined basis for the subsequent problem-solving process would be created. If an iteration between the system model and the standardized procedure of Lean Systems Engineering was planned, so that

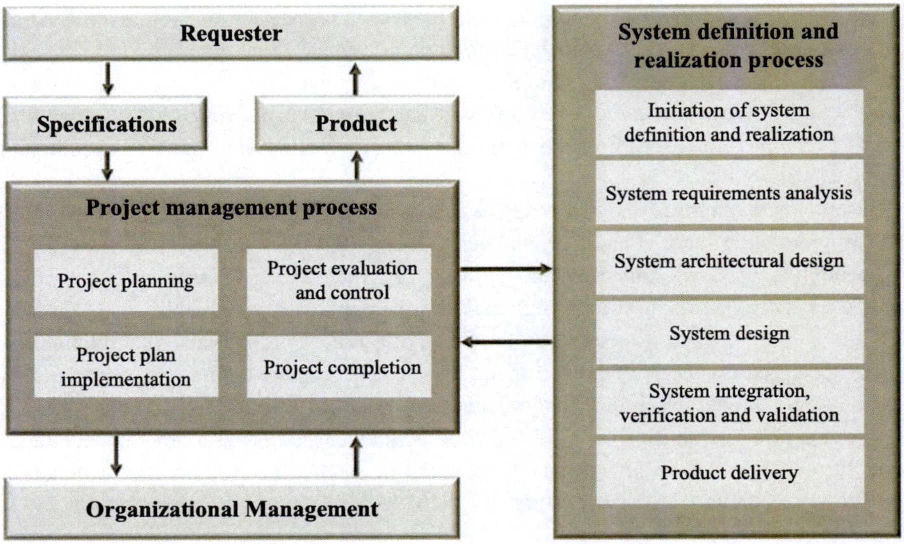

Fig. 2.11 Interaction of project and organizational management with SE. (After Gaupp et al. 2014, p. 355)

the problem-solving process is designed transparently and traceably, then Lean Systems Engineering could unite all universal Systems Engineering approaches in itself. However, the data artifacts of the subsystems or the various models of the system model must also be linked. These so-called trace links allow traceability, which is indispensable for tracing the lifecycle of the system. Basically, the rigidity between the procedural concept (problem-solving cycle) and the creation, maintenance, and documentation of the system model in the universal Systems Engineering approaches must be eliminated. A connection between SE, project and organizational management, as shown in Fig. 2.11, does not solve the outlined problem because the product system is constantly refined or improved.

All the SE approaches briefly characterized here have in common that they are universal, i.e., they can be applied in any scientific discipline. Thus, they are quite suitable for a transdisciplinary problem-solving approach. Each of the aforementioned SE approaches claims to be universally valid and to be based on standardized modules. This is exactly where the problem lies for the users of SE. Which procedural concept should an interdisciplinary team apply if it wants to prevent the sliding of subway trains on autumn leaves? To what extent these SE approaches fundamentally make their respective problem-solving algorithm traceable and transparent is not apparent. None of the analyzed solution approaches explicitly points to a connection with the basic principles of systemic thinking and action. Some of the presented universal SE approaches connect the system model with the procedural concept sequentially, although actually the system model must be supplemented, refined, maintained, and documented iteratively over the problem-solving process (Weiss 2013).

This problem was also considered in the ReMaiN project, which was funded by the German Research Foundation (DFG) (ReMaiN 2020). In this project, a methodological approach to requirements management in cross-company networks (ReMaiN) was developed. For this purpose, requirements management and requirements management were standardized and linked with each other. According to MISTLER, Requirements Engineering (RE) represents a process interdisciplinary to system development for specifying requirements for a system (Mistler 2021). According to MISTLER, Requirements Management (RM) forms a supporting process to Requirements Engineering to ensure the continuous networking of relevant information within system development and at the same time ensures consequence analysis and traceability in case of requirement changes (Mistler 2021). It should be noted that RE&RM have a connection to Systems Engineering and project management (Arnaut et al. 2016). Therefore, RE&M is connected with Systems Engineering throughout the entire system development process, because as GAUSEMEIER et al. (Gausemeier et al. 2013a) show, a system is considered holistically from the definition of requirements to validation throughout the development process (Gausemeier et al. 2013a). This means that RE&M must be designed as universally applicable as Systems Engineering. Thus, the DFG project resulted in the ReMaiN approach shown in Fig. 2.12.

The ReMaiN approach outlined in Fig. 2.12 arose from NICKLAS's demand to standardize the different discipline-specific RE&M approaches and make them universally usable with SE (Nicklas 2016). In addition, Mistler (2021) points out that RE&M supports SE and project management by forming an interdisciplinary process for the collection, structuring, weighting, validation, and management of requirements in system development" (Mistler 2021, p. 19). Accordingly, as shown in Fig. 1.12, RE&M encompasses all activities of Requirements Engineering (x-axis) and Requirements Management (y-axis). Due to the modular structure of the ReMaiN approach with the cubes shown, the activities of RE&M can be assigned to the respective companies in the corporate network (z-axis) (Schlueter et al. 2019).

With the ReMaiN approach, according to Schlueter et al. (2019), a uniform procedural concept with the corporate partners for the joint development of products can be worked out. This contributes to reducing product and organizational complexity in corporate networks. The sensitization phase shown in Fig. 2.12 serves as the basis for developing the procedural concept. In this phase, a procedural concept can be developed by the corporate partners using the ReMaiN cubes. The respective cubes contain methods and tools, the selection of which is problem-specific (Schlueter et al. 2019). The application of the ReMaiN approach primarily focuses on product development according to SCHLUETER et al. However, MISTLER shows the universal applicability of the ReMaiN approach by using it for the development of organizational systems by involving SE. However, MISTLER's work also makes it clear that the ReMaiN approach lacks the explicit connection to SE and the connection to project management. The use of

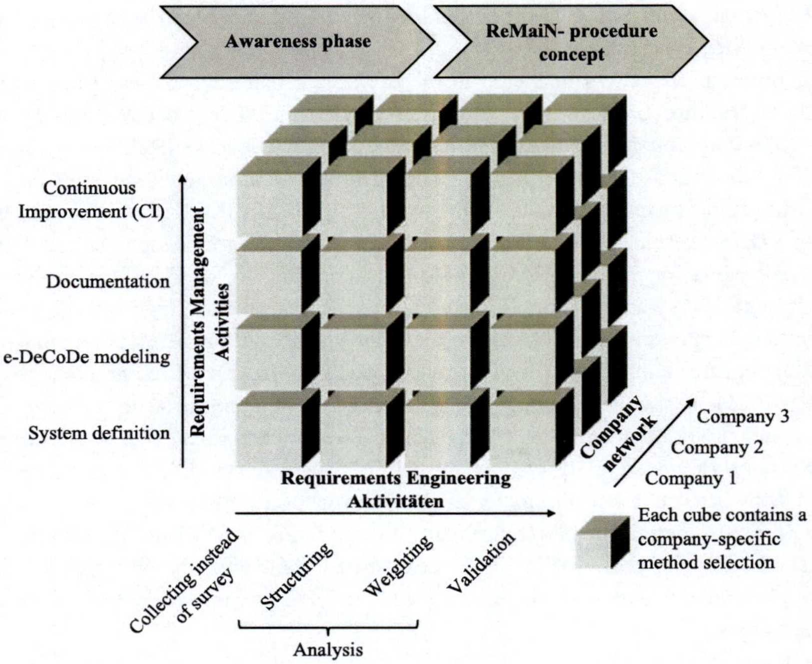

Fig. 2.12 ReMaiN approach. (After Schlueter et al. 2019)

SE and project management in the ReMaiN approach is only implicit due to the standardization of requirements management, which is why MISTLER was able to make the ReMaiN approach universally usable by expanding it to include Systems Engineering (see also Chap. 4).

However, none of the specific SE approaches presented here describes how a cross-disciplinary system model is created and how it can be designed iteratively with a problem-solving process that is controlled via project management and supported by methods. Therefore, the following will examine specific problem-solving approaches of SE in more detail to possibly find corresponding solution approaches.

2.4.2 Discipline-Specific Approaches to SE

The approaches to SE described in the previous chapter are so general that they can be transferred to all types of systems, such as technical systems, socio-technical systems, sociological systems, or biological systems. Their cross-disciplinary application is possible. However, SE is also applied in various disciplines, such as computer science and safety engineering, and further developed according to specific concerns. In addition to this disciplinary view, specific approaches to SE can also consider certain types of systems, such as the product system (technical system) or the company (sociotechnical system). They will be examined selectively in the following to check how thinking in systems is implemented, how the mental model is interlocked with the special procedural concept, and how the respective discipline-specific approaches to SE are connected with systematic thinking and action. Possibly, ideas for the redesign of the universally valid SE approach of the past can be derived from the discipline-specific approaches to SE in the sense of "Best Practice".

In an initial selection phase, the specific SE approaches are examined in terms of their contribution to managing the complexity that arises from the trends of globalization, individualization, miniaturization, dynamization, and sustainability. However, these aspects must be taken into account when developing new product systems. As these change very quickly, innovation cycles and thus the phase of product development are dramatically shortened. To counteract this, attempts are made to develop products simultaneously using international division of labor and distribute them globally. Consequently, it can be assumed that the various procedural concepts used in product development (Haberfellner et al. 2019; Bender and Gericke 2021; Haberfellner and Daenzer 1999) already meet the new challenges of complexity, but initially only from the specific professional perspective, i.e., that of mechanical engineering. The same is probably true for software development and its procedural concepts, which also build on the basic principles of SE (Balzert 1998, 1996; Böhm and Fuchs 2002; Rink 2002). It also takes place globally and is subject to very short innovation cycles.

Both, i.e., product and software development, take place in organizations. Thus, modern SE approaches to business modeling are actually closely related to product or software development, depending on the purpose of the respective company. Nevertheless, special procedural concepts for business modeling (also referred to as factory design) are identified in the literature (Grundig 2018; Haberfellner et al. 2019; Winzer 1997; Schenk 2004; Schuh 2007; Wiendahl et al. 2009; VDI 5200 2011), some of which are based on SE approaches. In this process, the process and organizational structure of the company are modeled. In product and software development, the focus is exclusively on the process of system development, i.e., on optimizing the development process of the respective system so that the desired product or software is created. In contrast, the perspective in the design of organizational systems (the company is considered as such) is expanded to include the organizational structure, which regulates responsibilities and authorities, as well as process optimization, which focuses on business processes (Haberfellner et al. 2019; Mistler 2021; Gausemeier et al. 2009). Therefore, selected approaches to business modeling, which are based on SE, are also analyzed in more detail in this section.

Requirements Engineering & Management (Danner 1996; Davis et al. 2006; Schlueter et al. 2019) focuses on managing the growing diversity of requirements and the increasing number of stakeholders. Both aspects result from the trend towards globalization, but are very closely interrelated with the trends towards individualization and miniaturization. For this reason, the approaches of **Requirements Engineering** & Management based on SE will also be considered in the following.

In both product and software development, as well as in business modeling and requirements engineering, ensuring safety plays a very special role. This specific aspect is addressed by Safety Engineering (Rink 2002). The approaches used here also partly based on SE. Therefore, they will also be included in the subsequent analysis. They focus on ensuring compliance with legal requirements ("must requirements"), which are particularly difficult to manage in the course of globalization due to the contradictions of various national and international laws and their overwhelming diversity (Mistler 2021; Winzer et al. 2010).

As a result of the first selection phase, it can be stated that there are approaches from specialist disciplines that make use of the intellectual world of SE from their respective professional perspective. They make a specific contribution to managing the trends of complexity in the present. For this reason, they are included in a further, detailed consideration, i.e., in the second selection phase. Here, it is to be analyzed to what extent these discipline-specific approaches of SE, which focus on the consideration of a special object from a professional perspective, use the same or similar thinking models or procedural concepts and how both interact. This is intended to generate ideas for the development of a general SE approach. Furthermore, it is to be examined to what extent these discipline-specific approaches meet the requirements for a generalized SE approach identified in Sect. 2.3, such as:

- Transparency,
- Traceability,
- Transdisciplinarity as well as
- Consideration of the basic principles of systematic thinking and action.

The following results of the second selection phase are presented in a structured manner. They refer to:

- specific types of systems, such as product or software or organizational systems (company/factory) (see Sects. A to C)
- specific aspects in system analysis and design, such as managing the diversity of requirements using Requirements Engineering or the safety-oriented design of systems within the framework of Safety Engineering (see Sects. D and E).

A) Use of SE in product development or product design

The multitude of methodological approaches used in product development seems almost infinite. Pahl/Beitz (see Pahl et al. 2005) created a chronological overview of these methods up to the year 2002 and came up with about 130 different methodological approaches recommended for product design. The following will discuss those that are strongly oriented towards the SE approach and that take into account the new dimensions of complexity. First, to the methodological approaches of product design that are very closely aligned with SE.

The approach of Pahl/Beitz (Pahl et al. 2005, see Fig. 2.13) is fundamentally based on SE. It forms the foundation for construction methodology in mechanical engineering and describes the procedure for designing technical systems.

The procedure of PAHL/BEITZ clearly characterizes the various iteration loops, i.e., the possibility of a renewed specification of the goals after system analysis, evaluation, or after system decision. The system definition, i.e., the description of the system in demarcation from its environment, implicitly goes into the system analysis. There is no explicit reference to the development of a system model, which can form the unified basis for the system synthesis process and the system analysis. The system synthesis process contains ways to find design solutions. The principles of systematic thinking and action are pointed out. It can be assumed that the use of this methodological sequence in product development contributes to transparency and traceability. The sequence of steps

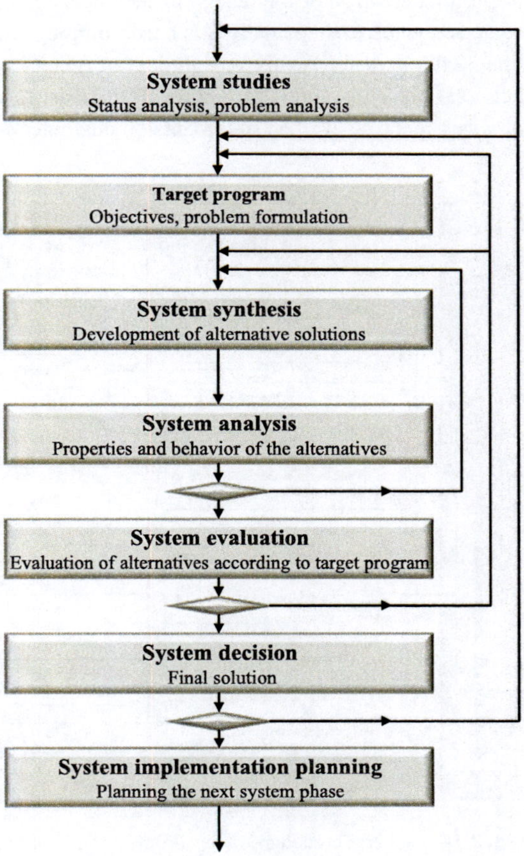

Fig. 2.13 Sequence of steps for product development. (After Pahl et al. 2005, p. 19)

by Pahl/Beitz includes a very general procedure that is obviously transferable to any type of system and not just to the technical systems of mechanical engineering assumed by them. It is quite conceivable that socio-technical or technological systems can also be designed using this procedure, i.e., with the steps of system study, target program, system synthesis, system analysis evaluation, and system decision.

The procedure for product design presented in VDI guideline 2221 is also valid for wide areas of application. It is also essentially based on the principles of SE, but also on the procedure of Pahl et al. 2005, as illustrated in Fig. 2.14.

It describes the requirement-oriented development of products, technologies, and processes under the condition that these can be represented as technical or technological systems. From requirements, functions are first derived and then solution variants. These are to be evaluated, modified, and ultimately implemented in practice according to the objective. However, this approach lacks an explicit emphasis on system definition and demarcation.

The VDI guideline 2221 and the VDI 2247 are closely interrelated with the process of product creation or the design of technical systems. While the VDI 2221 describes the basic sequence of steps, which is based on the SE approach of INCOSE, the VDI guideline 2247 characterizes the assignment of quality science methods to the phases of product creation. Thus, the latter approach links aspects of product safety and reliability with quality aspects as well as statements on customer satisfaction and costs.

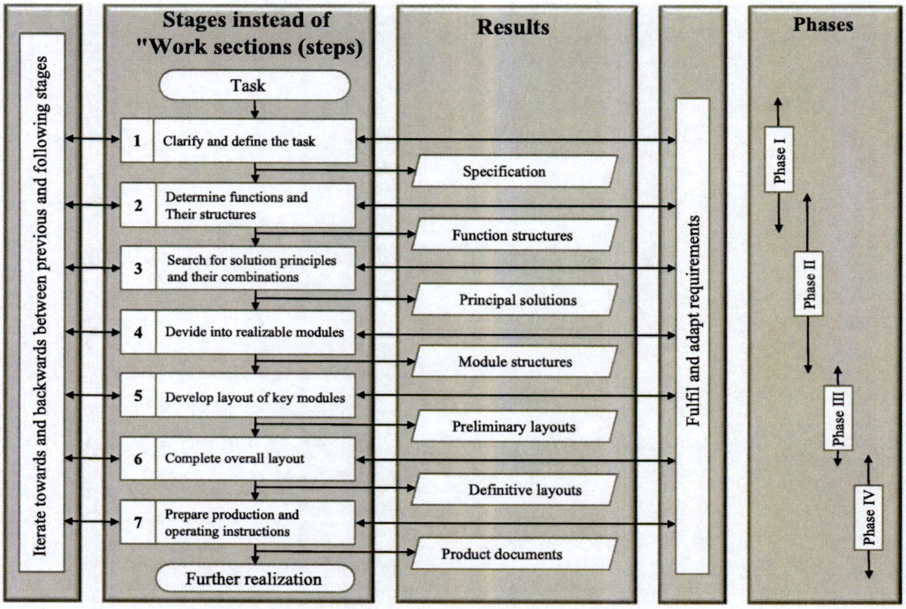

Fig. 2.14 General procedure in development and construction. (After VDI 2221 1993, p. 9)

In this context, Quality Function Deployment (QFD) systematically captures customer wishes. Value analysis evaluates them. Subsequently, the Failure Mode and Effects Analysis (FMEA) and the Failure Tree Analysis (FTA) estimate the risks and finally the statistical experimental design checks the design. However, it should be noted that these methods in the respective phases are different in precision and not connected to each other due to the system definition. In the case of a very large system—such as the car—both a QFD and an FMEA only provide rough results. However, if the injection valve of the engine of this car is examined, the mentioned methods lead to highly detailed findings. The relationship between the overall system "car", the subsystem "engine" and the component "injection valve" is not automatically established. Since the overall system is connected with these mentioned subsystems, it is important to consider the various methods and their interactions, for which the mentioned procedural model does not provide any hints.

The procedure according to PLATZECK considers the life cycle of product systems. It is characterized by RINK as an interdisciplinary, or technical approach to requirement-oriented problem solving (Rink 2002).

This includes system needs analysis, system planning, system control, system use, system shutdown, and system destruction. Over these individual phases, PLATZECK lays out a problem-solving process with the following steps:

- Situation analysis,
- Problem definition,
- Problem analysis,
- Concept,
- Concept analysis,
- Structuring,
- Design analysis,
- Compatibility analysis and
- Realization.

This procedural model contrasts the system view, here represented as the life cycle of the systems, i.e., from planning to realization to the passing of the system, with the problem-solving cycle. This allows the problem-solving cycle to be linked in each life phase of the system. Questions about an exact system demarcation or system environment description or about the derivation of goals and variants remain open in the procedural approach presented by PLATZECK. Nevertheless, some basic principles of systematic thinking and action are applied, such as the basic principle of structuring or the basic principle from the whole to the detail.

The phase model according to SCHNIEDER (Schnieder and Schnieder 2013) corresponds to the basic principle from the abstract to the concrete, here specifically applied to the development of automated systems (see Fig. 2.15). It also applies the basic principle from the whole to the detail in system design. In general, goals are set in the planning

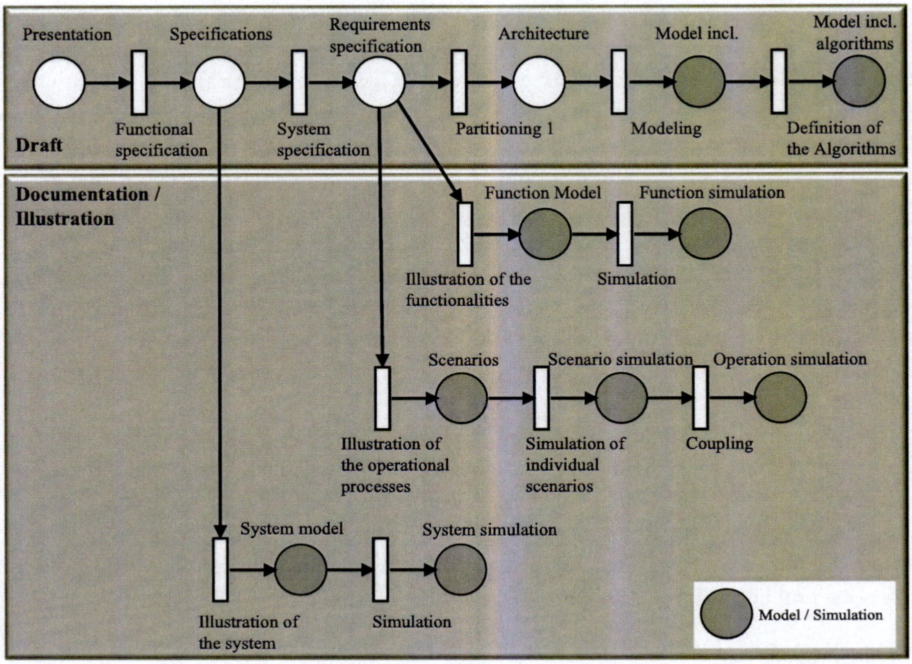

Fig. 2.15 Phase model of development. (After Schnieder and Schnieder 2013, p. 535)

phase, on the basis of which a system design is created. Here too, the corresponding iteration loops are provided. In the realization phase, the basic principles of discursive procedure, minimization of interfaces, and recurring reflection are particularly used. The latter also comes into play in the test phase and in the application of the system (usage phase).

Although this approach by SCHNIEDER is supposed to be valid exclusively for product development, it can certainly be used for the development of various systems, such as software systems, security systems, and technological systems, to name just a few. However, the delimitation of the system and its description are not explicitly addressed.

Due to the individualization of products, innovation cycles and product life cycles are becoming increasingly shorter. Simultaneous engineering, cooperative product engineering, smart engineering, and prototyping are responses to this trend towards complexity. Consequently, these are increasingly used in product design.

The basic idea of the prototyping approach is to develop a sample or a prototype of the final product to be designed relatively quickly. This gives the designers and developers, as well as the clients or potential users, the opportunity to quickly agree on whether the product actually meets the requirements set. The developed sample is the result of implementing one or more solution alternatives. It forms the basis for discussing the product system with the client or potential user and further improving it. The prototyping approach was continued with CAD support in the form of re-prototyping, by immediately converting CAD drawings into corresponding samples. Such an approach allows a product design to be held in the hands in real terms quickly.

Since not every potential client or user is able to recognize the potential product exactly from the drawings due to a lack of spatial imagination, the product sample of prototyping compensates for this deficit. The prototype helps all participants to evaluate the found solution variant in terms of the degree of fulfillment of the requirements. This allows a faster and more accurate decision on whether it makes sense to continue pursuing the favored solution idea, embodied in the prototype, and implement it as the final solution. In principle, prototyping corresponds to the SE approach. Its procedure refers exclusively to technical systems and supports the process of finding solutions. Thus, prototyping, which is currently being further developed by adaptive technologies, is directly to be classified in the solution finding process of SE.

The procedural model of simultaneous engineering according to Haberfellner (Haberfellner et al. 2019; Haberfellner and Daenzer 1999) was developed in the Nineties. It is an overlapping phase concept that is fundamentally based on the procedure of PAHL/BEITZ or VDI 2221 and thus on the SE approach. Simultaneous Engineering refers exclusively to technical systems. In the meantime, the variants of Simultaneous Engineering are proving to be increasingly diverse. It finds broad application possibilities especially in the automotive industry. Due to the globally distributed locations of companies that, for example, develop instruments, seats, air conditioning systems for the same type of car at the same time, it now depends on the engineers not only using the same procedural model, but also the systems and their structures being subject to a globally uniform characteristic or specification. Only in this way can a large number of errors be avoided in the early phases of product development at distributed manufacturing locations. The constant increase in recalls in the automotive industry (FAZ 2005; Pankow 2016) however, with a high probability, proves that the complex system "car" is not developed and built according to this knowledge.

As part of a joint research project on the topic of "Holistic Innovation Processes in Modular Corporate Networks" (GINA), a new problem-solving cycle was developed, which on the one hand promotes the innovation process and on the other hand better corresponds to the current conditions of globalization (Franke 2005). Basically, the authors assume that each phase of the systematic procedural model, shown in Fig. 2.16, is run through several times, or that a specific method can be assigned to each step. These are both methodologies with high universality, i.e., they are suitable for several steps, and methods that are characterized by a deep effect, i.e., they are only suitable for special steps. As a result, software was developed that supports the designer and developer in the targeted selection of methods. The basic principles of systematic thinking and action are embedded. Basically, the mentioned procedure is based on VDI 2221. Consequently, the

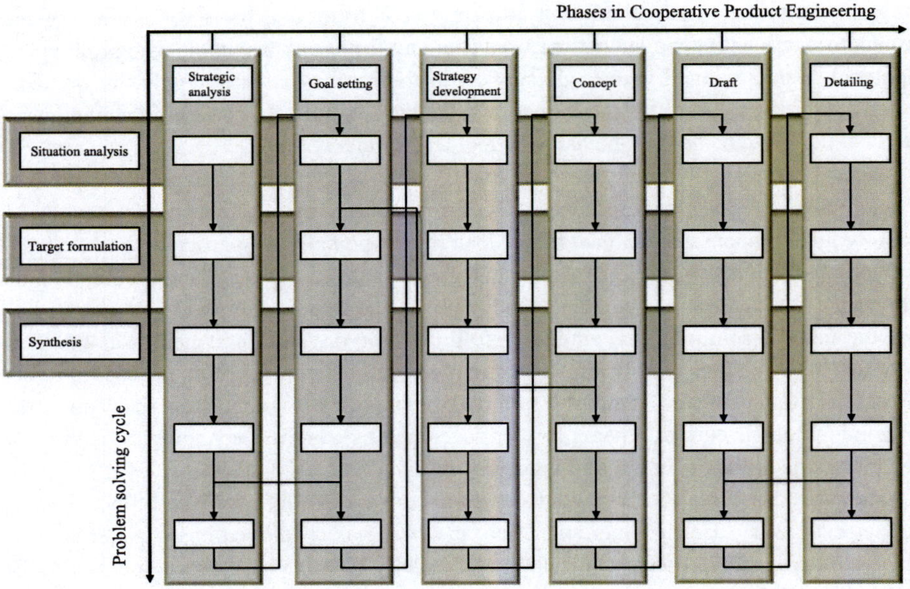

Fig. 2.16 Systematic procedural model. (Based on VDI 2221 1993, p. 3)

evaluations already made on this can be transferred to the procedural model of Fig. 1.15. However, thinking in systems is not explicitly required. Systems are a variable that goes into the problem-solving cycle. The selection of methods and their application in conjunction with the principles of systematic thinking and action are very advantageous for systematic problem solving.

Currently, a number of completely new procedural models are being developed in product development. Due to resource conservation, products are being miniaturized. This in turn means that the system boundaries in the product system are merging. An expression of this is mechatronics. Numerous new procedures have also been developed for the development of mechatronic systems. They serve the purpose of computer scientists, electrical engineers, and mechanical engineers developing a mechatronic product together, i.e., using a uniform procedure.

So the V model (see Fig. 2.17) according to VDI 2206 is a standardized approach to bundle the various procedures (Frank and Gausemeier 2004; Gausemeier and Bigl 2006; Friedrich et al. 2010) for the development of mechatronic systems.

The V-model uses the SE approach by proposing system design, model building and analysis with their domain-specific designs, and system integration as essential steps after requirement determination. However, how systems are to be delimited, how the model building is transdisciplinary, remains open in this standardized V model. For this

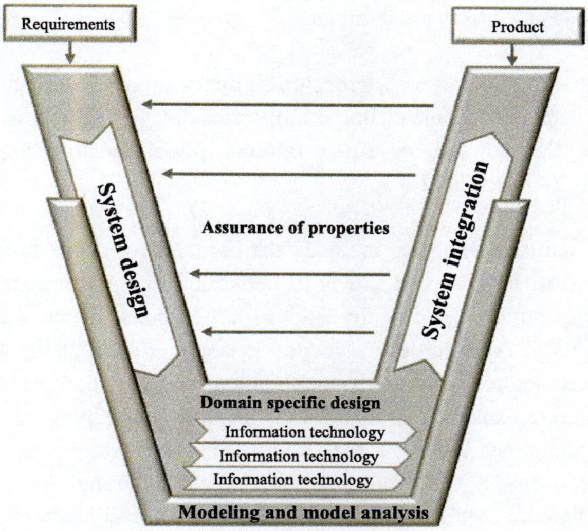

Fig. 2.17 Standardized V model according to VDI 2206. (After Ott 2009, p. 106)

reason, OTT developed a modified procedural model, which allows system thinking, traceability and transparency or a transdisciplinary approach and integrates the principles of systematic thinking and action in a differentiated way (Ott 2009). Parallel to this, BEIER developed a Systems Engineering process for mechatronic systems (Beier et al. 2012). This interlocks the steps of the V model according to VDI 2206 with those of Systems Engineering according to INCOSE (see Fig. 2.18). A product requirement model is created from the requirement analysis. Based on this, the function model is derived, which characterizes the most important functions and their interaction. Subsequently, the function carriers (components) are selected to develop the system architecture. This step is realized in parallel by the mechanics, electronics, and software development and subsequently the system verification is carried out in the multidisciplinary team. But this is exactly where the problem lies. On the one hand, three models, i.e., the requirement, function, and component models are created, which are not interlocked and are not demonstrably documented and improved over the system development process. On the other hand, the system architecture is not created in a transdisciplinary team but in a discipline-specific manner, which demonstrably leads to problems, as the many recalls in the automotive industry alone show. Smart Engineering tries to provide an answer to these questions (Anderl 2012). The term "Smart Engineering" stands for an interdisciplinary, networked, intelligent, clever approach in product development to successfully enable attractive innovations in future intelligent, networked products (Anderl 2012, p. 5). For this, ANDERL et al. demand a new cross-disciplinary development methodology, a corresponding model building, and corresponding IT tools.

Particularly essential are:

- the demand for a cross-disciplinary model building, which synergistically and transparently connects the multitude of discipline-specific models, and
- the demand for a new interdisciplinary method approach for product development (Anderl 2012).

Smart Engineering includes the demand for a systemic or integrated product development. Its core concern is to network strategic product planning with product development more specifically and to better consider the interactions between product and product system development (Gausemeier et al. 2014). Product development in this context includes: "the cross-disciplinary product conception, the discipline-specific design and the subsequent elaboration as well as the integration of the results of the individual disciplines into an overall solution" (Gausemeier et al. 2014, p. 216). In product system conception, "the four aspects: process planning, work equipment planning, workplace planning, and production logistics are to be considered integratively" (Gausemeier et al. 2014, p. 216).

In summary, it can be assessed that the procedural models of product development are indeed very diverse, but they refer to the product system without explicitly referring to the system thinking of the SE approach. Generalizable steps of SE can be found in the multitude of procedural models, especially in the development of mechatronic products, such as system design, model building, and system integration. Nevertheless, GAUSEMEIER (Gausemeier et al. 2014) states that the handover documents between Strategic

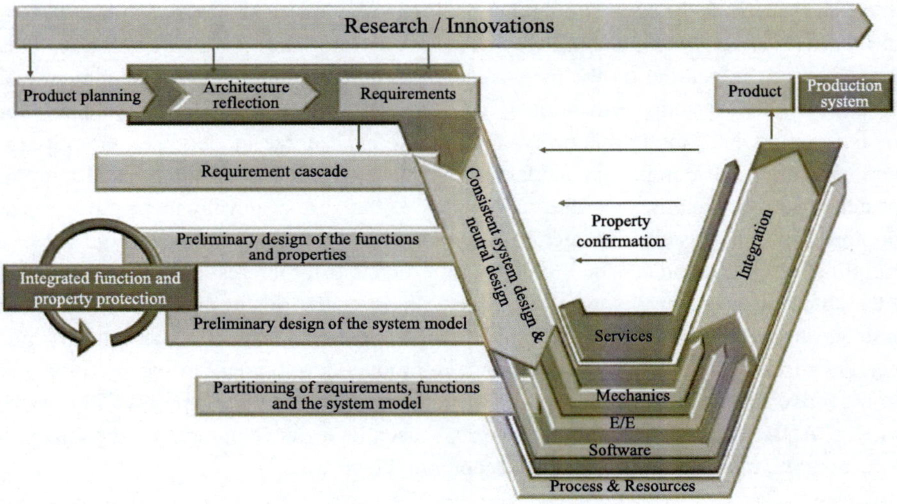

Fig. 2.18 Systems Engineering Process for mechatronic systems. (Based on Beier et al. 2012)

Product Planning and Product Development are not uniformly defined (see Fig. 2.19). A possible solution to this could be a standardized system model that is demonstrably maintained and documented. Almost all procedural models of product development refer in different ways to the observance of the basic principles of systematic thinking and action. Since software development is of great importance in the design of mechatronic systems, but its development is still domain-specific, the procedural models of software development will be examined more closely in the following in order to gain insights for a domain-overlapping approach, as is required for mechatronic systems.

B) Application of SE in Software Development

SE dominates in computer science and software engineering and is modified according to the specific field (Sommerville 2007). Evidently, problems in software development can be described and solved more easily using modified SE approaches. Thus, the IEEE 1220–1994 recommends the following phases for the development of computer systems:

- System definition,
- Preliminary design,
- Detailed design,
- Production,
- Introduction,
- Training,
- Maintenance and
- Decommissioning.

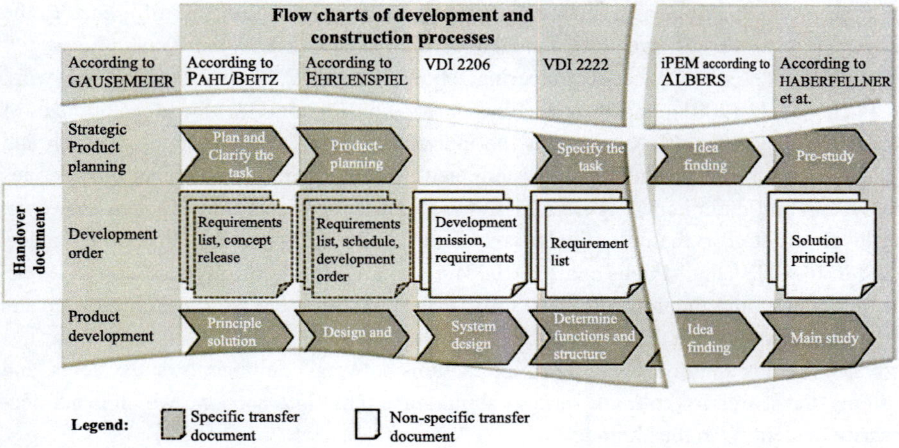

Fig. 2.19 Excerpt of analyzed procedural models in the context of product development of mechatronic systems. (Based on Gausemeier et al. 2014, p. 128)

Here too, the problem-solving cycle is applied to each phase. It thus includes the analysis and validation of requirements, the derivation of functions from the requirements, and their analysis and validation. While the SE approach includes a system definition, a connection between system delimitation and the problem-solving approach is implicit because the software itself is supposed to contribute to problem-solving for the customer. However, information on the nature and manner of system description is missing. Requirement determination takes place both in the system definition phase and in the phases of preliminary design, detailed design, production, etc. Moreover, each of these phases includes the development and validation of the function. Logically, a specification of the system definition would therefore belong in each problem-solving cycle phase. However, this is not explicitly stated.

The classic phase model according to SommerVille, which does not primarily refer to technical systems but exclusively to software development, is based on SE. Consequently, the software is referred to as a system. The procedure includes the following steps, which are linked iteratively:

- Requirement definition,
- System and software design,
- Implementation and testing of the components,
- Integration and system testing and
- Use and maintenance (Sommerville 2007).

The spiral model according to Balzert for software development, shown in Fig. 2.19, has only four phases. In contrast to SOMMERVILLE, BALZERT produces alternative solution variants. However, no system delimitation is made. The representation of the iteration process is done as a loop. Nevertheless, there is no significant difference to the classic SE or to the iterative phase model according to SOMMERVILLE.

The V-model of software engineering, shown in Fig. 2.20, is based on the work of Fuchs et al. (2001) and was the basis for the V-model for the development of mechatronic systems. It characterizes the individual phases of the system approach and links them with the system, the subsystem, and its components. This model thus defines the system and delimits it. Figure 2.21 shows the three levels of the V-model. However, the level of detail in the definition of a system and its subsystems is difficult to understand in FUCHS et al. (Fuchs et al. 2001). If a system consists of more than three levels, the V-model would theoretically need to be extended. But there is no description for the iteration loops between the levels or between the phases.

Since there is a high probability of interactions between components, subsystems, and systems, the search for solution variants should take these into account, which is not necessarily evident from the V-model.

In relation to software engineering and computer science, it can be summarized that both use system thinking, but developed a multitude of field-specific procedural models.

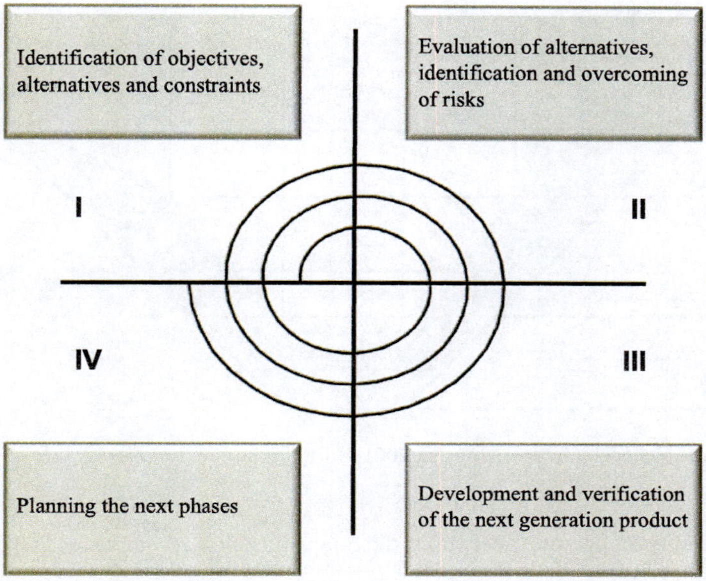

Fig. 2.20 The spiral model. (After Balzert 1998, 1996)

The interaction between the system model and the procedural concept is described. It is not clear from the literature to what extent these are transdisciplinary, traceable, transparent in shaping the problem-solving cycle, and observe the basic principles of systematic thinking and action. Especially due to the challenges of Industry 4.0, work is currently being done on how complex systems can be sensibly structured using, among other things, the Systems of Systems (SoS) approach (Warnecke 2015) and how data exchange between the various subsystems is guaranteed (Alt 2015).

C) Modified SE Approach in Enterprise Modeling

Manufacturing-Systems-Engineering focuses on the manufacturing factory (Meier et al. 2002). Consequently, Mertens also understands manufacturing engineering as the sub-discipline of SE that specifically deals with the production of goods, while he fundamentally shares the view of Adam and Scheer (see Adam 1997; Scheer 2001). HABERFELLNER, on the other hand, believes that SE must be suitable for the design of both product systems and factory systems (Haberfellner et al. 2019). Factory planning is increasingly focusing on SE (Wiendahl et al. 2009; Müller et al. 2012). In this context, ACKERMANN, NÄSER, and KING-KORDI present the factory system as a sociotechnical system (Ackermann 2007; Näser 2007; King-Kordi 2010). However, their views on the elements of this system and its determinacy to the system environment differ. For

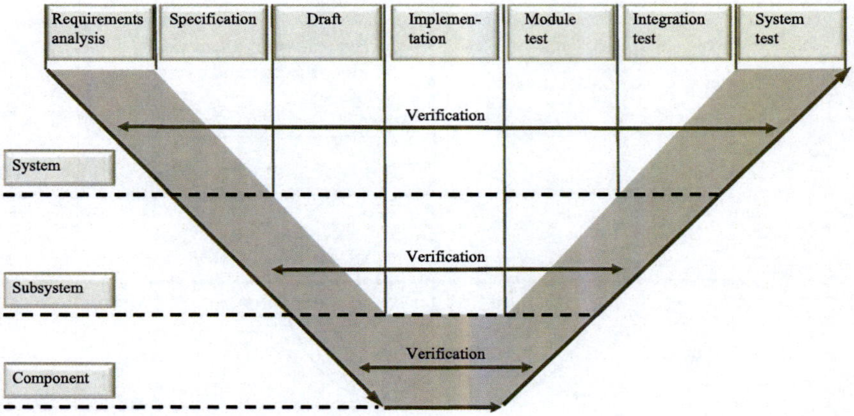

Fig. 2.21 The V-model. (After Fuchs et al. 2001)

the consideration of corporate networks, the SoS approach is recommended (Weiss 2013; Schlüter et al. 2019).

Schenk and WirtH (see Schenk 2004) very closely link the SE approach with the development and realization of product systems, as can be seen from Fig. 2.22.

They assume that a definition of the system already exists. The system is now to be designed according to the requirements, here referred to as project kick-off. This design process is time-determined and can be continued indefinitely.

Unlike other approaches, this approach explicitly points out the time determinacy both with regard to the system itself and with regard to the design cycle of the SE approach. However, how the delimitation, design, and introduction of the systems are carried out cannot be fully understood. The outlined procedures make the design process of the "company" or "factory" system partially transparent. To what extent this is traceable and observes the principles of systemic thinking and action is not indicated. Similar approaches are pursued by WIENDAHL, SCHUH, and KAMPKER (Schuh 2004; Wiendahl et al. 2009; Müller et al. 2012; Kampker et al. 2014).

D) Application of the SE Approach in Requirements Engineering & Management

Requirements management (Requirements Engineering and Requirements Management) (Hull et al. 2005; Pohl and Rupp 2015; Pohl 2016) refers exclusively to the requirement-compliant design of systems, i.e., the analysis of requirements, their comparative consideration, and their consideration in system design. Although requirements management is a general approach for the design of systems of any kind, it also primarily focuses on software development (see Hull et al. 2005; Sommerville 2005; Davis et al. 2007; Marques-Lucena et al. 2015; ISO/IEC/IEEE 29148 2018; Partsch 2010).

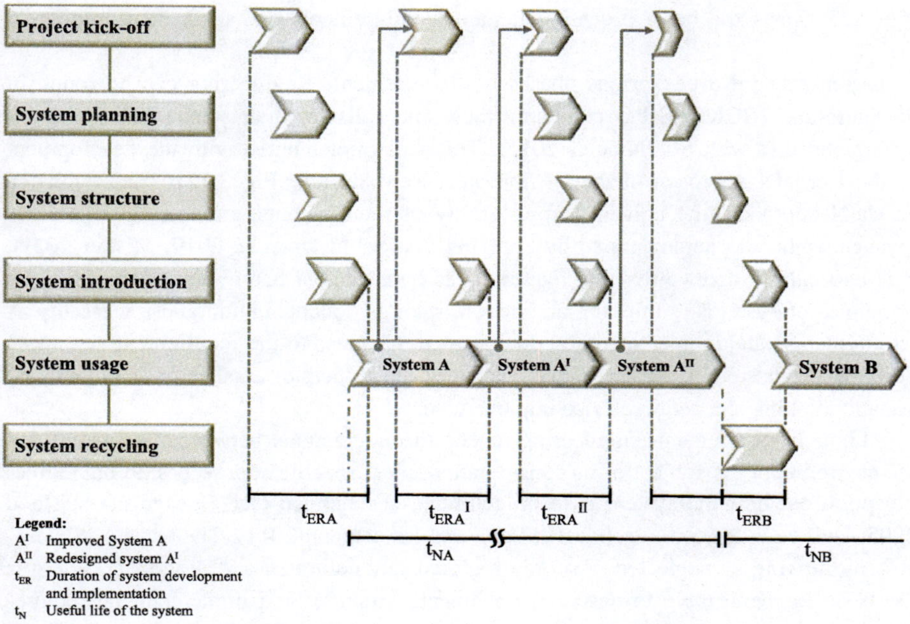

Fig. 2.22 Classification of the development, realization, and use of production systems. (After Schenk 2004, p. 120)

Product development is also unthinkable without the analysis and structuring, as well as the implementation of the requirements in a product solution concept (Verstegen 2004). Consequently, it is hardly explainable why requirements management developed as an independent discipline, detached from product development, software development, and SE.

In principle, requirements management focuses exclusively on the interaction of the system with its environment, thereby largely neglecting the elements within the system and their interaction, as SE in particular requires (see Rupp 2021; Pohl 2016; Pohl and Rupp 2015). It is seen as a link between the disciplines of the product life cycle, such as change, verification, risk, variant, and supplier management (Kreß et al. 2005).

Thus, according to OTT (Ott 2009), requirements management includes the following essential activities:

- Capture of requirements,
- Analysis and evaluation of requirements,
- Documentation,
- Verification and validation and
- Adjustment of requirements.

Fig. 2.23 shows the basic process schema of requirements management according to OTT.

In general, however, various phases of Requirements Engineering can be found in the literature. NICKLAS has compared these and calls for their standardization and a synergistic link with SE (Nicklas 2016). This was implemented with the development of the ReMaiN approach in the DFG project ReMaiN (see Fig. 2.12). Thus, with the ReMaiN approach, the universal applicability of requirements engineering as well as managemenbt was demonstrated by applying SE (Schlueter et al. 2019; Mistler 2021). It is particularly noteworthy that requirements management has implemented the basic principles of systematic thinking and action, such as documentation and traceability. A requirements catalog is created and documented. Changes to the requirements are to be described consistently and traceably. The interface principle, another principle of systematic thinking and action, is also implemented.

Although requirements engineering makes the interactions between the system and the environment the subject of its consideration, most specific approaches do not explicitly point out how the system is to be delimited and defined (see Goeken 2006; Hood 2005; Hull et al. 2005; Partsch 2010; Ponn and Lindemann 2011). The basis of SE, i.e., system thinking, is neglected. For only the precisely delimited and described system is the basis for being able to describe, document, structure, and implement the requirements placed on this system exactly. Therefore, in the ReMaiN approach, system delimitation and definition are explicitly required at the beginning.

E) Application of the SE Approach in Safety Engineering

Safety Engineering is understood as a sub-discipline of SE, which specifically deals with safety-relevant analyses of systems (cf. Leveson 2001). As a result, the S T A M P model was developed. The acronym STAMP stands for: Systems, Theory, Accident, Modelling

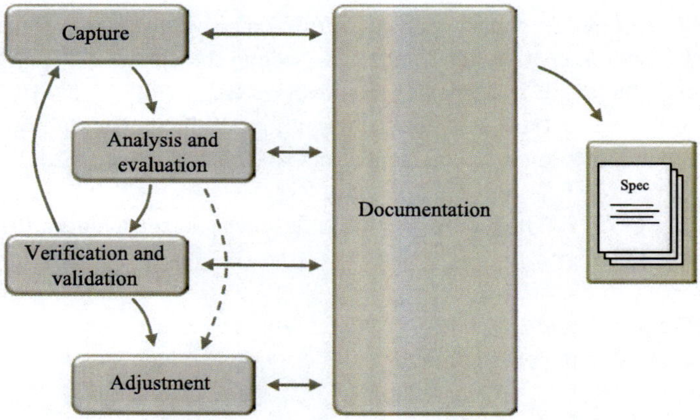

Fig. 2.23 Process schema of requirements management. (After Ott 2009, p. 96)

and Process, which refers to a method for modelling and analysing accidents using systems theory. The IEC 61508 focuses on the design of safety functions in electronic systems based on lifecycle models, as shown in Fig. 2.24.

This approach also does not rely on a precise system definition. The focus is on risk analysis. Although the procedure is based on SE, it primarily focuses on controlling the risks in the system.

While the IEC61508 focuses on the design of safety functions in electronic systems over the product lifecycle, the procedure model according to EN 954-1 includes safety-related components and controls. It mainly covers the steps shown in Fig. 2.25. Although there is no precise definition of system boundaries, the model begins with a hazard analysis and a risk assessment, which, however, change significantly with changing system boundaries. Thus, the hazard assessment of a car is likely to differ significantly from the hazard assessment of the same car's steering system. Safety-relevant requirements understandably focus first on protecting human health, animals, and the environment. However, a one-sided view of these requirements runs the risk of neglecting technical functions, social aspects, aesthetic aspects, and even efficiency.

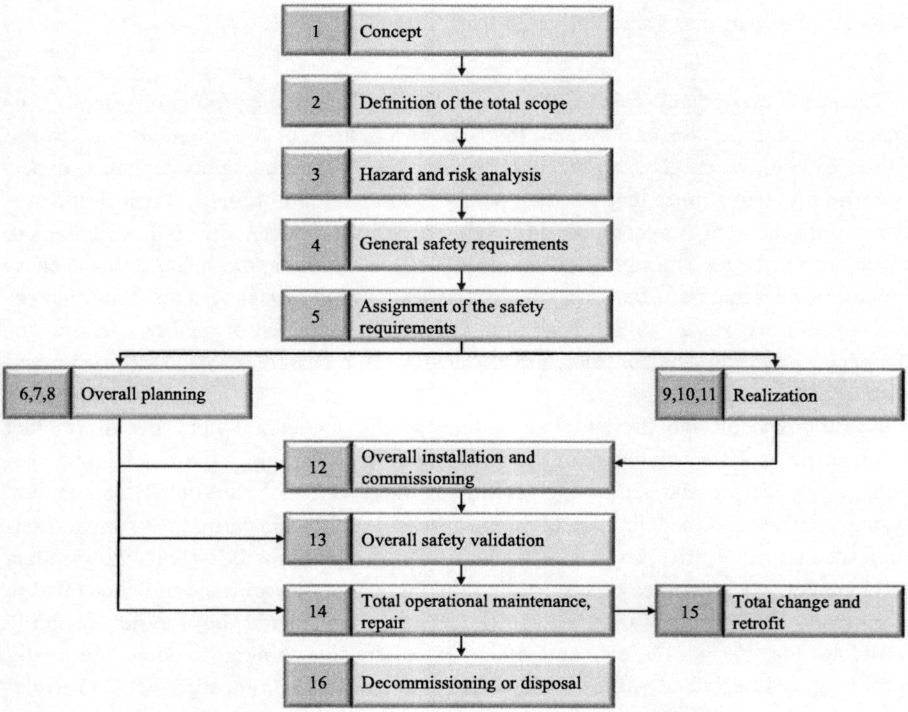

Fig. 2.24 Procedure model according to IEC 61508 (IEC 61508)

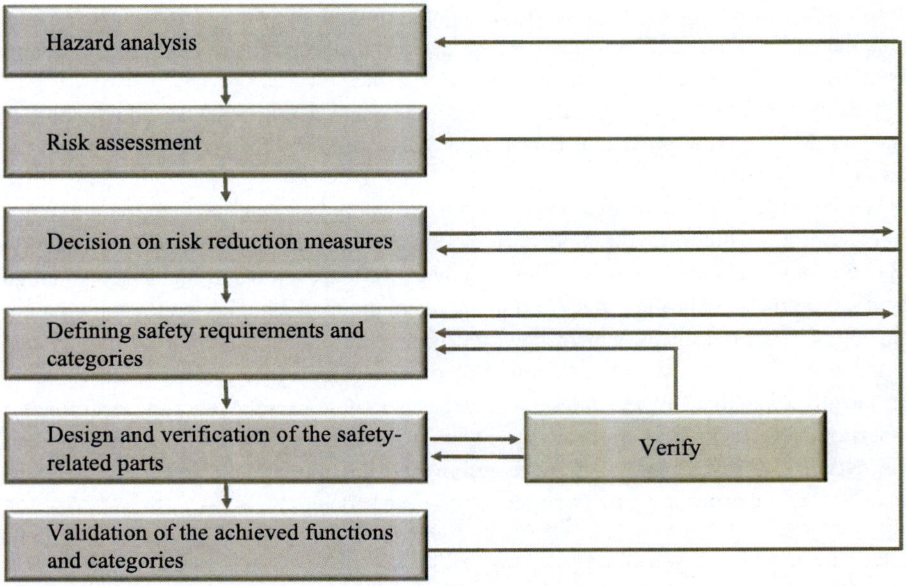

Fig. 2.25 Procedure model according to EN 954-1 (EN 954)

The procedure models in Safety Engineering also differ in the professional focus considered, i.e., the hazards of a system, the risks of a system, or the potentials for damage caused by the system. These procedure models require special safety technical expertise. The problem here is that the term safety is interpreted differently. As a result, further system-theoretical approaches to solving security or safety aspects have emerged, which need to be reconnected (Beterer et al. 2010). Regardless, the safety of a system is only one aspect that must be considered in the design of technical systems. Environmental compatibility, sustainability, economic efficiency, guarantee of quality, etc. are further aspects that must be considered in the analysis and design of systems (Beyerer and Winzer 2018).

In summary, all the presented discipline-specific procedural concepts of product development, software engineering, manufacturing engineering, requirements management & engineering and safety engineering are more or less based on SE approaches. System analysis and modeling are often mentioned, but they are carried out from a discipline-specific perspective, so their transdisciplinary use is limited. Especially in product development, the principles of systematic thinking and action are used to better manage complexity. Transparency and traceability are partly considered, but are not explicitly verifiable. For this reason, elements of individual specific thinking models and procedural concepts described earlier could well be suitable, in combination with other elements, for coping with the new dimensions of complexity.

2.4.3 Comparative Consideration of Universal and Special SE Approaches

The initial thesis was that SE could contribute to managing complexity in the new dimensions. The requirements for SE that resulted from this were:

- Thinking in systems,
- Presence of a thinking model, which can be used by representatives of all disciplines,
- Transdisciplinary applicability, transparency, and traceability of the procedural concept and
- Purposeful integration of the basic principles of systematic thinking and action into the procedural concept when creating the thinking model.

These requirements for SE served as the basis for comparing the various approaches. It had to be determined that a multitude of thinking models and procedural concepts of SE developed, which claim to be universal for developing solutions for any types of problems.

How they correspond to the new tendencies of complexity is shown in Table 2.2 at a glance.

In addition, specific thinking and procedural models based on SE have emerged. These are also reflected in the requirements in an overview in Table 2.3.

It can be observed that both the universal and the specific SE approaches make use of systems thinking to some extent (Balzert 1998; Fuchs et al. 2001; Sommerville 2007; Ott 2009). The majority of the considered universal SE approaches use thinking models and present their procedural concepts as transparent and transdisciplinary, and in part as traceable (Bahill and Gissing 1998; Haberfellner and Daenzer 1999; Lindemann 2016; IEEE 1220 2005; Sage and Rouse 2009; Haberfellner et al. 2019; Schlueter et al. 2019; Dumitrescu et al. 2021; Mistler 2021).

They partially also use basic principles of systemic thinking and action such as the basic principles of recurring reflection, from the whole to the detail, or of discursive action. This is also occasionally true for the special SE approaches (Schenk 2004; Sommerville 2007; Ott 2009; Pahl et al. 2005).

This multitude of thinking models and procedural concepts, which could only be presented here in an excerpt and without claim to completeness, make it more difficult for the potential user to make a selection. In addition, the new dimensions of complexity today and in the future require transdisciplinary thinking and action. This can only be done on a standardized basis.

For this, it must be possible for the teams, which today always consist of different specialists, to use a common thinking model when solving a problem. If the air conditioning system of a car is to be improved, the team must first have a common metamodel, i.e., a common image of the car. Only then can the specialist team turn to the

Table 2.2 Comparison of universal approaches based on SE

Requirement to SE / Source	Think in systems	Thinking model	Procedure concept: transparent	Procedure concept: traceable	Procedure concept: transdisciplinary	Basic principles of the systemic thinking and acting	Connection of thinking model and procedure concept: sequential	Connection of thinking model and procedure concept: iterative
Bahill and Gissing 1998	◐	○	●	◐	●	○	●	○
Sell 1989	◐	○	◐	○	◐	○	●	○
Haberfellner and Daenzer 1999	●	●	●	◐	●	○	○	◐
Wulf 2002	◐	●	●	◐	◐	○	●	○
IEEE 1220-2005	◐	●	●	◐	●	○	●	
Sage and Rouse 2009	●	●	●	◐	●	○	●	
Gausemeier et al. 2014	◐	◐	●	●	●	◐	○	◐
Lindemann 2016	◐	●	●	◐	●	●	●	
Ehrlenspiel and Meerkamm 2017	◐	●	●	◐	●	●	○	◐
Haberfellner et al. 2019	●	●	●	◐	●	◐	○	◐
Schlueter et al. 2019	●	●	●	●	●	●	●	
Dumitrescu et al. 2021	●	●						
Mistler 2021	●	●	●	●	●	●	●	●

legend: ○ not applicable ◐ partially applicable ● applicable

change of the subsystem "air conditioning", because it is now clear which other subsystems of the car the air conditioning is connected to. If this approach is taken, it cannot happen that in summer, when the air conditioning is running, the automatic handbrake, which is controlled by the engine speed, releases and the car, which was on a hill, starts rolling (Motor-Talk 2012).

In addition to the common thinking model, a standardized, modular approach to problem solving is also necessary. The change of the braking system and the climate system should be implemented in such a way that the various specialists first agree on how the problem analysis and basically the problem solution could look like. Only then does a discipline-specific search for solution alternatives seem sensible, while at the same time

Table 2.3 Comparison of specific approaches based on SE

Requirement to SE / Source	Thinking in Systems	Thinking model	Procedure concept: transparent	Procedure concept: traceable	Procedure concept: transdisciplinary	Basic principles of the systemic thinking and acting	Connection of thinking model and procedure concept: sequential	Connection of thinking model and procedure concept: iterative
Product development								
Pahl et al. 2005	◐	●	●	◐	●	●	●	○
VDI 2221	◐	◐	◐	◐	●	◐	○	○
Schnieder 2013	◐	◐	◐	◐	◐	◐	●	○
VDI 2206	◐	◐	◐	◐	●	◐	●	○
Software Engineering								
IEEE 1220	●	◐	◐	◐	◐	◐	◐	○
Sommerville 2007	●	◐	◐	◐	◐	◐	●	○
Fuchs et al. 2001	●	◐	◐	◐	○	◐	●	○
Balzert 1998	●	◐	◐	◐	◐	◐	●	○
Manufacturing Engineering								
Schenk et al. 2014	◐	◐	◐	◐	◐	◐	◐	◐
Wiendahl 2009	◐	◐	◐	◐	◐	◐	○	●
VDI 5200	◐	◐	●	◐	◐	◐	●	○
Grundig 2018	◐	◐	●	◐	◐	◐	●	◐
Requirements Engineering								
Ott 2009	●	●	●	●	●	●	●	○
Safety Engineering								
IEC 61508	○	◐	◐	◐	○	◐	●	○
EN ISO 13849-1	○	◐	◐	◐	○	◐	●	○

Legend: ○ not applicable ◐ partially applicable ● applicable

using the principles of systemic thinking and action. The corresponding transparency and traceability must be ensured so that at the "marriage" of the solution alternatives in the car, the entire team can think and act together. Consequently, a universal modular, standardizable, cross-disciplinary problem-solving approach is required. SE could be such an approach if it is possible to typify and bundle modules from the multitude of thinking models and procedural concepts in a standardized way through comparative considerations, so that a universal problem solution using a unified thinking model and procedural concept becomes possible for multidisciplinary teams. However, due to the diversity of SE approaches described here only in part, this has not yet been achieved.

Consequently, further demands on SE arise. It must be:

- modular,
- standardizable as well as
- universal and
- allow for specific problem solutions.

Can this claim already be achieved by modifying some SE approaches or does SE need to be fundamentally reformed? This question will be pursued in the following chapter.

2.5 SE and Possibilities of its Reformability

The great variety of SE approaches, which already existed in the past, is still present today, as illustrated in Fig. 2.26 at a glance. The potential users are more confused by this than encouraged to apply SE in today's times.

When solving problems, it is therefore difficult to decide which of the recommended approaches is most suitable.

Going further, selected SE approaches will be reconsidered with the aim of checking, through a comparative consideration, whether they contain modules that are used repeatedly. This can then be used to investigate whether the determined modules have the potential for generalization and standardization. This forms the basis for the restoration of a universal SE.

A similar approach was already applied by BAHILL and ARLT at the end of the nineties (Bahill and Gissing 1998; Arlt 1999). They were able to filter out core elements or similar steps of SE. ARLT subjected various differentiated methods of problem solving to a comparative consideration and came to the conclusion that the procedural concepts he considered basically show similar steps, as shown in Fig. 2.27.

Bahill and Gissing attempted something similar. Their comparison is based on about 15 methodological approaches to SE, which were, however, applied specifically, e.g., in product design, in shaping the product lifecycle, in business engineering, in scientific work and in re-engineering, as well as in the DEMING cycle (Plan, Do, Check, Act) and within the framework of the Business Excellence Model.

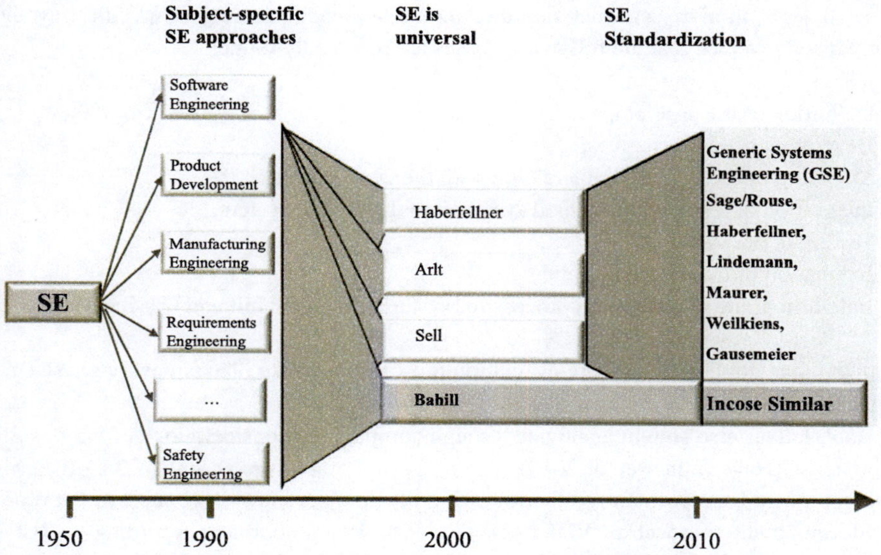

Fig. 2.26 SE over time. (After Mistler 2021, p. 22; Sitte and Winzer 2011)

Fig. 2.27 Overarching steps of a procedural model in SE. (After Arlt 1999)

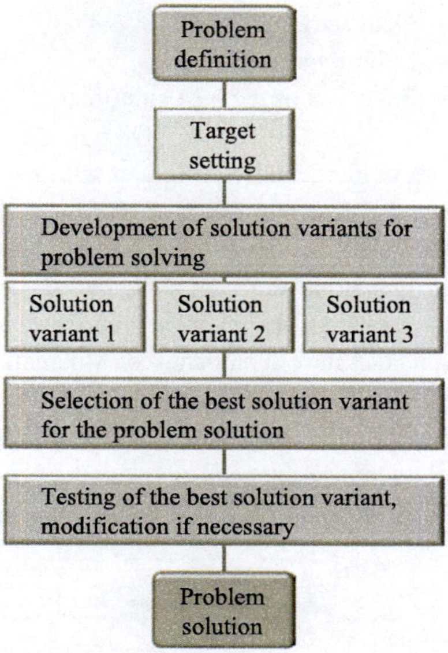

In all application cases, which modified the SE approach in a subject-specific way, the comparison revealed that the following phases are repeatedly found:

- Definition of the problem,
- Development of alternatives,
- Modeling of the system in interaction with the environment,
- Integration of the newly designed system into the overall system,
- Testing of the system,
- Evaluation of the test results and
- Initiation of the re-evaluation process for system design (Bahill and Gissing 1998).

Both authors emphasize that a re-evaluation must actually take place in every phase (see Fig. 2.28).

Haberfellner also sought analogiės through comparative considerations of various SE approaches (Haberfellner et al. 2019). The basis for comparison is, unlike BAHILL, his own approach. He presents it, as the following illustrations show, in contrast to the product development approach of VDI 2221, the work design approach according to REFA and the value analysis (Fig. 2.29, Tables 2.4 and 2.5).

The comparative considerations lead Haberfellner to the realization that the problem-solving cycle, obers consisting of:

- Goal setting,
- Solution search and
- Selection of the best solution,

can fundamentally be linked with systems thinking (Haberfellner et al. 2019; Haberfellner and Daenzer 2002). This is a tool to narrow down problems, to order and structure the facts.

If systems thinking is neglected, i.e. if the problem is not clearly assigned to a system, then efficient problem solving is difficult. For example, the problem "O-Bus does not run" can have many causes. A demarcation and designation of the various subsystems of the O-Bus can contribute to initially assigning the problem to one or more subsystems. In this case, it was the drive system. It is closely interrelated with other subsystems via plug connections and screw connections (interaction of the system with its environment).

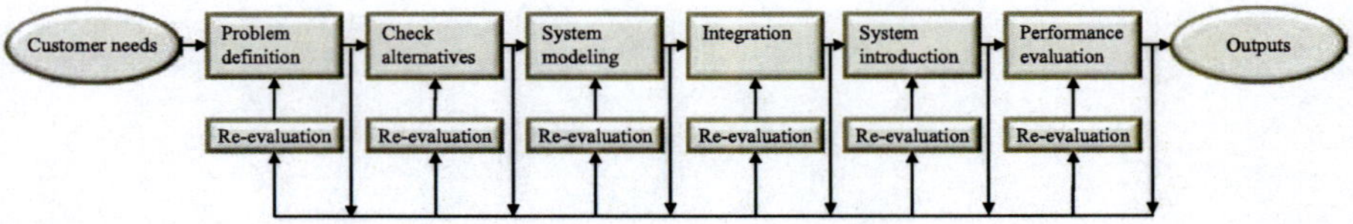

Fig. 2.28 Similar Process. (After Bahill and Gissing 1998)

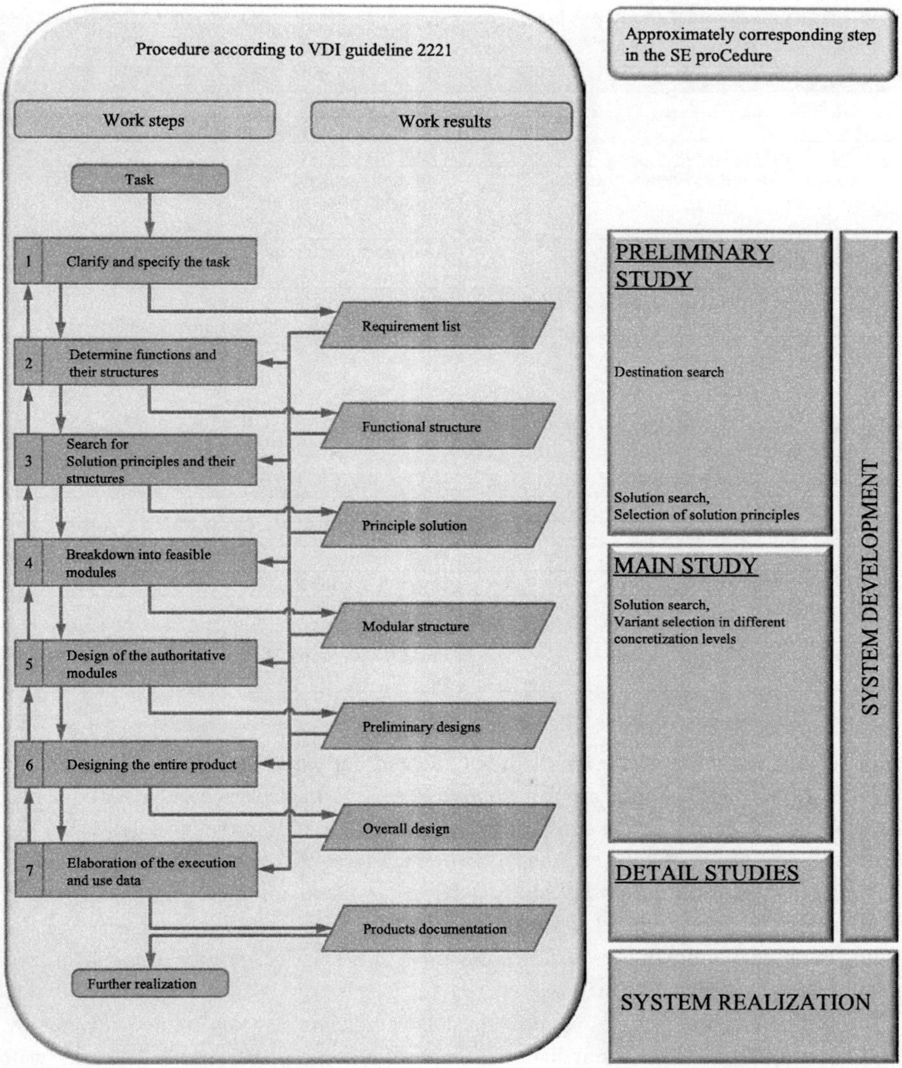

Fig. 2.29 Comparison of the procedure of the VDI guideline 2221 and SE procedure (Haberfellner et al. 2019, p. 65)

Consequently, these must first be checked before the drive itself is removed and examined on a special test stand. In the example described here, unfortunately, the approach was not based on systems thinking, but a mechanic immediately removed the drive. When the drive was on the test stand, it had to be determined that it worked flawlessly.

Table 2.4 Comparison of the REFA 6-step method with the classic SE concept. (After Haberfellner and Daenzer 2002, p. 62)

Steps of the REFA 6-step method	Corresponding steps in the SE concept
1. Set goals	Goal
2. Delimit task	Project order possibly partial situation analysis
3. Search ideal solution	Synthesis of solutions
4. Collect data and develop workable solutions	Situation analysis Synthesis of solutions Analysis of solutions
5. Select optimal solutions	Analysis of solutions Evaluation
6. Control solutions and target fulfillment	Phases: System construction, introduction, completion, incl. project management

Table 2.5 Comparison of the WA work plan according to DIN 69 910 with the classic SE concept. (After Haberfellner and Daenzer 2002, p. 62)

Steps in the WA work plan according to DIN 69910	Corresponding steps in the SE concept
1. Prepare project	Project planning
2. Analyze object situation	Situation analysis
3. Describe target state	Goal formulation
4. Develop solution ideas	Concept synthesis
5. Set solutions	Concept analysis (possibly additional synthesis steps), evaluation, decision
6. Realize solutions	Phases: System construction, introduction, completion, incl. project management

Consequently, the drive was reinstalled, but still did not work, because, as it turned out much later, the plug connections that connect the drive to the electrical control of the O-Bus were loose and thus the current flow was disturbed. If an effective solution for the sustainable avoidance of such errors is to be found, systems thinking, as HABERFELLNER emphasizes, must be closely linked with the problem-solving cycle. Furthermore, a standardized thinking model must exist for the system, in this specific case for the O-Bus, in order to quickly narrow down the error. WEILKIENS speaks of a generalized thinking model (Weilkiens 2007).

The problem-solving cycle, in turn, must be guided in a goal-oriented manner by project management. This is emphatically underscored by SAGE and ROUSE as well as HABERFELLNER (Haberfellner et al. 2019). These authors particularly emphasize that SE must be guided, steered and controlled, otherwise it does not lead to success in a goal-oriented manner. Haberfellner's approach clearly structures SE as a kind of universal concept for coping with complexity. It is divided into:

- the basic philosophy of systematic thinking and action, which is underpinned and supported by
- systems thinking and
- the problem-solving oriented procedual concept.

This basic structure of SE into the SE philosophy, systems thinking and the problem-solving procedure, as well as the coupling with the techniques of system design and project management, proposed by HABERFELLNER already in the 90s, is taken up by WEILKIENS, SAGE and ROUSE and further developed by him in 2012 (Haberfellner et al. 2019). The aforementioned are interested in restoring the lost universal character of SE. Their effort is based on the realization that the current challenges in the development of new products and technologies could be better solved by a universal approach to SE. WEILKIENS further developed this idea by building a modular concept for SE, which is illustrated in Fig. 2.30.

His modules can optionally be used to solve specific problems, whether for the new development of a drive or the redesign of a logistical facility or the development of a service concept, i.e. these modules contribute to supporting the problem-solving algorithm in a goal-oriented manner. The modular system according to WEILKIENS does not necessarily prescribe when which module must be used for problem solving. However, he emphasizes that the individual modules must be applied problem-specifically. This modular concept contributes to universally adapting SE to any kind of problems to be solved, in order to develop efficient solutions in a goal-oriented manner.

In summary, it can be stated that the SE approach is quite suitable for coping with the new dimensions of complexity. Due to systems thinking, it is possible to structure and simplify complex facts in a transdisciplinary way. Because all disciplines depict the system in a system description language, i.e. develop a thinking model and use it during the problem-solving process, transdisciplinary work is possible.

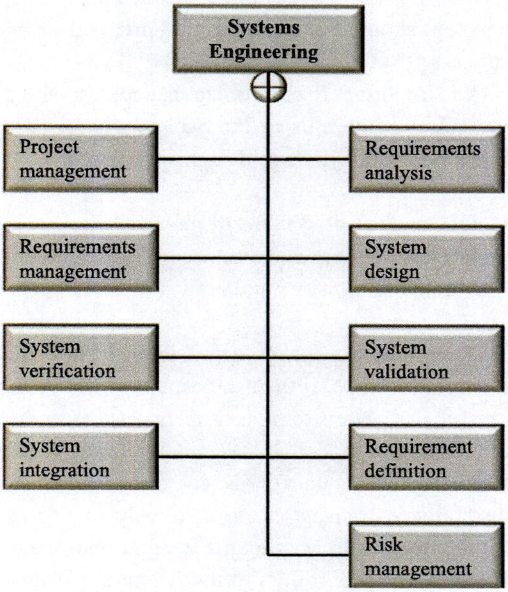

Fig. 2.30 Modular concept of SE. (After Weilkiens 2007, p. 9)

Thus, the basis can be created to transparently and traceably realize and understand the process of designing the system together for each of these disciplines. Transferred to the example of the logistical facility, this means: If it is successful in the design process to depict the logistical facility as a total system in a common language—i.e. everyone in the team designates the subsystems, the components, the elements, the functions and their interrelationships in the same way -, then the logistician, the computer scientist, the mechanical engineer, the business economist and the electrical engineer, to name just a few disciplines, can work together on a meta-thinking model. The resulting thinking model of the logistical facility is now understood in the same way by all team members. It can be used for the discipline-specific development of solution variants. Thus, on this basis, various solution concepts for the logistical facility can be selected, tested and the best solution implemented in a transparent and traceable manner for all participants of the interdisciplinary team. However, there are different approach concepts in SE for this.

This must be changed and a universal approach concept must be created, as demanded by BAHILL and GISSING. However, this also requires that the principles of systematic thinking and action must be integrated into the SE procedural concept in a goal-oriented manner. The principle "from the rough to the detail" can help, for example, to get an overall view of the logistical facility and subsequently to work out a multi-level model as a system image of it. The integration of further principles and corresponding methodsand procedures into the problem-solving algorithm depends on the problem to be solved. Haberfellner emphasizes that the SE procedural concept can be designed more systematically if it is coupled with the methods and procedures of modern project management. This view must definitely be followed. Consequently, the design process of the logistical facility is to be planned and controlled as a project. In doing so, methods of project management should be used in a goal-oriented manner, as Haberfellner, SAGE and ROUSE suggest.

It is important to emphasize that coping with the new dimensions of complexity of the present and future using the SE approach necessarily requires the realization of the following demands in its re-design:

- Preservation of systems thinking,
- Interdisciplinary uniform description of the thinking model and
- Modularization or standardization of the procedure in the problem-solving process.

Systems thinking makes it possible to make the systems with their elements and their interrelationships to themselves and to the system environment transparent. If it is successful that interdisciplinary teams use systems thinking together and develop a common thinking model (meta-model) of the considered system, for example the logistical facility, which is understood by every specialist of this team, then the basis for a transdisciplinary, transparent problem-solving process is given. This can be supported in a discipline-specific way by the integration of simulation methods, appropriate testing procedures etc. This results in the demand that there must be methods and procedures that support systems thinking, the development of the thinking model and the approach concept in a problem-specific way.

The re-design of the thinking model in SE must therefore implement the following goals:

- The system model must make all elements and their interrelationships transparent.
- It must be able to be created and changed transdisciplinarily.
- It should consist of a few standardized views.
- Appropriate methods and procedures are to be linked with systems thinking.

The principles of systematic thinking and action are to be included in the creation of the thinking model.

The changes to the thinking model, which arise during the problem-solving process, must be traceable.

The re-design of the SE approach concept must implement the following demands:

- It must be universal.
- It is to be built modularly.
- It is to be designed in a standardized way.
- The principles of systematic thinking and action are to be integrated.
- It must be controlled by project management.
- It is to be modified in a problem-oriented way by coupling with specific methods, procedures and (IT) tools.

If it is successful to implement these demands on the thinking model and on the procedural concept in the reform of the SE approach, a basis could be created to guarantee the complexity of the present and future in the new dimensions.

References

acatech (2021): Verantwortung in Unternehmen und Institutionen für eine nachhaltige Technikentwicklung. München: (acatech POSITION).

acatech; Körber-Stiftung; Universität Stuttgart (2021): TechnikRadar 2021. Stakeholderperspektiven. Langenhagen: Gutenberg Beuys Feindruckerei GmbH.

Ackermann, Jörg (2007): Modellierung, Planung und Gestaltung der Logistikstrukturen kompetenzzellenbasierter Netze. Techn. Univ., Diss.-Chemnitz, 2007. Chemnitz: IBF (Wissenschaftliche Schriftenreihe des Institutes für Betriebswissenschaften und Fabriksysteme, 59).

Adam, Dietrich (1997): Planung und Entscheidung. Modelle – Ziele – Methoden; mit Fallstudien und Lösungen. 4., vollst. überarb. und wesentlich erw. Aufl., Nachdr. Wiesbaden: Gabler (Gabler-Lehrbuch). Online verfügbar unter http://www.gbv.de/dms/ilmenau/toc/239223624.PDF.

Albers, Albert; Matthiesen, Sven; Bursac, Nikola; Moeser, Georg; Schmidt, Sebastian; Lüdcke, Robert et al. (2014): Abstraktionsgrade der Systemmodellierung-von der Sprache zur Anwendung. In: *Tag des Systems Engineering.*, pp. 183–192.

Alt, U. (2015): Modellbasiertes Systems Engineering und seine Technologien als Schlüssel für Industrie 4.0. In: Maik S. Maurer, Jutta Abulawi und Sven-Olaf Schulze (Eds.): Tag des Systems Engineering. Bremen, 12.–14. November 2014; [TdSE]. München: Hanser, pp. 3–10.

Anderl, Reiner (2012): Smart Engineering. Interdisziplinäre Produktentstehung. Berlin: Springer Vieweg (acatech DISKUSSION).

Arlt, Gregor (1999): Systemansatz eines produkt- und ablauforientierten Qualitätsmanagements durch Integration der Systemtechnik. Univ., Diss.-Duisburg, 1999. Als Ms. gedr. Düsseldorf: VDI-Verl. (Fortschritt-Berichte VDIReihe 16, Technik und Wirtschaft 109).

Arnaut, Bruno M.; Ferrari, Denise B.; Oliveira e Souza, Marcelo Lopes de (2016): A requirements engineering and management process in concept phase of complex systems. In: 2016 IEEE International Symposium on Systems Engineering (ISSE). 2016 IEEE International Symposium on Systems Engineering (ISSE). Edinburgh, United Kingdom, 03.10.2016–05.10.2016: IEEE, pp. 1–6.

Badke-Schaub, Petra; Frankenberger, Eckart (2004): Management kritischer situationen: Springer.

Bahill, A. Terry; Gissing, Bruce (1998): Re-evaluating systems engineering concepts using systems thinking. In: *IEEE Transactions on Systems, Man, and Cybernetics, Part C (Applications and Reviews)* 28 (4), pp. 516–527.

Balzert, Helmut (1998 // 1996): Lehrbuch der Software-Technik. [1.–2. Aufl.] // [Ausg. in 2 Bd.]. 2 Bände. Heidelberg: Spektrum Akad. Verl.; Spektrum Akad. Verl (Lehrbücher der Informatik, 2).

Bauer, Horst (2003): Kraftfahrtechnisches Taschenbuch. 25., überarb. und erw. Aufl. Wiesbaden: Vieweg.

Bauernhansl, Thomas (Hg.) (2014): Industrie 4.0 in Produktion, Automatisierung und Logistik. Anwendung, Technologien und Migration. Wiesbaden: Springer Vieweg.

Baumann, Georg; Erlenspiel, Klaus (1981): Entwicklung einer Methode zur Erarbeitung von Kostenfrühermittlungssystemen beim Maschinenentwurf. Für die Zeit vom 01.01.1979 bis 31.12.1981; Bericht zum DFG-Forschungsvorhaben Nr. 46/15. München.

Beier, G.; Rothenburg, U.; Woll, R.; Stark, R. (2012): Durchgängige Entwicklung mit erlebbaren Prototypen. Modellbasiertes Systems Engineering. In: *Digital Engineering* 3 (2012), pp. 14–17.

Bender, Beate; Gericke, Kilian (2021): Pahl/Beitz Konstruktionslehre. Methoden und Anwendung erfolgreicher Produktentwicklung. 9th ed. 2021. Berlin, Heidelberg: Springer Berlin Heidelberg; Imprint: Springer Vieweg.

Benno, Stützel; Borchart, Laura; Illa, Thomas; Gerling, Christoph (2018): Systems Engineering in Deutschland. Die deutsche Unternehmenslandschaft im Vergleich. Studie. Frankfurt am Main: Dialogistiker GmbH.

Beterer, Jürgen; Geisler, Jürgen; Dahlem, Anna; Winzer, Petra (2010): Sicherheit: Systemanalyse und-design. In: Petra Winzer, Friedrich-Wilhelm Bach und Eckehard Schnieder (Eds.): Sicherheitsforschung. Chancen und Perspektiven // Chancen und Perspektiven (acatech DISKUTIERT). 1st ed. Berlin, Heidelberg: Springer; Springer-Verlag, pp. 39–72.

Beyerer, Jürgen; Winzer, Petra (Eds.) (2018): Beiträge zu einer Systemtheorie Sicherheit. Deutsche Akademie der Technikwissenschaften; Herbert Utz Verlag GmbH. München: Herbert Utz Verlag GmbH (acatech DISKUSSION). Online verfügbar unter http://web.archive.org/web/20181117000045/http://www.acatech.de/wp-content/uploads/2018/10/acatech_DISKUSSION_Systemtheorie_WEB.pdf.

Bielefeld, Ovidiu (2020): Entwicklung einer Methodik für eine modellbasierte und ganzheitliche Fehleranalyse. Dissertation. Wuppertal: Bergische Universität Wuppertal.

Bing, Torsten (2001): Zeitduplexbasierte Mobilkommunikation, untersucht am Beispiel eines TD-CDMA-Mobilfunksystems.

Böhm, R.; Fuchs, E. (2002): System-Entwicklung in der Wirtschaftsinformatik. Zürich: vdf Hochschulverlag AG.

Boyd, Drew; Goldenberg, Jacob (2019): Inside the Box. Berlin, Heidelberg: Springer.

Bruijn, Hans de; Herder, Paulien M. (2009): System and actor perspectives on sociotechnical systems. In: *IEEE Transactions on systems, man, and cybernetics-part A: Systems and Humans* 39 (5), p. 981–992.

colourbox (2022): Innovation communication. Online verfügbar unter http://de.fotolia.com/Download, zuletzt geprüft am 07.11.2022.

DAG (2010): Defense Acquisition Guidebook 2010. Online verfügbar unter https://dag.dau.mil, zuletzt geprüft am 11.03.2013.

Dalhöfer, Jörg; Rall, Klaus (2009): Komplexitätsbewertung indirekter Geschäftsprozesse. Techn. Univ., Institut für Werkzeugmaschinen, Roboter und Montageanlagen, Diss.-Hamburg-Harburg, 2009. 1. Aufl. Aachen: Shaker (Schriftenreihe des Instituts für Werkzeugmaschinen, Roboter und Montageanlagen der Technischen Universität Hamburg-Harburg, 19).

Danner, S. (1996): Ganzheitliches Anforderungsmanagement für marktorientierte Entwicklungsprozesse. München, Wien: Hanser.

Davis, A.; Dieste, O.; Hickey, A.; Juristo, N.; Moreno, A. M. (2006): Effectiveness of Requirements Elicitation Techniques: Empirical Results Derived from a Systematic Review. In: 14th IEEE International Requirements Engineering Conference (RE'06). 14th IEEE International Requirements Engineering Conference. Minneapolis/St. Paul, MN, 11.09.2006–15.09.2006: IEEE, pp. 179–188.

Davis, Alan M.; Hickey, Ann M.; Zweig, Ann S. (2007): Requirements Management in a Project Management Context. In: Peter W. G. Morris und Jeffrey K. Pinto (Eds.): The Wiley Guide to Project Technology, Supply Chain, and Procurement Management. Hoboken, N.J: John Wiley & Sons, Inc., pp. 1–31.

Detel, Wolfgang (2005): Aristoteles. 1st ed. Leipzig: Reclam.

Detel, Wolfgang; Wildberger, Jula; others (2009): Aristoteles, Metaphysik, Bücher VII und VIII: [Text und Kommentar; Griechisch-Deutsch]. In: *Suhrkamp-Studienbibliothek*.

Dimario, M. J. (2010): Systems of Systems Collaboration Formation. In: *Systems Research Seriells* VOL. 1.1 Singapur.

Domenico, Manlio de; Sayama, Hiroki (2019): Complexity Explained.

Dörner, Dietrich (2007): Die Logik des Misslingens. Strategisches Denken in komplexen Situationen. 6th ed. Reinbek bei Hamburg: Rowohlt-Taschenbuch-Verl.

Dumitrescu, Roman; Fechtelpeter, Christian; Kühn, Arno (2014): Systematische Berücksichtigung von Fertigungsanforderungen im Model-Based Systems Engineering. In: *Tag des Systems Engineering*, p. 21.

Dumitrescu, Roman; Riedel, Oliver; Gausemeier, Jürgen; Albers, Albert; Stark, Rainer (2021): Engineering in Deutschland – Status quo in Wirtschaft und Wissenschaft. Ein Beitrag zum Advanced Systems Engineering. Padaborn. Online verfügbar unter www.advanced-systems-engineering.de, zuletzt geprüft am 06.05.2021.

Ebert, Christof (2019): Systematisches Requirements Engineering. Anforderungen ermitteln, dokumentieren, analysieren und verwalten. 6., überarbeitete und erweiterte Auflage. Heidelberg: dpunkt.verlag.

Ehrlenspiel, Klaus; Meerkamm, Harald (2017): Integrierte Produktentwicklung. Denkabläufe, Methodeneinsatz, Zusammenarbeit. 6., vollständig überarbeitete und erweiterte Auflage. München, Wien: Hanser. Online verfügbar unter http://www.hanser-fachbuch.de/buch/Integrierte+Produktentwicklung/9783446440890.

Eigner, Martin; Stelzer, Ralph (2009): Product Lifecycle Management. Ein Leitfaden für Product Development und Life Cycle Management. 2nd ed. Berlin u. a.: Springer.

Enzyklopädie der Wirtschaft (1982). Berlin: Die Wirtschaft.

DIN EN 954-1:1997-03 (zurückgezogen) Sicherheit von Maschinen – Sicherheitsbezogene Teile von Steuerungen – Teil 1: Allgemeine Gestaltungsleitsätze; Deutsche Fassung, Beuth Verlag.

FAZ (2005): Der deutschen Autoindustrie droht ein Imageschaden. In: *Frankfurter Allgemeine Zeitung (FAZ)*, 2005. Online verfügbar unter https://www.faz.net/aktuell/wirtschaft/unternehmen/pannenserie-der-deutschen-autoindustrie-droht-ein-imageschaden-1226944.html, zuletzt geprüft am 05.07.2021.

Flückiger, Matthias; Rauterberg, Matthias (1995): Komplexität und Messung von Komplexität. In: *ETH Zürich-Institut für Arbeitspsychologie, Technical Report IfAP/ETH/CC-01/95.*

Focus Online (2011): Toyota Rückruf für 400.000 Hybrid-Autos. Online verfügbar unter http://www.focus.de/auto/news/toyota-rueckruf-fuer-400-000-hybrid-autos_aid:478330.html, zuletzt geprüft am 26.07.2011.

Foerster, Heinz von; Köck, Wolfram K.; Schmidt, Siegfried J. (1993): Wissen und Gewissen. Versuch einer Brücke. Frankfurt am Main: Suhrkamp.

Frank, U.; Gausemeier, Jürgen (2004): Selbstoptimierende Systeme des Maschinenbaus. Definitionen und Konzepte. Paderborn: Heinz Nixdorf Institut.

Franke, Hans-Joachim (2005): Kooperationsorientiertes Innovationsmanagement. Ergebnisse des BMBF-Verbundprojektes GINA, „Ganzheitliche Innovationsprozesse in modularen Unternehmensnetzwerken". Berlin: Logos-Verl.

Friedrich, M.; Schmidt-Colinet, J.; Naß, A. (2010): Funktionsorientierte Modellierung von Wirkzusammenhängen zur Beherrschung von Veränderungen mechatronischer Produkte. In: E. Schnieder, U. Jumar und C. Diedrich (Eds.): Entwurf komplexer Automatisierungssysteme. Beschreibungsmittel, Methoden, Werkzeuge und Anwendungen; 11. Fachtagung mit Tutorium, 25. bis 27. Mai 2010 in Magdeburg, Denkfabrik im Wissenschaftshafen. Magdeburg: Ifak.

Fuchs, M.; Lersch, F.; Pollehen, D. (Hg.) (2001): Neues Rollenverständnis für die Entwicklung verteilter Systemverbunde in der Karosserie- und Sicherheitstechnik. VDI. Düsseldorf: VDI-Berichte (1646).

Gaupp, Franz; Schulze, Sven-Olaf; Steffen, Daniel (2014): Die ISO29110-Möglichkeiten und Chancen für den Mittelstand. In: *Tag des Systems Engineering*, pp. 351–359.

Gausemeier, J.; Dumitrescu, R.; Steffen, D.; Czaja, A.; Wiederkehr, O.; Tschirner, C. (2013a): Systems Engineering in der industriellen Praxis. Universität Paderborn: Heinz Nixdorf Institut.

Gausemeier, Jürgen; Gaukstern, Tobias; Tschirner, Christian (2013b): Systems Engineering Management Based on a Discipline-Spanning System Model. In: Conference on Systems Engineering Research (CSER 13) (Ed.): Proceedings, 19.–22. März 2013: Elsevier B.V.

Gausemeier, J.; Wiederkehr, O.; Dumitrescu, R. (2014): Der Entwicklungsauftrag als Basis für eine vorausschauende und systemorientierte Produktentstehung. In: *Tag des Systems Engineering, Hrsg.: Maurer, M.; Schulze, S.-O., Carl Hanser Verlag, München.*

Gausemeier, Jürgen (2007): Strategisches Produktionsmanagement. 1st ed. München: Hanser.

Gausemeier, Jürgen; Bigl, Thomas (2006): Integrative Entwicklung räumlicher elektronischer Baugruppen. München [u. a.]: Hanser.

Gausemeier, Jürgen; Lanza, Gisela; Lindemann, Udo (2012): Produkte und Produktionssysteme integrativ konzipieren. Modellbildung und Analyse in der frühen Phase der Produktentstehung. München: Carl Hanser Verlag. Online verfügbar unter http://www.hanser-elibrary.com/action/showBook%3Fdoi=10.3139/9783446429857.

Gausemeier, Jürgen; Plass, Christoph; Wenzelmann, Christoph (Eds.) (2009): Zukunftsorientierte Unternehmensgestaltung. Strategien, Geschäftsprozesse und IT-Systeme für die Produktion von morgen. München, Wien: Hanser.

Goeken, Matthias (2006): Entwicklung von data-warehouse-systemen: Anforderungsmanagement, modellierung, implementierung: Springer-Verlag.

Grundig, Claus-Gerold (2018): Fabrikplanung. Planungssystematik – Methoden – Anwendungen. 6., neu bearbeitete Auflage. München: Hanser. Online verfügbar unter http://www.hanser-fachbuch.de/9783446454002.

Haberfellner, Reinhard; Daenzer, Walter F. (1999): Systems engineering. Methodik und Praxis. 10., durchges. Zürich: Verl. Industrielle Organisation.

Haberfellner, Reinhard; Daenzer, Walter F. (2002): Systems engineering. Methodik und Praxis. 11., durchges. Aufl. Zürich: Verl. Industrielle Organisation.

Haberfellner, Reinhard; Weck, Olivier L. de; Fricke, Ernst; Vössner, Siegfried (2019): Systems engineering. Fundamentals and applications. 14. überarbeitete Auflage. Cham: Springer International Publishing; Birkhäuser.

Hall, Arthur David (1965): Systems Engineering from an Engineering Viewpoint. In: *IEEE Transactions on Systems Science and Cybernetics* 1, 1965 (1), pp. 4–8.

Hanenkamp, Nico (2004): Entwicklung des Geschäftsprozesses Komplexitätsmanagement in der kundenindividuellen Serienfertigung. Ein Beitrag zum Informationsmanagement in mehrdimensional modellierten Produktionssystemen. Univ., Diss. -Bochum, 2004. Aachen: Shaker (Schriftenreihe des Lehrstuhls für Produktionssysteme/Institut für Automatisierungstechnik, Lehrstuhl für Produktionssysteme, 2004,5).

Haskin, C.; Krueger, M.; Forsberg, K.; Walden, D.; Hameling, R. D. (2011): Systems Engineering Handbook. V3.2.2. San Diego, USA: INCOSE (TP-2002-002-03.2.2).

Heinrich, Harald (2015): Systemisches Projektmanagement. Grundlagen, Umsetzung, Erfolgskriterien. München: Hanser.

Heinrichsmeyer, Marius (2020): Entwicklung eines zielgerichteten Fehlerursachensuch- und Lösungsalgorithmus [FusLa]. Dissertation. Wuppertal: Bergische Universität Wuppertal.

Heinrichsmeyer, Marius; Ansari, Amirbabak; Schlueter, Nadine; Boehmer, Christian (2020): Investigation of Problems with High Initial and Update Efforts in the Modeling of Production Systems. A Review on System Modeling Approaches. In: *International Journal on Advances in Systems and Measurements* 13 (3 & 4), pp. 264–274. Online verfügbar unter http://www.iariajournals.org/systems_and_measurements/, zuletzt geprüft am 30.04.2021.

Hinrichsen, Diederich; Pritchard, Anthony J. (2005): Mathematical Systems Theory I. Modelling, State Space Analysis, Stability and Robustness. Berlin, Heidelberg: Springer-Verlag Berlin Heidelberg (Springer-11649/Dig. Serial], 48). Online verfügbar unter https://doi.org/10.1007/b137541/http://www.zentralblatt-math.org/zmath/en/search/%3Fan=1074.93003.

Hood, Colin (2005): iX-Studie zum Thema Anforderungsmanagement. Hannover: Heise-Zeitschriften-Verl.

Huber, Miriam (2014): Ansatz zur Nutzung vernetzter virtueller Produktmodelle für die kundenintegrierte Produktentwicklung. Bochum: Ruhr-Universität Bochum (Schriftenreihe des Institut Product and Service Engineering, Ruhr-Universität Bochum).

Hull, Elizabeth; Jackson, Ken; Dick, Jeremy (2005): Requirements engineering. 2nd ed. London: Springer. Online verfügbar unter http://www.knovel.com/knovel2/Toc.jsp?BookID=1805/http://www.zentralblatt-math.org/zmath/en/search/%3Fan=1058.68104.

IEC 61508 (2010): Functional Safety: safety related systems parts 1 to 7.

IEEE 1220 (2005): IEEE-STD-1220-2005. IEEE standard for application and management of the systems engineering process.

INCOSE (Hg.) (2015): Systems engineering handbook. A guide for system life cycle processes and activities. Unter Mitarbeit von David D. Walden, Garry J. Roedler, Kevin Forsberg, R. Douglas Hamelin und Thomas M. Shortell. International Council on Systems Engineering. 4. edition. Hoboken, NJ: Wiley.

INCOSE, T. (2007): Systems engineering vision 2020. INCOSE-TP-2004-004-02. Version/Revision: 2.03. In: *INCOSE, San Diego, CA, accessed Jan 26.*

ISO/IEC/IEEE 29148 (2018): Systems and software engineering – Life cycle processes – Requirements engineering: IEEE.

Jackson, Michael C. (2000): Systems approaches to management. New York, NY: Kluwer Academic/Plenum.

Jamshidi, Mo (2009): Systems of systems engineering. Innovations for the 21st century. Hoboken, NJ: Wiley (Wiley series in systems engineering and management).

Jenkins, Gwilym Meirion; Youle, Philip V. (1971): Systems Engineering. A unifying approach in industry and society. London: Watts (The new thinker's Library).

Jing, Z, Ming-Yang, W; Wie, L.; Jan, H.; Li-Qun, Y.; Ze-Min, L.; Qin-Zhang, Y. (2013): Enormal Approach auf SOS Modelling und Comprehensive Evaluation based Ontology. In: *Proceedings of the 8th international conference on Systems of Systems Engineering, Mauwi, Hawaii, USA.*

JuraForum.de (2021): Denken. Definition, Begriff und Erklärung im JuraForum.de. Online verfügbar unter https://www.juraforum.de/lexikon/denken, zuletzt aktualisiert am 31.08.2021, zuletzt geprüft am 31.08.2021.

Kampker, Achim; Burggräf, Peter; Deutskens, Christoph; Maue, Andreas; Förstmann, Ruben (2014): Integrated product and process development: Modular production architectures based on process requirements. In: *Procedia Cirp* 20, pp. 109–114.

Keating, C.; Roger, R.; Ruault, R.; Dreier, D.; Sousa-Poza, A.; Safford, R. et al. (2003): Systems of Systems Engineering. In: *Engineering Management Journal (ENJ)* (VOL 15, No. 3).

King-Kordi, Amal (2010): Methodik zur bausteinbasierten Planung und Organisation von verfahrenstechnischen Produktionssystemen. Techn. Univ., Diss.-Chemnitz, 2010. Chemnitz: IBF (Wissenschaftliche Schriftenreihe des Institutes für Betriebswissenschaften und Fabriksysteme, 88).

Kreß, A.; Hood, C.; Stevenson, R.; Versteegen, G.; Wiebel, R. (2005): iX-Studie zum Thema Anforderungsmanagement: Heise Zeitschriftenverlag GmbH & Co. KG.

Lamm, J. G.; Weilkiens, T. (2015): Method for Deriving Functional Architectures from Use Cases. In: Maik S. Maurer, Jutta Abulawi und Sven-Olaf Schulze (Eds.): Tag des Systems Engineering. Bremen, 12.–14. November 2014; [TdSE]. München: Hanser, pp. 225–236.

Lanza, Gisela; Nyhuis, Peter; Fisel Johannes; Jacob, Alexander; Nielsen, Lars; Schmidt, Matthias; Stricker, Nicole (2018): Wandlungsfähige, menschzentrierte Strukturen in Fabriken und Netzwerken der Industrie 4.0. München: Herbert Utz Verlage (acatech STUDIE). Online verfügbar unter https://elibrary.vahlen.de/10.15358/0935-0381-2015-8-9-455.pdf?download_full_pdf=1.

Leveson, Nancy G. (2001): Safeware. System safety and computers: [a guide to preventing accidents and losses caused by technology]. 5th printing. Boston, Mass.: Addison-Wesley.

Lim, S. L.; Ncube, C. (2013): Social Networks and Oursourcing for Stakeholder Analysis for Systems of System Projects. In: *Proceedings of the 8th Conference of Systems of Systems Engineering, Mauwi, Hawaii.*

Lindemann, U. (2005): Methodische Entwicklung technischer Produkte. Methoden flexibel und situationsgerecht anwenden. Berlin: Springer.

Lindemann, Udo (Ed.) (2016): Handbuch Produktentwicklung. München: Carl Hanser Verlag.

Ludwig, Björn (2001): Management komplexer Systeme. Der Umgang mit Komplexität bei unvollkommener Information: Methoden, Prinzipien, Potentiale. Techn. Univ., Habil.-Schr.-Clausthal, 2000. Berlin: Ed. Sigma (Technik – Gesellschaft – Natur, 4).

Luhmann, N. (1980): Komplexität. Enzyklopädie der Betriebswirtschaftslehre. Stuttgart: Poeschel.

Luhmann, Niklas (2000): Organisation und Entscheidung. Opladen [u. a.]: Westdt. Verl.

Luzeaux, D.; Ruault, R. (Eds.) (2010): System of Systems. Hoboken, NJ: John Wiley & Sons, Inc.

Maier, M. W. (2005): Research Challenges for Systems of Systems in Systems. In: *Man and Cybernetics, 2005, IEEE international Conference* (Vol. 4), pp. 3149–3157.

Mamrot, Michel (2014): Entwicklung eines Ansatzes zur modellbasierten Felddatenrückführung in die Produktentwicklung. 1. Aufl. Herzogenrath: Shaker (Berichte zum Generic-Management, 2014,1).

Marques-Lucena, Catarina; Agostinho, Carlos; Marcelino-Jesus, Elsa; Sarraipa, Joao; Jardim-Goncalves, Ricardo (2015): Collaborative Management of Requirements Using Semantic Wiki Modules. In: Ioan Dumitrache (Ed.): 20th International Conference on Control Systems and Computer Science (CSCS 2015). Bucharest, Romania: IEEE, pp. 665–672.

Maurer, Maik; Schulze, Sven-Olaf (2014): Tag des Systems Engineering. Proceedings Systems Engineering Konferenz, Bremen, 12.–14. November 2014: Carl Hanser Verlag GmbH Co KG.

Meier, Marco; Sinzig, Werner; Mertens, Peter (2002): SAP Strategic Enterprise Management/ Business Analytics. Integration von strategischer und operativer Unternehmensführung. Berlin: Springer (SAP kompetent). Online verfügbar unter http://www.gbv.de/dms/bsz/toc/bsz-095165223inh.pdf.

Mistler, Marian (2021): Entwicklung eines Vorgehenskonzeptes zum modellbasierten agilen Anforderungsmanagement (Requirements Engineering und Requirements Management) für Organisationen – REMOt. Dissertation. Wuppertal: Bergische Universität Wuppertal.

Mistler, Marian; Schlueter, Nadine; Löwer, Manuel (2021): Analysis of software tools for model-based Generic Systems Engineering for organizations based on e-DeCoDe. In: 2021 IEEE International Systems Conference (SysCon): IEEE.

Motor-Talk (2012): Klimakompressor defekt nach 52.000 km Astra J 2.0 CDTI. Feststellbremse löst bei Gas im Leerlauf. Online verfügbar unter http://www.motor-talk.de/forum/klimakompressor-defekt-nach-52-000kmastra- j-2-0-cdti-feststellbremse-loest-bei-gas-im-leerlauf-t3830394.html, zuletzt geprüft am 20.09.2012.

Müller, Egon; Jentsch, David; Trommler, Ullrich; Horbach, Sebastian; Ackermann, Jörg (2012): Entwicklung eines interaktiven Planungshandbuches für kompetenzzellenbasierte Produktionsnetze. In: E. Müller und A.C Bullinger (Hg.): Fachtagung "Vernetzt Planen und Produzieren – VPP2012" sowie Symposium "Wissenschaft und Praxis – W&P". Intelligent vernetzte Arbeits- und Fabriksysteme VPP 2012. Fachtagung Vernetzt Planen und Produzieren und Symposium Wissenschaft und Praxis. Chemnitz. Chemnitz: Eigenverlag TU Chemnitz, ISSN: 0947–2495 (Wissenschaftliche Schriftenreihe des IBF, Sonderheft 18), S. 53–62.

Näser, Peggy (2007): Methode zur Entwicklung und kontinuierlichen Verbesserung des Anlaufmanagements komplexer Montagesysteme. Techn. Univ., Diss.-Chemnitz, 2007. Chemnitz: Inst. für Betriebswiss. und Fabriksysteme (Wissenschaftliche Schriftenreihe/Technische Universität Chemnitz-Zwickau, Institut für Betriebswissenschaften und Fabriksysteme, 56).

Ncube, Cornelius; Lim, Soo Ling; Dogan, Huseyin (2013): Identifying top challenges for international research on requirements engineering for systems of systems engineering. In: 2013 21st IEEE International Requirements Engineering Conference (RE). IEEE, pp. 342–344.

Nicklas, Jan-Peter (2016): Ansatz für ein modellbasiertes Anforderungsmanagement für Unternehmensnetzwerke. Dissertation. Aachen: Shaker Verlag (Berichte zum Generic-Management, Band 2016, 2).

Oppenheim, Bohdan W. (2011): Lean for systems engineering with lean enablers for systems engineering: John Wiley & Sons (82).

Ott, Stefan (2009): Konzept zur methodischen System-Modellierung in der anforderungsgerechten Produktentwicklung. Aachen: Shaker.

Padilla, Jose J.; Logan, Bradford; Sousa-Poza, Andres; Keating, Charles B. (2008): A system of systems engineering environment to deal with complex situations. In: 2008 IEEE International Conference on System of Systems Engineering. IEEE, pp. 1–5.

Pahl, Gerhard; Beitz, Wolfgang; Feldhusen, Jörg; Grote, Karl-H (2005): Pahl/Beitz Konstruktionslehre. Grundlagen erfolgreicher Produktentwicklung Methoden und Anwendung. 6. Auflage. Berlin, Heidelberg: Springer-Verlag Berlin Heidelberg (Springer-11774/Dig. Serial]). Online verfügbar unter https://doi.org/10.1007/b137606.

Pankow, Gabriel (2016): Rückrufe und kein Ende – Qualitätsmangel in der Autoindustrie? In: *Produktion*, 2016 (50), pp. 6–7.

Partsch, Helmuth (2010): Requirements-Engineering systematisch. Modellbildung für softwaregestützte Systeme. 2nd ed. Berlin, Heidelberg: Springer.

Pohl, Klaus (2016): Requirements Engineering. Fundamentals, principles, and techniques. Berlin, Heidelberg: Springer-Verlag.

Pohl, Klaus; Rupp, Chris (2015): Requirements engineering fundamentals. A study guide for the certified professional for requirements engineering exam, foundation level. IREB compliant. 2nd edition. San Rafael, CA: Rocky Nook.

Ponn, Josef; Lindemann, Udo (2011): Konzeptentwicklung und Gestaltung technischer Produkte: systematisch von Anforderungen zu Konzepten und Gestaltlösungen: Springer-Verlag.

ReMaiN (2020): DFG – GEPRIS – Entwicklung eines methodischen Ansatzes zum Requirements Management in unternehmensübergreifenden Netzwerken (ReMaiN). Hg. v. Deutsche Forschungsgemeinschaft (DFG) und Petra Winzer (Projektnummer 320394109). Online verfügbar unter https://gepris.dfg.de/gepris/projekt/320394109%3Fcontext=projekt%26task=showDetail%26id=320394109, zuletzt aktualisiert am 06.07.2021, zuletzt geprüft am 06.07.2021.

Rink, Anton W. (2002): Entwicklung einer Methode für die systemtechnische Auslegung verteilter und sicherheitskritischer Führungsfunktionen für Fahrzeugantriebe. Dissertation. Bergische Universität Wuppertal, Wuppertal.

Ropohl, Günter (2012): Allgemeine Systemtheorie: Einführung in transdisziplinäres Denken: edition sigma.

Rupp, Chris (2021): Requirements-Engineering und -Management. Das Handbuch für Anforderungen in jeder Situation. 7., aktualisierte und erweiterte Auflage. München: Hanser.

Sage, Andrew P.; Rouse, William B. (Hg.) (2009): Handbook of systems engineering and management. 2. ed. Hoboken, NJ: Wiley (Wiley series in systems engineering and management).

Sahin, Ferhad; Jamshidi, Mo.; Sridhar, Prasanna (2009): A system-of-systems simulation framework and its applications. In: Mohammad Jamshidi (Ed.): Systems of systems engineering. Principles and applications. Boca Raton, FL: CRC Press/Taylor & Francis Group, pp. 107–132.

Scheer, August-Wilhelm (2001): ARIS – Modellierungsmethoden, Metamodelle, Anwendungen. 4th ed. Berlin: Springer. Online verfügbar unter http://www.gbv.de/dms/hebis-darmstadt/toc/96055979.pdf.

Schenk, Michael (2004): Fabrikplanung und Fabrikbetrieb. Methoden für die wandlungsfähige und vernetzte Fabrik. Berlin, Heidelberg: Springer-Verlag Berlin Heidelberg (Springer-11774/Dig. Serial]). Online verfügbar unter https://doi.org/10.1007/3-540-35046-2.

Schenk, Michael; Wirth, Siegfried; Müller, Egon (2014): Fabrikplanung und Fabrikbetrieb. Methoden für die wandlungsfähige, vernetzte und ressourceneffiziente Fabrik. 2., vollständig überarbeitete und erweiterte Auflage 2014. Heidelberg, Berlin: Springer Vieweg.

Schlueter, Nadine (2016): Der DyNamic-Ansatz. Entwicklung eines Ansatzes zur verlässlichen Gestaltung von Unternehmensnetzwerken und ihren Produkt-Service-Systemen. Habilitationsschrift. Wuppertal. Online verfügbar unter http://elpub.bib.uni-wuppertal.de/edocs/dokumente/fbd/maschinenbau/habi2016/schlueter; http://nbn-resolving.de/urn/resolver.pl%253Furn=urn%253Anbn%253Ade%253Ahbz%253A468-20171009-111216-2.

Schlueter, Nadine; Heinrichsmeyer, Marius; Bielefeld, Ovidiu; Mistler, Marian; Ansari, Amirbabak (2019): ReMaiN-Concept for Requirements Management and Engineering in R&D business networks in Germany. In: Proceedings of 14th International Conference on System of Systems Engineering. Anchorage, Alaska: IEEE, pp. 308–312.

Schlueter, Nadine; Winzer, Petra; Ansari, Amirbabak; Bielefeld, Ovidiu; Dransfeld, Hendrik; Heinrichsmeyer, Marius (2018): KAUSAL. A New Methodological Approach for Model Based Analysis of Complex Failure Chains by Example of an Electromobility Concept. In: 2018 IEEE International Conference on Systems, Man, and Cybernetics (SMC). Miyazaki, Japan: IEEE, pp. 947–952.

Schmitt, Robert; Amini, P.; Bergholz, M.; Falk, B.; Humphrey, S.; Müller, C. et al. (2014): Anforderungsmanagement 4.0. Robuste Spezifikation in turbulentem Umfeld. In: Christian Brecher, Fritz Klocke, Robert Schmitt und Günther Schuh (Ed.): Integrative Produktion. Industrie 4.0 – Aachener Perspektiven; AWK, Aachener Werkzeugmaschinen-Kolloquium. 1st ed. Aachen: Shaker.

Schnieder, Eckehard; Schnieder, Lars (2013): Verkehrssicherheit. Maße und Modelle, Methoden und Maßnahmen für den Straßen- und Schienenverkehr. Berlin, Heidelberg: Springer Vieweg (VDI-Buch).

Schuh, Günther (2004): Komplexitätsmanagement in der produzierenden Industrie. Fortschritte – zukünftige Anforderungen – Handlungsbedarf. WZL; RWTH Aachen; GPS Schuh & co. GmbH.

Schuh, Günther (2007): Effizient, schnell und erfolgreich. Strategien im Maschinen- und Anlagenbau. Frankfurt, M: VDMA-Verl.

Schuh, Günther; Anderl, Reiner; Dumitrescu, Roman; Krüger, Antonio; Hompel, Michael ten (2020): Industrie 4.0 Maturity Index. Die digitale Transformation von Unternehmen gestalten. Update 2020: acatech STUDIE.

Schuh, Günther; Riesener, Michael (2018): Produktkomplexität managen. Strategien – Methoden – Tools. Unter Mitarbeit von Stefan Breunig, Christian Dölle, Manuel Ebi, Michael Gerrit Schiffer, Sebastian Schloesser und Elisabeth Schrey. 3., vollständig überarbeitete Auflage. München: Hanser. Online verfügbar unter http://www.hanser-fachbuch.de/9783446452251.

Schuldt, Christian (2003): Systemtheorie. Hamburg: EVA Europäische Verlagsanstalt (Wissen 3000).

Sell, Robert (2013): Angewandtes Problemlösungsverhalten: Denken und Handeln in komplexen Zusammenhängen. 2nd ed.: Springer-Verlag.

Senge, Peter M. (1999): The dance of change. The challenges of sustaining momentum in learning organizations: a fifth discipline resource. 1st ed. New York: Currency/Doubleday.

Sitte, J.; Winzer, P. (2011): Systemmodellierung im Fokus von Generic Systems Engineering. In: Gesellschaft für Systems Engineering e. V. (Ed.), Tag des Systems Engineering.

Sommerville, J. (2005): Integrated Requirements Engineering: A Tutorial. Version 22 no.1: IEEE.

Sommerville, J. (2007): Software Engineering. München VDI-Richtlinie 2221 – Methodik zum Entwickeln und Konstruieren technischer Systeme und Produkte. Berlin: Beuth Verlag.

ISO/IEC/IEEE 29148:2018-11: System- und Software-Engineering – Lebenszyklus-Prozesse – Requirements Engineering, Beuth Verlag.

VDI 2221 (1993): VDI-Richtlinie 2221 – Methodik zum Entwickeln und Konstruieren technischer Systeme und Produkte. Berlin: Beuth Verlag.

VDI 2247 (Hg.) (1994): VDI-Richtlinie 2247 – Qualitätsmanagement in der Produktentwicklung. VDI-Gesellschaft Produkt- und Prozessgestaltung. Berlin: Beuth Verlag.

VDI 5200 (Hg.) (2011): Fabrikplanung. Planungsvorgehen. VDI-Gesellschaft Produktion und Logistik. Berlin: Beuth Verlag.

Verstegen, G. (2004): Anforderungsmanagement. Formale Prozesse, Praxiserfahrungen, Einführungsstrategien und Toolauswahl. Berlin, Heidelberg: Springer.

Vester, Frederic (2003): Die Kunst vernetzt zu denken. Ideen und Werkzeuge für einen neuen Umgang mit Komplexität; ein Bericht an den Club of Rome. 3rd ed. München: Deutscher Taschenbuchverlag.

Warnecke, T. (2015): Sichere Integration von Teilsystemen zu Systems-of-Systems durch formale Verifikation. In: Maik S. Maurer, Jutta Abulawi und Sven-Olaf Schulze (Eds.): Tag des Systems Engineering. Bremen, 12.–14. November 2014; [TdSE]. München: Hanser, pp. 299–310.

Weilkiens, T. (2007): Systems engineering with SysML. Modeling, analysis, design. Amsterdam: Morgan Kaufmann OMG Press/Elsevier.

Weilkiens, Tim (2019): Systems Engineering mit SysML/UML. Anforderungen, Analyse, Architektur. 3., überarb. und aktualisierte Aufl. Heidelberg: dpunkt.verlag.

Weiss, Stanley I. (2013): Product and systems development: a value approach: John Wiley & Sons.

Westphal, Jan R.; Kummer, Sebastian (2001): Komplexitätsmanagement in der Produktionslogistik. Ein Ansatz zur flussorientierten Gestaltung und Lenkung heterogener Produktionssysteme. 1st ed. Wiesbaden: Dt. Univ.-Verl.

Wiendahl, Hans-Peter; Reichardt, Jürgen; Nyhuis, Peter (2009): Handbuch Fabrikplanung. Konzept, Gestaltung und Umsetzung wandlungsfähiger Produktionsstätten. München: Hanser.

Wiener, Norbert (1994): Cybernetics or control and communication in the animal and the machine. 2nd ed., 7. print. Cambridge, Mass: MIT Press.

Wildemann, Horst (2004): Komplexitätsmanagement. Leitfaden zur Einführung eines durchgängigen Komplexitätsmanagement. München: TCW Transfer-Centrum.

Winzer, Petra (1997): Chancen zur umfassenden Unternehmensgestaltung. Methodischer Ansatz zur qualitäts-, human- und ökologieorientierten Gestaltung von Arbeits- und Fabriksystemen. Techn. Univ., Habil.-Schr.-Berlin, 1996. Frankfurt am Main: Lang (Europäische Hochschulschriften Reihe 5, Volks- und Betriebswirtschaft, Bd. 2189).

Winzer, Petra (2015): Generic System Description and Problem Solving in Systems Engineering. In: *IEEE Systems Journal* 11 (4), pp. 2052–2061. https://doi.org/10.1109/JSYST.2015.2428811.

Winzer, Petra; Bach, Friedrich-Wilhelm; Schnieder, Eckehard (Hg.) (2010): Sicherheitsforschung. Chancen und Perspektiven // Chancen und Perspektiven (acatech DISKUTIERT). 1st ed. Berlin, Heidelberg: Springer; Springer-Verlag.

Woll, R.; Hayka, H.; Stark, R. (2015): Ontologiebasierte Datenintegration für das modellbasierte Systems Engineering. In: Maik S. Maurer, Jutta Abulawi und Sven-Olaf Schulze (Eds.): Tag des Systems Engineering. Bremen, 12.–14. November 2014; [TdSE]. München: Hanser, pp. 33–42.

Wulf, Joachim Erich (2002): Elementarmethoden zur Lösungssuche. Technische Universität München.

Zilahi-Szabó, M. (2002): Prinzipien des Software-Engineering. Vorlesungsunterlage. SS 2002, Giessen. Online verfügbar unter http://www.informatik.uni-giessen.de/Sommersemester/SE/Software_Engineering_Prinzipien.ppt, zuletzt geprüft am 21.09.2012.

Züst, Rainer (2004): Einstieg ins Systems Engineering. Optimale, nachhaltige Lösungen entwickeln und umsetzen. 3., überarb. Aufl. Zürich: Verl. Industrielle Organisation.

Generic Systems Engineering—An Approach to Mastering Complexity in a New Dimension

3

The conclusion of the second chapter proves that SE is particularly suitable for systematically and efficiently addressing the problems arising from the new trends of complexity through the targeted application of its methodological approach.

However, it also becomes apparent that SE has lost its original, universal approach (cf. Sect. 2.2) due to the constantly growing number of new problem-specific or subject-specific thinking models and procedural concepts (cf. Sect. 2.4). Therefore, despite overarching SE approaches (cf. Sect. 2.5), it seems questionable to develop a general and generic, i.e., a ***Generic Systems Engineeringapproach (GSE approach),*** according to the new requirements. Nevertheless, HABERFELLNER, as one of the few representatives of universal SE in Germany, emphasizes that the basic principles of SE fundamentally require a system thinking and a generalizable procedural concept for an industry-independent problem solution (cf. Haberfellner and Daenzer 2002; Haberfellner et al. 2018). The latter is subsequently referred to as a procedural concept, because the term model is used for an image of reality, but in this context, a planned approach for problem-solving is meant. Concepts are mental representations that can connect procedures and models (Seghezzi 2007). According to WEILKIENS, system thinking must generate a generalized conceptual model regardless of the respective discipline (Weilkiens 2007, 2019). Consequently, in a further to be developed ***generic SE,*** the basic building blocks, i.e., system thinking, the conceptual model, and the procedural concept, must be connected in such a way that a transdisciplinary, general use for problem solutions of any kind in any industry becomes possible. This, in turn, requires a ***synergetic connection between the conceptual thinking model and the procedural concept,*** which the approaches of SE presented in Sect. 2.4 do not have. Why exactly this requirement must be implemented in the new GSE approach will be considered in Sect. 3.1.

How this can be achieved will be examined in more detail in Sect. 3.2. Thus, a conceptual thinking model with corresponding tools needs to be developed, which illustrates

N. Schlüter, *Generic Systems Engineering*, https://doi.org/10.1007/978-3-662-67994-4_3

the complexity of systems in a generally understandable way. The further development of the system itself, i.e., its genesis, must be recognizable in the new general conceptual model, i.e., the ***GSE thinking model***. Recognition is the prerequisite for change. This should follow simple rules and must be considered in the SE procedural concept. But also transdisciplinarity, transparency, and traceability pose specific demands on the SE procedural concept to be changed.

How can the SE procedural concept be developed or modularized in a generalist or generic way into a ***GSE procedural concept***? These and other questions will be examined in more detail in Sect. 3.3. Only on this basis can the question be answered as to what a ***GSE approach*** could look like (cf. Sect. 3.4).

3.1 The Synergy between the Thinking Model and the Procedural Concept—A Necessary Condition in GSE

The genesis between the thinking model and the procedural concept in SE has often been neglected in the different approaches. However, it is a necessary condition for coping with the new dimensions of complexity of today and in the future. The following example illustrates what can happen when system thinking is neglected in the problem-solving process:

A few years ago in autumn, there were significant delays in rail traffic in North Rhine-Westphalia (3Sat 2010). The newspapers wrote that the causes of the delays were due to wet autumn leaves. Of course, such a reason sounds rather paradoxical, especially since autumn as a leaf-intensive season is not a phenomenon of modern times. What happened?

The very light train slid on the wet leaves during all its braking operations, while the traditional, heavy trains crushed the leaves on the tracks. The wheels sliding on the tracks caused one-sided wear, which led to square wheels after several braking operations. These, in turn, worsened the running properties and had effects on the entire undercarriage. Consequently, the undercarriages, which were originally supposed to be avoided, needed frequent repair. In addition, due to the originally forecast low maintenance effort, the responsible parties no longer stored undercarriages to minimize storage costs for the company. As a result, the necessary undercarriage changes could not be carried out. A large number of trains were in the depot and were missing in public local transport in North Rhine-Westphalia (NRW). Regardless of this, the responsible parties were striving for quick solutions such as:

a) the re-use of heavier trains (via a chargeable loan from other local transport associations) and
b) the development of new, constructive solutions for driving light trains in autumn (including the installation of sandblasting blowers in the front of the train or a leaf blower).

Solution variant a) initially led to a rapid improvement in the situation in local transport. On the other hand, solution variant b) required more extensive investigations, which, however, did not yield any significant results. While the sand from variant b) again led to increased wear of the wheels and thus an impairment of both the track and the wheel, the leaf blower merely swirled the leaves. In contrast to the situation in NRW, the leaf blower worked in the local transport associations in Bavaria, who also saw their future in the light trains. The causes, however, were not with the constructors of the leaf blowers from NRW, but required the expansion of the solution space. Comparing the various local transport systems of Bavaria and North Rhine-Westphalia, it turned out that in NRW, unlike in Bavaria, almost 80% of the rail system is limited by noise barriers. On routes without a noise barrier, as is predominantly the case in Bavaria, the leaf blower blew the leaves to the side. On the routes in NRW, the leaves, due to the noise barriers, did not fly far enough into the environment. Consequently, the development of the leaf blower in Bavaria was successful because the missing noise barriers created completely different air flow conditions than in North Rhine-Westphalia. The example illustrates, among other things, the close interrelationship between the thinking model and the procedural concept. If the thinking model sets very narrow boundaries, e.g., the development of a lighter, resource-saving, fast, low-maintenance train, the solution space for the procedural concept is also limited, i.e., to the interaction of train and track. If the solution space, or the system, were delimited so that the transport system in Bavaria would be considered exclusively, the light train with the blower would have become the preferred solution. If the system were to be expanded further, e.g., to the rail system in Bavaria and NRW, the interaction of the moving train with its environment could be recognized by the constructors and developers, that the heavy trains, which crush the leaves, would be the better solution for the rail systems in NRW and Bavaria.

The development of "an all-weather train" for the transport networks of North Rhine-Westphalia and Bavaria requires a different, expanded approach than the task "development of a resource-saving train". Basically, thinking in systems is a simplification. The human being himself, according to HÄUSLEIN, can perceive his environment in its complexity in a system-theoretical way (Häuslein 2004). He divides it into systems, separates the systems from each other, and can thus describe the interaction of systems with each other and with their environment. ***Systems as a delimited unit do not exist in reality.*** They are formed to represent complex relationships, as the example of the local transport system just described illustrates. Consequently, images, i.e., a ***system model,*** can be created to design such fixed systems using system thinking, which can be specified and/or refined as required and serves a specific purpose. ***Thinking in systems helps humans recognize a structure in their environment.*** As a result, a ***conceptual model,*** which describes the cognitive way of creating images of reality, is created. If system thinking is used for this, the result of this process is a system model. Simple rules are required for this. They can, for example, help to break down the system "local transport" into further subsystems such as "rail system", "train system", "noise protection systems" etc. and to describe their interrelationships with each other or with the environment.

Only through this special approach is man better able to optimize the interaction between the system and its environment, as in the example sketched above. As a logical consequence, the ***thinking model andthe procedural conceptmust form a generic unit.***

When a product, as in this example the train, is to be described, a model is needed. This represents an image of reality and abstractly illustrates, for example, the relationships of the components or the functions of the product "train". Through a specific approach, which is oriented towards the problem to be solved, in the above-mentioned example it was the sliding of the wheels on the tracks in the wet leaves, the interactions between the components "wheels", "track" or "wheel drive" or "wheel axle", to name just a few, are now analyzed and designed by special methods and procedures. The design results in turn lead to changes in the product model "train" (conceptual model). On the one hand, the initially thought interactions between the components can be described more precisely, on the other hand, function simulations are also possible, which, for example, lead to a different design of the drives themselves or the axles in the case of the wheel drives. A renewed analysis of the interactions of the system "train" with its environment leads in the example just described to the addition of a blower to the train to remove the leaves. The system "train" is thus extended by a component "blower". For this new component "blower", on the one hand, the interactions with already defined components or subsystems of the system "train" must be established, and on the other hand, already defined relationships must be questioned and possibly specified or newly designed. The added blower, for example, can lead to changed solution concepts, such as the reinforcement of the drive, new dimensioning of the wheel axle, to name just a few. The discussion of this example is intended to illustrate the interaction between the thinking model and the procedural concept. This can be efficiently influenced by the basic principles of systematic thinking and action, which were explained in Sect. 2.4. The implementation of the basic principle from the whole to the detail, related to the example of public local transport, could have resulted, for example, in the constructors first dealing with the process "train travels nationwide in all weather conditions". This would have brought the system "public local transport", which is superior to the system "train", into the constructors' view. The application of the basic principle of recurring reflection, coupled with the basic principle of discursive action, would then result in the solution variant "leaf blower" being examined on the one hand for necessary changes to the system "train" but on the other hand also for the interaction of the modified train with its environment. These results in turn lead, as here in this example, to the specification of objectives not to develop lighter, but rather heavier trains again and to critically view already found solutions, i.e., e.g., the train with blower at noise barriers. The implementation of these selected basic principles of systematic thinking and action require thinking models and procedural concepts that allow this.

While the procedural concept only has to "implement" the steps sketched above in the logical temporal sequence, special requirements are placed on the thinking model. It must be understood by all potential users, i.e., e.g., the railway engineer, the computer scientist, the electrical engineer (basic principle of comprehensibility and basic principle of standardization) and can be further processed in a discipline-specific way (basic principle of multiple usability). The thinking model should be structured in such a way that it allows both the view of the whole and the detail. A connection is required between these views. If the model is to be critically reflected repeatedly (basic principle of recurring reflection) in order to derive new goals from it (implementation of the basic principle of discursive action), the model of the system, here in the special case "train", must be specified in each step of the procedure. But exactly this step of ***permanent model specification*** is often neglected or not explicitly stated in the implementation of the procedural concept, as proven in Sect. 2.4. Accordingly, the individual steps of the procedural concept always result in a change of the thinking model. The thinking model must fundamentally allow this, preferably traceable, for the individual steps of product development "train", here referred to as procedural concept.

SAGE and ROUSE demand that technical systems be managed (Sage and Rouse 2009). This includes the interaction between the system's thinking model and its project-driven design process. The thinking model itself, as well as its changes, which it experiences purposefully during its design process through the procedural concept, must be documented transparently so that it is generally understandable and not just subject-specific. This means that the model of the "train" system must be understood by the mechanical engineer as well as by the electrical engineer or computer scientist, who are also involved in the design of the "train" system. Each of these specialists brings new ideas into the system design process, which result in changes to the "train" thinking model. These must in turn be inserted into the thinking model in a timely and traceable manner for the other participants in the design process. How this should be done must be described in more detail within the framework of the GSE. HABERFELLNER's suggestion that this should be controlled via project management is certainly helpful (Haberfellner and Daenzer 2002; Haberfellner et al. 2018). However, it is not sufficiently detailed because it does not precisely describe how the results achieved per milestone are to be integrated into the thinking model or contribute to its precision. Only if this genesis of the thinking model and the procedural concept is consistently recognized, considered, described, methodically supported and implemented, can the problem-solving process be designed transparently and traceably. If the problem-solving process of the mentioned example for the development of an "all-weather train" is to be generalized, i.e. universally designed, then further principles of systematic thinking and action, such as the principle of standardization, the principle of multiple use of modules, the principle of minimizing interfaces and the principle of proceeding from the abstract to the concrete must be implemented both in the model and in the procedure.

WEINBERG has clearly worked out that a ***general thinking model,*** i.e. a general cognitive way of creating images of reality, is needed (Weinberg 2001). He understands ***general*** in the sense of a generalistic, universally or cross-disciplinary usable model. On the other hand, Weinberg also associates his demand for a general system thinking with the ***genesis*** of the thinking model itself. The thinking model can also consist of several

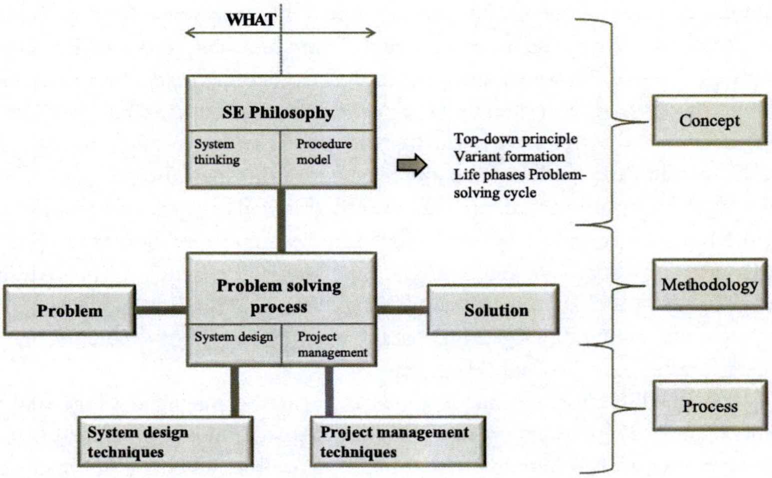

Fig. 3.1 The SE concept according to. (After Haberfellner et al. 2018)

sub-models. A genesis must then also be established between them (Schnieder and Schnieder 2013).

The generalized system thinking is also demanded by SAGE and ROUSE (Sage and Rouse 2009). They justified their vision of generalized system thinking on the one hand with the necessity of complexity management and on the other hand with the philosophical basic idea that complex relationships can only be solved by generalistic, simplified and standardized explanatory patterns. Systems themselves develop or change. The same applies to the system environment and consequently to the interactions between the two. This change process must be taken into account when developing the problem-solving approach, i.e. the procedural concept. It is conceptually implemented by HABERFELLNER (Haberfellner and Daenzer 2002; Haberfellner et al. 2018) by connecting the system thinking with the procedural model for the problem-solving process. This in turn can be supplemented by appropriate methods and procedures, as Fig. 3.1 illustrates.

HABERFELLNER does not use the term thinking model. However, this should be retained in the GSE because ***system thinking and the image of the system in the form of a thinking model have been proven to be an important basis of SE*** (see Sects. 2.2 and 2.4). It must be generally, i.e. for all types of systems, creatable, as demanded by WEINBERG, SAGE and ROUSE. But it should also represent the development of the system, which can also be a result of the step-by-step implementation of the problem-solving cycle, in a generic way. But what consequences does this have for the thinking model, or the procedural concept? This will be examined in more detail in the following Sects. 3.2 and 3.3.

3.2 The Demands on the Thinking Model of the GSE

The ***GSE thinking model*** must be based on system theory. According to DÖRNER, humans are not capable of recognizing complex relationships without failures (Dörner 2007). Through ***thinking in systems*** (see Häuslein 2004; Lindemann 2016; Bender and Gericke 2021), complexity can first be made comprehensible and then made transparent in the next step. HABERFELLNER characterizes system thinking as a "way of thinking… that makes it possible to better understand and shape complex phenomena (systems)…" (Haberfellner et al. 2012, p. 31). System thinking uses explanations derived from system theory, such as iterative, structured, relation-oriented, alternative thinking (Simon 2006). Thus, it is excellently suited to better control the complexity and dynamics of the problem-solving process. The system as a mental construct and its environment are a result of system thinking (Heinrich 2015). LUHMANN holds the view that the system itself must reduce complexity. By working out the difference and the interactions between system and environment, a starting point for dealing with complexity is thus created (Luhmann 1980). As a result, a modular image of real systems based on standardized views is to be developed as a thinking model for the GSE. The basic requirement that it must be usable as a ***metamodel,*** i.e. as a cross-disciplinary thinking model, for various scientific disciplines, had already been worked out and justified. But what other requirements must the thinking model of the GSE to be developed meet?

WEINBERG holds the view that everything can be described as a system and refers to this as a "general system" (see Weinberg 2001, p. 28). ***Regardless of what is being considered, it can be described as a system,*** whether it is a person, politics, a natural event, a product, a company, etc. Also SIMON clearly points out that systems can describe facts and things abstractly and are not bound to specific contents. Thus, they can be used for problem solving in almost all scientific disciplines (Simon 2006).

Following these views, another essential requirement for the thinking model of the GSE emerges. It must generally apply to many scientific disciplines and for all types of systems. The further specification of this requirement includes a consideration of the ***types of systems*** and the search for cross-system characteristics with the aim of being able to describe systems universally and standardized. The character or type of system is described by its elements and their interrelationships with each other or with the system environment.

Consequently, ***categories of system types*** can be formed, among others, according to:

a) the type of system elements,
b) the relationship of the system with its environment,
c) the location reference of the system and
d) their properties.

to a) System types characterized by their elements

a1) technical systems

Technical systemsrefer to product systems with which humans interact, but are not themselves part of the system. These include, for example, machines, facilities, software and hardware. Technical systems can be subdivided at will into sub- or partial systems or elements, which in turn interact with each other. The transformation of the input into the output is carried out via the function of the system. Thus, the function describes the purpose of the system. Consequently, the function of a lathe is to remove chips through rotational movements, or the function of a wind turbine is to convert wind into electrical energy. PAHL and BEITZ (Bender and Gericke 2021) advocate considering such technical structures as systems that are connected to their environment via their input (inputs) and output variables (outputs). What belongs to the considered system is determined by the respective system boundaries.

a2) sociotechnical systems

In contrast to the technical, ***sociotechnical systems*** consist of the relationships between humans and machines or facilities. The human being is an inherent and determining part of the sociotechnical system (Ehrlenspiel and Meerkamm 2017; Bender and Gericke 2021; Mistler 2021). Thus, people (social systems) and technical systems are components (elements) of the sociotechnical system (Maucher et al. 2002; Bender and Gericke 2021). These interact with each other to create products and/or services (Mamrot et al. 2012; Mistler 2021). In the case of defining a work system as a sociotechnical system, the division of functions between human and machine significantly determines the character and level of the sociotechnical system (Winzer 1997). In this context, sociotechnical systems can also be understood as organizational systems with an implicit process and structural organization according to MISTLER (Mistler 2021). An organizational system is "to be understood as a set of interconnected and mutually influencing elements to achieve a common goal (function) via input and output. The input of an organization is customer requirements and resources, which are transformed into fulfilled customer requirements and consumed resources as output. The fulfilled customer requirements refer to the developed product or service (performance)" (Mistler 2021, p. 41). Basically, the human being, the operating resource and the work object as the essential system elements belong to the work system and the way they interact determines the work process, the organization and are influencing factors on the quality-compliant production (Westkämper 1991; Braunholz 2006; Mamrot 2014; Mistler 2021).

a3) social systems

If only humans interact with each other, it is a matter of pure ***social systems.*** (However, the cohabitation of animals also falls into the category of a social system.) Social systems are of particular interest to sociology, psychology and pedagogy.

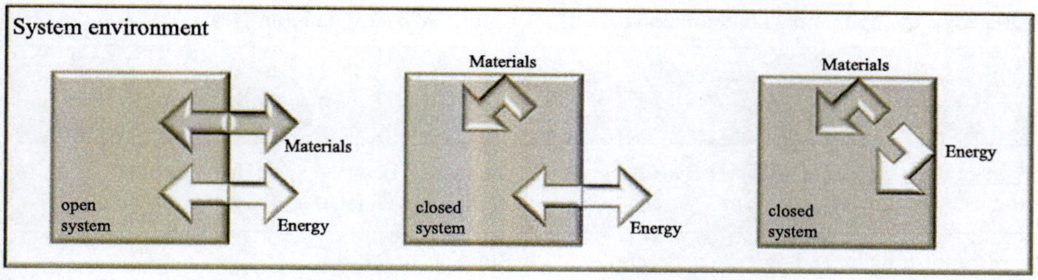

Fig. 3.2 Example of system types. (After Atkins and Paula 2006)

to b) System types that differ in their relationship to the environment
In this context, ***open, closedand isolated systems*** are distinguished. They can be described by the type of their material and energy exchange with the environment, as Fig. 3.2 illustrates. This can also be extended to the system's exchange of information with its environment. In open systems, there is an exchange with the system environment, but not in isolated ones. The description of the system's interaction with the environment can only take place if the reference system is defined, i.e. the system boundary has been determined. As a result, input and output variables can be fixed and their flow direction can be described (Schnieder and Schnieder 2013).

to c) System types that differ in their location reference
Social or sociotechnical systems that refer to a region are referred to as ***regional systems***. If these systems apply to various regions, they are national, and if they extend worldwide, they are ***international systems.***

d) Types of systems distinguished by their properties
However, it is also possible to categorize types of systems based on their properties. Such ***system properties*** can be:

- State (e.g., stable, unstable, dynamic),
- Function (i.e., the goal of the system, such as a replacement system, compensation system),
- Structure (e.g., hierarchical, chaotic) or
- Behavior (e.g., stochastic, discontinuous) (see Jumar et al. 2010).

LINDEMANN derives the following ***characteristics of the system*** which he exemplifies as shown in Table 3.1.

In principle, the view of HÄUSLEIN (Häuslein 2004) should be followed, ***to describe systems universally, regardless of the type of system, using the following characteristics:***

Table 3.1 Components and characteristics of a system. (After Lindemann 2005, p. 10)

Components and characteristics of a system are:	Example: Car engine
System boundary: demarcation between a system and its environment	Distinction between basic machine and auxiliary units (starter, alternator, oil pump)
Structure: structure and order of a system	Building structure
Elements: (different) building blocks of the system	Piston, valve, cylinder
Relations: (different) connections between elements and systems	Type of connection, heat transfer
Input/Output: Relations of the system to an environment (open systems)	Air, fuel, electrical energy - mechanical power, thermal energy, exhaust gases
Environment: systems and elements outside the system boundary	Engine compartment, car occupants
System behavior: dynamic behavior of the system	Elasticity

- System boundaries,
- System input and output,
- System relations,
- System elements and
- System environment (see Fig. 3.3).

In addition, HABERFELLNER (Haberfellner and Daenzer 2002; Haberfellner et al. 2018) state that systems can be described in their complexity by:

- Elements:
  - Type and diversity of elements,
  - Number of elements and.
  - Inequality of the distribution of elements,
- Relations:
  - Type and diversity of relations,
  - Number of relations and
  - Inequality of the distribution of relations and
- Dynamics, also the system's own dynamics
  - Type and number of possible states of the system,

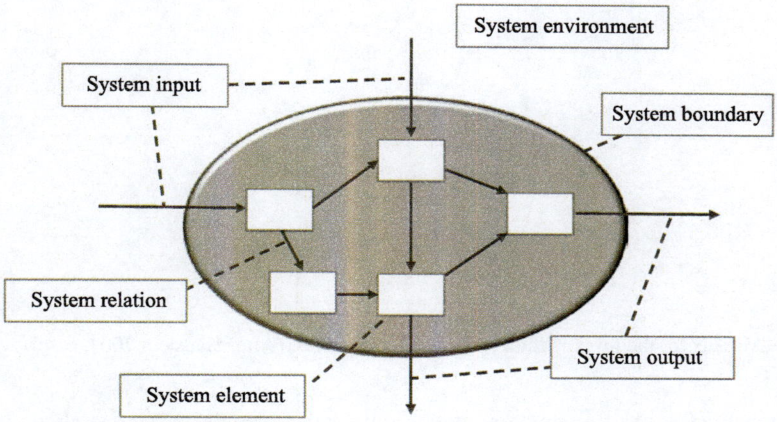

Fig. 3.3 Characteristics of systems. (After Häuslein 2004, p. 29)

are further generally describable (Haberfellner and Daenzer 2002; Haberfellner et al. 2018).

Delineating the system from its environment is a significant contribution to reducing complexity. Thus, the system boundary has a special significance in the system consideration (Simon 2006; Schnieder and Schnieder 2013). The authors point out that the representation of the system's interaction with its environment can be concretized by defining superior and subordinate systems, as shown in Fig. 3.4. The system environment can also remain undifferentiated and general (Simon 2006). However, the interaction between the system and its environment must be considered.

Fig. 3.4 System—Subsystem—System elements. (After Haberfellner et al. 2018, p. 17)

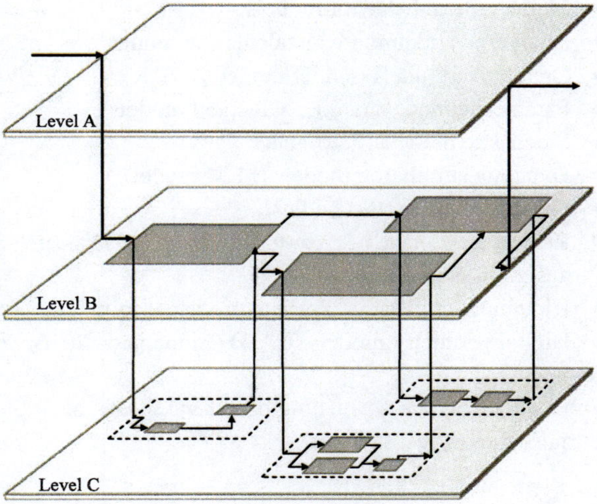

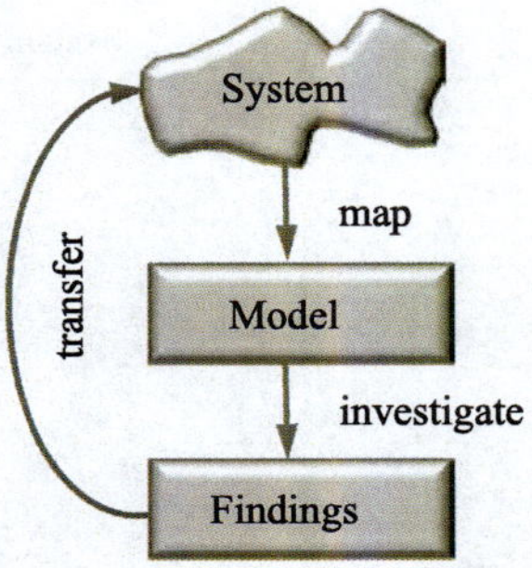

Fig. 3.5 Models for the investigation and design of systems. (After Häuslein 2004, p. 40)

After evaluating the literature and defining the characteristics of a generalistic system to be described, these can now be transferred into a model. The ***model is understood here as a representation of the system***. It is therefore not a representation of reality, but only a representation of the previously outlined system and its system characteristics. Models are purpose-bound representations of systems that serve general knowledge gain. They are the basis for system designs and carry a certain uncertainty (Viertl and Yeganeh 2013). Furthermore, EBERT (Ebert 2019) points out that models are not perfect and it is not sensible to strive for perfection in modeling. In modeling, it is much more crucial to represent the essential content, omit unimportant content, and not to deal with or get caught up in minor details (Ebert 2019). Models have various addressees and subject areas. Thus, a variety of model types arise (Schnieder and Schnieder 2013) (Fig. 3.5).

Consequently, there are a variety of ***model types*** such as:

- Thought models,
- Mathematical behavior models,
- Analytical and numerical calculation models,
- Qualitative models (ladder model, ACT-R),
- Parametric models (static, statistical models),
- Stochastic models (Petri nets),
- Dynamic simulation models (FEM model),
- Physical models (test bodies),
- Real models made of easy-to-process materials, such as wood and hard foam, to make design models tangible,
- Kinematic models, in physical or virtual form, and
- Digital geometry models (CAD) (Lindemann 2016; Schnieder and Schnieder 2013).

Other model types are distinguished based on the subject matter they represent or the purpose they serve, such as:

- Object model,
- Function model,
- Product model,
- Process model,
- Information model and
- Computer-integrated model (Bender and Gericke 2021).

These various model types serve, among other things, due to the high complexity of the products, to better understand both the designers and developers among themselves, but also the designers and developers in the communication process with the customer. Thus, VR technology is increasingly used to better and faster transform customer wishes into a concrete product (FQS 2016).

Accordingly, the models or model types were developed to:

- have a better understanding of the system to be developed or existing,
- better understand the complex relationships through abstraction and reduced images,
- distinguish the essential from the non-essential,
- enable interdisciplinary communication in the development process,
- recognize the specification, the structure, and the specific behavior in advance and simulate it, as well as
- document the findings about the system,

to name just a few. This corresponds in analogy to the demands that a model and thus also the ***GSE thinking model*** must fulfill.

The ***thinking model*** of the GSE approach should initially only be the cognitive way of creating images of reality using system thinking. It serves as a ***metamodel,*** i.e., it is a transdisciplinary thinking model. However, there are also very differentiated views on the metamodel. They range from the transdisciplinary thinking model, the transdisciplinary system model, the networked system model of subsystems to the model that integrates various IT models (Jeckle 2000; Tabeling 2002; Bender and Gericke 2021; Gausemeier et al. 2013; Pohl 2016; Krcmar 2015). Despite these different views on the metamodel, interestingly, there are similar requirements for it, such as:

- The basis should contain few basic terms.
- Transparency and clarity are desired.
- The networking of various models must be possible.
- Changes to the submodels are to be incorporated into the metamodel and vice versa.
- The various system worlds or models should be integrated.
- The transparency of the link to the other models must be guaranteed (Jeckle 2000; Tabeling 2002; Bender and Gericke 2021; Gausemeier et al. 2013; Bernhard and Jahn 2009; Krcmar 2015; Pohl 2016).

Accordingly, the thinking model of the GSE approach has to represent the system with its characteristics in such a way that it is understood transdisciplinarily by all participants and meets the previously mentioned requirements for a metamodel. Thus, the GSE thinking model is a concept for the simplified representation of complex systems and must also meet the general requirements for models. The GSE thinking model serves to represent the most important properties of the system and to disregard the secondary properties. The basic principles of system thinking, such as the basic principle of minimal models, the basic principle of structuring, and the basic principle of standardization, must be observed and implemented. It is essential that the internal relations, i.e., the interrelationships between the elements, and the external relations, i.e., the relation between the system and the environment, are described. ***The model is thus understood as a simple mental image of the system and its interaction with its environment*** (see Imboden and Koch 2005).

Models, according to HÄUSLEIN, are not wrong or right. They are exclusively purpose-bound. He characterizes different ***usage categoriesof models:***

- "***Descriptive models***serve to document the depicted systems, so that their essential properties become comprehensible and communicable.
- ***Explanatory models*** represent explicit assumptions about the depicted systems, with which phenomena of the systems, whose reasons cannot be directly determined, are to be explained, especially their behavior.
- ***Forecast models*** serve to predict the future behavior of the depicted systems. They are often realizable as simulation models, in which the state of a model is gradually changed and thus its behavior can be determined.
- ***Design models*** serve the design of new or the modification of existing systems and
- ***Optimization models*** serve to improve the systems. This can affect both the structure and the behavior of the system" (according to Häuslein 2004, p. 40).

SCHNIEDER adds the following additional model usage categories:

- ***Decision models,*** which help to select the preferred solution from the developed variety of variants.
- ***Cognition models,*** which contribute to improving system understanding and can reveal specification gaps.
- ***Knowledge models,*** which serve to store knowledge (cf. Schnieder and Schnieder 2013).

The usage categories of SCHNIEDER can be coupled with those of HÄUSLEIN through a corresponding function extension. These are:

- The ***descriptive model***serves not only for the ***documentation of systems,*** but also for the ***storage of knowledge.***

- The ***design model*** not only enables the ***design of new and/or the modification of existing systems,*** but supports system understanding and ***reveals system specification gaps.***
- The ***optimization model improves*** not only the ***systems,*** but ***supports*** necessary ***decision-making processes.***

The GSE thinking model must correspond to all these extended usage categories, i.e., it must be ***both a descriptive, explanatory, forecast, design, and optimization model at the same time.*** This means that the GSE thinking model initially represents, for example, the system "logistical facility". By a suitable, yet to be chosen description of the interactions between the system elements or between the system and its environment, certain behaviors of the logistical facility can be explained. Thus, the descriptive model also serves as an explanatory model. If it is apparent from the model that the interactions, for example, between the system elements rollers, belt, and drive are not optimal and this leads to an unfavorable efficiency of the drive, the model can also be used as a design or optimization model. Consequently, the GSE thinking model must correspond to all five usage categories of models, i.e., it must allow general use.

The ***models are*** also distinguishable by their properties ***in addition to their usage***, such as:

a) the black-box model,
b) the dynamic model and
c) the structured thinking model.

They are first briefly characterized below to then describe their possible applications.

a) **The black-box model:**
The ***black-box model*** is a greatly simplified image of reality. The system's characteristics are determined by defining system boundaries, which are defined via input and output. It focuses on the essentials of a system. The advantage of the GSE system approach by WINZER over the system approach by, for example, HABERFELLNER et al. is on the one hand the more precise demarcation of the system from its environment and on the other hand the bundling and filtering of the requirements for a system (Winzer 1997; Haberfellner et al. 2018). This allows the viewer of the system to specify problems more precisely and to develop a more precise idea of how far the system boundaries should be drawn (Mistler 2021). The black-box model models the system with its interactions with the environment, while the other system characteristics are ignored. This means that the interior of a system is not considered, i.e., it is temporarily irrelevant or unknown (Haberfellner et al. 2018; Parnell et al. 2008). This model approach is particularly suitable for simplifying complex problems and quickly generating solutions. However, it completely neglects the interrelationships of the elements with each other and the structure of the system—which of course also

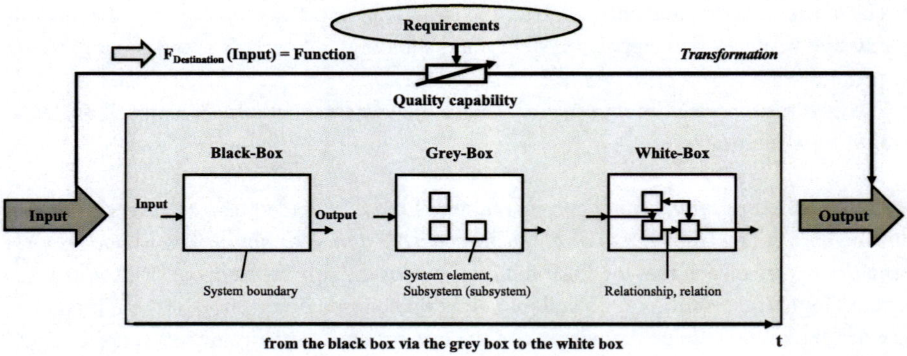

Fig. 3.6 Black-Box: System design from the black-box to the white-box. (After Mistler 2021)

significantly determine input and output. As MISTLER points out, not only new systems are developed, but also existing systems are modified and incrementally further developed. This requires the use of the grey- and white-box models (Mistler 2021). These represent an extension of the black-box model according to HABERFELLNER et al. and PARNELL et al. to detail the system view (see Fig. 3.6).

As Fig. 3.6 outlines, the black-box model represents the coarsest detailing of a system model, as it does not show any system elements and relationships. Only the system inputs and the system outputs are defined. Based on the black-box model, the grey-box model can be formed. It lies between the black- and white-box model view. The grey-box model shows a rough or partial structuring of the system elements. In the white-box model, the specific relationships between the system elements are additionally described. They are either of high importance, for example, to carry out simulations, or the internal relationships must be defined to allow a more concrete view of the system (Parnell et al. 2008; Haberfellner et al. 2018). MISTLER adapts the idea of the black-, grey- and white-box models according to PARNELL et al. and HABERFELLNER et al. to the GSE system approach according to WINZER (Winzer 1997), as can be seen in Fig. 3.6. This means that a system is always defined via input and output. The system input determines the function of the system. The function of the system is influenced by requirements that are placed on a system. The degree of their fulfillment is an expression of the system's quality capability. If requirements can be assigned to subsystems, a determination of the quality capability of the subsystems is also possible (Mistler 2021).

b) **The dynamic model:**

The ***dynamic model approach*** particularly takes into account that humans and the world around them are subject to constant changes or that the system elements are constantly in motion. Such a time-dependent, dynamic model describes the temporal change of system characteristics "discretely" for state changes at a time jump and

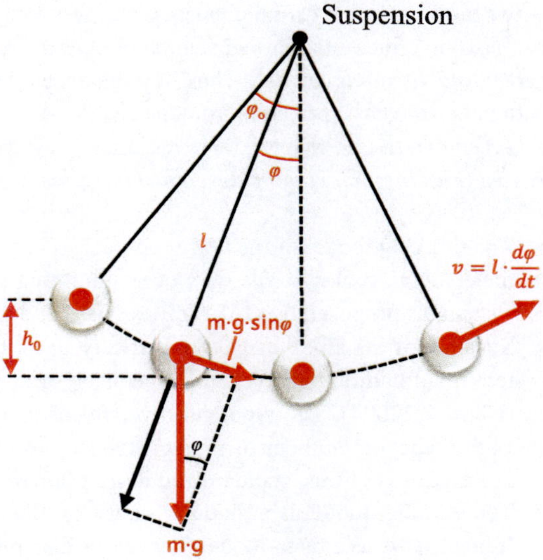

Fig. 3.7 The representation of the pendulum as a dynamic model[1]

"continuously" for state changes over any period of time. Today, modern mathematics in particular offers a multitude of methods for creating dynamic models (Fig. 3.7).

c) **The structured model:**
The ***structure model approach*** focuses on the subdivision of a system into subsystems as a hierarchy structure, as a network structure, as chaos, or as a cluster. Structured models thereby allow on the one hand a reduction of complex entities by summarizing similar states and on the other hand their refinement by a initially rough structuring subsequently experiencing an increasing detailing. Through this model approach, relationships in the system can be recognized and represented.

In principle, the individual mentioned model types can of course also be merged with each other in order to analyze a system from different aspects. For example, it makes sense to first consider a problem via a black-box model at time t_0. This way, it can be assigned to a system and its interactions with the system environment can be recognized. Subsequently, the system can be characterized more precisely in the form of a dynamic model at time t_n -***coupling of black-box with dynamic model***. A desired projection of the development process from the system to time t_{n+1} then supports a model that simulates these state changes. If detailed information about the effects of system-internal processes on the interactions of the system with its environment is required, the black-box model approach additionally needs, for example, a hierarchical image of the system. For deriving the relationships

[1] After http://www.ullala.at/experiments/movement/images/pendel1.gif

between the systems or the elements, matrices or network structures are suitable, which can also be represented in their temporal change—***coupling of black-box with the structured and dynamic model***. Thus, a fundamental principle of systematic thinking and action, i.e., the basic principle from the abstract to the concrete, can be implemented.

As shown in this chapter, there are many ways to derive images of systems, so a ***generalist thinking model to be developed must meet the following requirements:***

- The thinking model of the GSE approach is based on systems theory. It is a generalist understanding of systems, i.e., everything that surrounds us can be represented as a system, as proposed by DAENZER et al. (Lindemann 2016).
- Systems of all kinds can be universally described through system boundaries, system input and output, system elements, system relations and the system environment (Häuslein 2004). The type and diversity of the elements and their relations, as well as the type and number of system states (system changes), also belong to the system description. For this, standardized description procedures need to be developed.
- The thinking model is a modular image of real systems, which is subject to changes. It must serve as a meta-model for various disciplines.
- The dynamic behavior and development of the system must be transparent and comprehensible through the thinking model.
- By coupling the black-box model with the implicit grey- and white-box model, as well as the hierarchical and dynamic models of the system, the complexity of systems can be systematically limited or expanded again.
- The image of this generalist system essentially serves in the GSE as a thinking model(cf. Lindemann 2016), which is intended to support discursive thinking.
- The modeling must represent the circularity of cause-effect relationships in their temporality as well as the connection between function and behavior (Schnieder and Schnieder 2013).
- Furthermore, the thinking model of the GSE approach to be developed should serve as a ***description,explanation, forecast, design, and optimization model*** (cf. Häuslein 2004; Schnieder and Schnieder 2013). The basic principles of systematic thinking and action should be included. This should result in a procedure for building and handling the thinking model.

The thinking model must also meet the detailed ***requirements for a system model*** set by SCHNIEDER (Schnieder 1999; Schnieder and Schnieder 2013), such as:

- the modularization or
- the structure of models,
- the representation of scaling and relations,
- the description of properties,
- the conditions of objects, the dynamics of the model and
- the causality, or determinacy of systems with its environment.

These detailed requirements for a system model resulted from a comprehensive literature review and will be tested for their feasibility in Chap. 4. They serve as a framework for the development of a generalist thinking model approach.

Consequently, the goal is to develop a ***GSE thinking model that modularly maps real systems through standardized views.***

This simultaneously requires a ***description procedure*** for the number and diversity of the elements and their relations, as well as the system changes. If, for example, an interdisciplinary team succeeds in describing the components, functions and processes of the robot, as well as their interactions with each other, in a common language and representing them as a system model in the form of a ***meta-model***, the basis for common thinking and action is thus created. How the robot is specifically designed can be described by the procedural concept. The same applies to the design appropriate to requirements of organizations, which represent sociotechnical systems. However, since the universal character of the SE procedural concept was lost, it must now be examined in the following chapter which requirements arise from the trends of complexity for the development of a procedural concept within the framework of the GSE.

3.3 The Possibilities of Restoring a General Procedural Concept within the Framework of the GSE

The demand for the restoration of the generalist character of the procedure concept of the SE can in principle be achieved in two ways:

1. An existing universal procedural concept is selected for the GSE approach, structured into standardized modules, and coupled with problem-specific methods.
2. Through a detailed comparison of the universal and specific procedural concepts, ideas for the development of a new procedure concept for the GSE are generated.

The first path seems very simple at first glance. BAHILL, HABERFELLNER, ARLT, LINDEMANN, WEILKIENS, SAGE, and ROUSE (Bahill and Gissing 1998; Arlt 1999; Züst 2004; Lindemann 2016; Weilkiens 2019; Haberfellner et al. 2018) developed universal procedural concepts. But which of these procedural concepts should be selected, how should it be modified, and with which methods and procedures should it be combined? The basis for this could be a purely pragmatic approach, i.e., the sole reduction of the procedural concepts to some overarching steps, as implemented by JENKINS, BAHILL, and RINK (cf. Jenkins and Youle 1971; Bahill and Gissing 1998; Rink 2002). However, there is little prospect that such an approach alone will prove to be successful. On the one hand, a reduced yet practical, solution-leading procedural concept would have already established itself in practice and would now be available. On the other hand, practice has shown that a meaningful problem solution often required additional steps due to subject- or problem-specific peculiarities. As a result, further specific and universal procedural concepts were created. This development continues to this day, as demonstrated in Sect. 2.5.

For this reason, the second path, i.e., the detailed comparison of the universal and the special procedural concepts, will be used to search for possibilities to restore its generalist character.

If the demand for the restoration of a universally valid procedural concept in SE is to be implemented, it initially requires:

1. conceptual clarity about what a procedural concept of SE is,
2. the uncoupling of demands on the general solution approach based on the basic principles of systematic thinking and action,
3. the detailed analysis of existing, universal elements in the various procedural concepts of SE, and
4. the combination of these aspects into a generalist procedure concept of the GSE.

to 1. The procedural concept—conceptual clarification

PAHL/BEITZ (cf. Bender and Gericke 2021), LINDEMANN (Lindemann 2016) and HABERFELLNER (Haberfellner et al. 2018) speak in SE of procedure models. The procedure models, according to LINDEMANN, support the planning of processes and the navigation within processes for the targeted determination of the next steps (cf. Lindemann 2016). Human actions, according to DAENZER, are based on certain patterns (cf. Lindemann 2016). If these lead to success and are to be made transparent, they correspond in an abstracted, generalized form to a model that describes the idealized procedure (cf. Lindemann 2016). In SE, procedure models are understood as a guide to achieving a certain goal. In procedure models, LINDEMANN continues, important elements of a sequence of actions are depicted, which can serve as aids for planning and controlling processes. HABERFELLNER emphasizes that the procedure model consists of "various basic principles and components that are supposed to help subdivide the development and realization of a solution into manageable sub-steps" (cf. Haberfellner et al. 2012, p. 27). SEGHEZZI refers to mental representations that can be implemented with the help of models as concepts (Seghezzi 2007).

Since the concept of the model (cf. Sect. 3.2) is understood as an image of the reality of systems, the concept of the procedural concept will be introduced. The ***procedure concept*** includes the temporally logical sequence of actions, which are supported by models and can be standardized and modularized in a certain way. The conceptual idea is based on that of SCHNIEDER (cf. Schnieder 1999), where concepts "also need to be adequately modularized and then also represented, or formalized" (Schnieder 1999, p. 200). A module is a chapter to be formed on the basis of fixed values, which can be combined arbitrarily to generate new solutions. It is to be designed for reuse by fulfilling a certain function and thus being clearly delimited. Consequently, a procedural concept can enable modularity[2] and agility[3] (cf. Mistler 2021).

to 2. Derivation of Criteria for Comparing Prodecural Concepts
However, before it comes to exploring or developing new modules that meet the generalist claim, it is of course obvious to first identify existing, universal potentials, concepts and elements of the existing SE approaches. The main difficulty here is to look for commonalities in the approach concepts, which are hard to recognize due to the many specializations of engineering sciences and their correspondingly very different technical terms.

It is therefore about finding and understanding overarching, engineering contexts or procedures, which could serve as the basis of a general, modular procedural concept of SE. However, such cross-disciplinary methodological approaches should be oriented towards the basic principles of systemic thinking and action. In addition to existing, so to speak visible, universal approaches of SE as well as previously rather hidden commonalities of individual discipline-specific procedures of SE, methods and procedures are particularly needed that ensure a practicable problem solution. Ultimately, only realistic system descriptions and problem-solving methods guarantee the ability to master and shape the increasing diversity of complex tasks. Universally manageable and applicable solution principles can only be mastered if a systematic ***limitation of the solution spaceis based on a universal system model.*** However, practice shows that such solution concepts quickly become self-sufficient if they do not immediately capture changes occurring simultaneously in reality. Here it is necessary to implement a meaningful problem-solving-oriented time determination on the one hand and to adapt the level of detail of the system model to the real problem description on the other hand. In principle, the procedural concept should apply the essential basic principles of systematic thinking and action, such as the basic principle from the whole to the detail, the basic principle from the abstract to the concrete, the basic principle of modality change as well as the basic principle of thinking in alternatives or the basic principle of problem decomposition, depending on the type of problem to be solved, in extension of HAFERFELLNER (cf. Haberfellner et al. 2018).

This is possible if the views of HABERFELLNER (cf. Haberfellner and Daenzer 2002; Haberfellner et al. 2018) and ZÜST and SCHREGENBERGER (cf. Züst and Schregenberger 2003) are followed and ***the procedural concept of the GSEis linked with project management.*** By coupling the basic ideas of SE with project management,

[2] In relation to the problem-solving process, modularity means that it is possible to omit steps or modules or to let them take place in a varying order (Mistler 2021).

[3] Agility in the problem-solving process means ensuring adaptability through modularity, so that new requirements or changes in requirements can be flexibly absorbed and implemented (Mistler 2021).

it is possible to implement the basic principles of systematic thinking and action in the approach concept depending on the problem to be solved. HABERFELLNER considers project management as an organizational component in the SE problem-solving process and recommends its use in large projects (Haberfellner et al. 2018).

Although there are again different views on project management, it essentially pursues the effective realization of projects or tasks or problem situations within a certain time, with corresponding resources (Hoffmann and Mörike 2008; Welge et al. 2015). This chronological sequence of actions and the use of resources are planned, realized, and the status of success is controlled using project management.

The phases of project management are on the one hand so universal that they can be applied to any type of task to be solved and on the other hand these phases offer the possibility to implement the basic principles of systematic thinking and action individually. As a rule, the way of project management is often differentiated into classic and agile, although there are also approaches that combine classic and agile project management. These are then referred to as hybrid approaches. Classic project management is characterized by a sequential or waterfall-like approach. Agile project management, on the other hand, aims for an iterative and incremental approach with stakeholders (cf. Habermann 2013; Kuster et al. 2019; Preußig 2015; Heinke and Mistler 2019). Agile approaches are currently gaining more and more importance, as the increasing dynamics of the market require new or constantly changing requirements in system development to be flexibly adapted (Schuh et al. 2020; Dumitrescu et al. 2021; Rupp 2021).

In summary, this means: if it were possible to extract universal modules from the multitude of procedural concepts discussed in Sect. 2.4 that must be used in a problem-solving algorithm, and if these modules could be standardized and simplified so that they are applicable to various disciplines, then these modules could be coupled temporally, logically, and individually according to the specification of the problem to be solved via the corresponding project management. This would enable efficient problem solving and an procedural concept could emerge that is applicable to problems of all kinds, i.e. generally, and that can be adapted again and again to the specific problem situation or that can continue to develop itself according to the existing requirements, i.e. is generic.

The ***developing GSE approach concept*** must therefore allow the following through ***interaction with the GSE thinking model:***

- the problem definition with its assignment to the considered system,
- the goal formation and specification,
- the analysis of the system and its environment,
- the development of solution alternatives,
- the evaluation of the various solutions,
- the testing and improvement of solutions,
- the planning, control, and monitoring of the problem-solving process and the necessary resources via project management, and

- the integration of various methods, procedures, and (IT) tools that require the specificity of the problem to be solved.

These are further requirements for the generalist procedural concept of GSE, in addition to those already fixed in Sect. 2.4. They serve as comparison criteria for the various universal and specific procedural concepts of SE.

to 3. DetailedAnalysisof Selected Universal and Specific SE Procedural Concepts

The following are already preselected, universal and specific approach concepts of SE. They are comparatively analyzed on the basis of the previously mentioned criteria. The goal is to find modules for the new procedural concept of GSE to be developed.

First, ***universal SE procedural concepts are considered in more detail than in Sect.1.4, which are based on system thinking, partly incorporate the basic principles of systematic thinking and action, and have a transparent procedural concept.***

Table 3.2 clearly shows in detail for the procedural concepts that the reference to system thinking is established directly or indirectly via the problem definition.

Basically, in all universal approaches, goal setting is at the center. This includes goal specifications based on determined requirements and the weighting of goals. The analysis of the problem or the system and its environment is also seen as an important step in

Table 3.2 Comparative consideration of universal SE procedural concepts

Request / Source	Problem definition	goal setting	Analysis	Development of alternative solutions	Solutions evaluation	Testing and improving solutions	Project management	Integrating methods and procedures
BAHILL 1998	●	●	○	●	●	●	○	○
SELL 1989	○	●	●	○	●	○	◐	○
WULF 2002	●	●	○	●	○	●	◐	○
IEEE 1220 2005	◐	●	●	●	◐	◐	◐	○
SAGE u. ROUSE 2009	●	◐	●	●	●	●	●	◐
LINDEMANN 2016	◐	●	◐	●	◐	◐	○	●
EHRLENSPIEL u. MEERKAMM 2017	○	●	◐	●	●	●	◐	●
WEILKIENS 2019	○	◐	◐	●	●	●	●	◐
HABERFELLNER et al. 2019	●	●	◐	●	●	◐	●	●
SCHLUETER et al. 2019	●	●	●	●	●	●	◐	●
MISTLER 2021	●	●	●	●	●	●	●	●

Legend: ○ not applicable ◐ partially applicable ● applicable the majority of procedural concepts. This is often followed by the development of solution alternatives. While some authors include the evaluation of solutions, their testing, and improvement in this step, others list this separately. The same can be observed with regard to project management. As HEINKE and MISTLER point out, Systems Engineering is often considered separately from Project Management (Heinke and Mistler 2019). To integrate Project Management into Systems Engineering, MISTLER structures the procedural concept in a modular way. The application of the modules can then determine the way of Project Management. For example, it can be determined whether a classic or agile approach is carried out (Mistler 2021).

Thus, MISTLER meets the emphatic demand of HABERFELLNER, SAGE, and ROUSE as well as WEILKIENS (Haberfellner et al. 2018; Weilkiens 2019; Sage and Rouse 2009) to control the ***problem-solving process*** via project management. Other authors standardize the problem-solving process by fixing cycles to be run through in the approach concept (cf. among others Sell 1989; Bahill and Gissing 1998; Rink 2002; Ehrlenspiel and Meerkamm 2017; IEEE Computer Society (IEEE) 2005). Of course, such a view neglects the consideration of the many special features of complex problems and their solutions. Therefore, it is not surprising that almost all procedural concepts, in addition to additional phases or methods that serve the specificity of the problem or the discipline, also contain these basic phases of a problem solution. With the help of Table 3.3, further ideas for the development of the generalist approach are to be collected through a deeper comparison of specific procedural concepts of SE approaches than in Sect. 2.4.

Also in the comparison of the special procedural concepts that make use of SE, it becomes clear that goal setting and analysis are also necessary in a specific problem-solving process. This also applies to the development of solution alternatives, the evaluation of solutions, and the testing and improvement of solutions. The latter is summarized in VDI 2221 under the term design. In product development but also in software engineering, it is emphasized that the product life cycle should be taken into account. Especially in product development, but also in manufacturing engineering, the integration of further problem-specific methods and procedures into the approach is pointed out. The chronological control of the problem-solving steps via project management is particularly in focus in the approach concept of software engineering. However, other specific sequences of steps emphasize that the problem-solving cycle is to be planned, controlled, and monitored.

If the respective comparison results are reconsidered in summary, it becomes clear that in all procedural concepts there are modules that contribute to setting goals, analyzing the system, and designing it. The design process includes the search for solutions as well as variant development including implementation. The problem-specific, chronological sequence of individual steps from these phases can be controlled via project management. The planning, implementation, and control of the problem-solving process must be ensured. Thus, the demand for a coupling of SE with project management (Weilkiens 2019; Sage and Rouse 2009; Haberfellner et al. 2018) is implemented.

Table 3.3 Comparative consideration of specific SE procedural concepts

Source \ Request	Problem definition	Goal setting	Analysis	Development of alternative solutions	Solutions evaluation	Testing and improving solutions	Project management	Integrating methods and procedures
Product development								
BENDER u. GERICKE 2021	●	●	●	●	●	●	◐	◐
VDI 2221	●	○	●	●	●	●	◐	◐
FEY 1998	◐	●	◐	●	●	●	●	○
VDI 2206	◐	●	●	●	●	●	◐	◐
GAUSEMEIER et al. 2014	●	●	●	●	●	●	◐	◐
Software development								
IEEE 1220	○	●	●	●	●	●	◐	○
SOMMERVILLE 2007	○	●	●	●	●	●	◐	○
FUCHS 2001	○	●	●	●	●	●	◐	○
Requirements Engineering								
OTT 2014	●	●	●	●	●	●	◐	◐
Manufacturing Engineering								
ICE 61508	◐	●	●	●	●	●	○	○
EN 954-1	◐	●	●	●	●	●	○	○
SCHENK 2004	●	●	●	●	●	●	◐	◐
VDI 5200	●	●	●	●	●	●	●	◐
GRUNDIG 2018	●	●	●	●	●	●	●	◐

Legend: ○ not applicable ◐ partially applicable ● applicable

Consequently, a universal problem-solving cycle, i.e. a ***generalapproach concept, can be limited to the four modules, i.e.:***

- the goal setting,
- the analysis,
- the design, and
- the project management,

which can be combined with methods and procedures according to a specific problem situation.

4. Combining the comparison results into a generalist approach of the GSE
Fundamentally, a generalist approach should naturally ensure the universal solving of very specific problems using analysis, goal formation and design modules, coupled with specific methods and procedures, and controlled via project management. Of course, the focus of the approach is on a permanent identification of the progress of problem solving, which is why the methodical approach should allow a systematic or iterative scanning of the problem-solving space, both as a top-down and bottom-up approach. ***The interaction between the GSE thinking model and the modules of the GSE approachmust be given at all times.*** The same applies to the integration of the basic principles of systematic thinking and action. In addition to these general, overarching demands on a general procedural conceptkann to problem solving, the modules goal formation, analysisand design each characterize specific properties.

In particular, the increasing dynamics of many processes in the context of value creation shape ***the goal formation.*** Accordingly, the demand to precisely define the goals at the beginning of a problem-solving cycle is to be questioned. With the increase in the complexity of the problems, not only the duration and scope of problem solving increase. At the same time, the problem itself and thus also the goal formation are subject to a noticeable change in dynamics. Therefore, iterations between goal formation, analysis, and design are needed to concretize the goal setting of system development. This means that a permanent goal specification or a constant goal correction should be possible in the course of such a problem-solving cycle. Consequently, the beginning of problem solving does not necessarily require the exact definition of medium-term goals for the next two to three years, as is traditionally envisaged in project management (cf. Lehner 2001; Burghardt 2002; Boy et al. 2003; Mistler 2021).

In addition, it is one of the current imperatives of our time to take into account the existing interests and regulations, both on the part of those affected by the solution and those involved in the solution, largely already in the context of goal formation. In particular, it is about capturing special requirements to be implemented from the perspective of value creation, such as regarding occupational safety, quality assurance with respect to individual wishes of the participants or changing market trends, etc., from the very beginning in the goal formation. Ideally, it is possible to recognize all necessities, commandments, and inclinations in the context of goal description. This can lead to contradictions and conflicts. To detect these, the classification into must, may, and should goals recommended by HABERFELLNER is as important as their comparative consideration, evaluation, and weighting (cf. Haberfellner et al. 2018). Must-goals are mandatory, while may-goals are to be pursued but not necessarily implemented. HABERFELLNER recommends should-goals as an additional category, which stands between must and may goals (Haberfellner et al. 2018). However, in the context of complex systems, it is initially sufficient to formulate the goals very generally and durably and only over time to detail or specify them systematically as needed. Due to the enormous change dynamics of today's and future value creation processes, a universal approach in goal formation thus allows a permanent comparison of the goals.

The analysis requires the system to be made transparent and comprehensible, both in the hierarchies of its subsystems or the elements and their relationships with each other, and in its interrelationships with the environment. HABERFELLNER refers to this as "situation analysis", the purpose of which is:

- to make the situation transparent,
- to structure the problem field,
- to delimit the area for the search for solutions, and
- to collect the information for the search for solutions (cf. Haberfellner et al. 2018).

There is always the danger that the realistic problem or problem-solving description will fail due to too complex data volumes. This danger exists even when implementing Industry 4.0 concepts, where efficient solutions are being sought (Kaufmann 2014; Hoppe 2014; acatech 2021). Here, it is particularly important to design the scope and content of the analysis data in such a way that a universal meta-model is developed and systematically specified with a reasonable effort. ***This meta-model is to be synergistically linked and updated with the subject-specific models,*** in order to increasingly meet the basic demand for reality. In addition to HABERFELLNER, it may well happen that market analyses are carried out in the development of solution variants or that transfer possibilities into other sectors are analyzed after the successful implementation of the solution. Consequently, the analysis can take place several times in the course of the problem-solving cycle and is to be planned and controlled via project management (Mistler 2021).

This is consequently followed by ***the design***. It has the claim to develop adapted design solutions. After all, the aim is to achieve realistic or practice-relevant, i.e. ***successful problem solutions with the help of the thinking model and the developed design variants.*** Comparatively quickly developed, realistic solution variants not only reduce the implementation effort, but also offer additional competitive advantages in the context of value creation, for example, against the background of constantly shortening product development times. The declared design goal is to generate solution variants with a reasonable resource expenditure, to view them comparatively, to test, improve, and realize them. Due to the enormous change dynamics of today's and future value creation processes, a generalist approach in the design phase, in addition to demand-oriented goal specifications, must also constantly integrate varying analysis data into the design process (Mistler 2021). Consequently, the design solutions are subject to a permanent improvement process, which can also extend over their product or system life cycle.

In summary, it can be stated that it is quite possible to develop a generalist procedural concept, which is part of the GSE approach. This presupposes that it is modularly structured. The basic principles of systematic thinking and action, in particular the basic principles of standardization, thinking in alternatives, and multiple usability, must be consistently implemented. Possible ***modules of the GSE approach*** are:

- the analysis module,
- the goal (setting) module,
- the design module and
- the project management module.

These ***modules*** should be designed and standardized in such a way that they can be used generally for any type of problem solving. Furthermore, they should be able to modify themselves problem-specifically through their combination with methods and procedures, i.e. develop generically and reusably. The interfaces between the modules are to be minimized. They ***must interact with the thinking model,*** so that, for example, the design results, which can lead to changes in the thinking model, are available in a traceable manner. If the design solution is changed again, the changes to the various states of the thinking model can be tracked and re-examined. To enable this in detail, further development work is required both for the thinking model and for the procedural concept of the GSE approach. In principle, the temporal; logical sequence of linking the individual modules should be supported by the tools of project management, specifically the phases

- planning,
- implementation and
- controlling

in such a way that efficient problem solving is possible. In these phases, the basic principles of systematic thinking and action, such as the basic principle of discursive approach, the basic principle of recurring reflection, and the basic principle of thinking in alternatives, must be observed.

How the mentioned modules of the GSE approach, i.e. the analysis, the goal formation and the design module look in detail and how they can be connected with methods and procedures to then enable a specific problem solution via the phases of project management, will be explained in detail in Chap. 5. In this process, the connection to the thinking model must not be lost. The interlocking of the modules with the thinking model and a first draft of the GSE approach will be described in the following Sect. 3.4.

3.4 The First Draft of a GSE and Ideas for Its Further Development

The evaluation of the literature clearly showed that the demand for a universal, standardized SE approach is increasing (cf. Bahill and Gissing 1998; Haberfellner and Daenzer 2002; Sitte and Winzer 2011; Winzer 2015; Lindemann 2016; Weilkiens 2019;

Haberfellner et al. 2018; Mistler 2021). It is justified with the requirement to cope with the current tendencies of complexity. Their analysis resulted in further demands on the SE, which served as criteria for a comparison of the various SE approaches (cf. Sects. 2.3, 2.4 and 3.3). As a result, it had to be determined that a new, i.e. a GSE approach, needs to be developed. It must meet the following general requirements, such as:

- be universally (generally) applicable,
- fundamentally based on system thinking,
- use a thinking model and procedural concept,
- allow a constant update of the thinking model or procedural concept, and
- guarantee changes continuously and traceably.

Since these demands are not completely met by any of the currently known SE approaches, a first draft for a new GSE approach will be created below.

The historical thinking model of the SE is based on the system-theoretical approach, i.e. explaining different phenomena using systems or as a combination of systems. Thus, due to its universal applicability, this system approach has proven itself overall in the case of complex systems, even in the context of analysis and design of economic and engineering contexts. The basis of this universally valid system philosophy of the SE naturally forms a corresponding understanding of the system concept. As already determined in Sect. 3.2, systems of all kinds are to be described via the following characteristics: the system boundary, system input and output, system elements, system environment as well as the relations and the system states. Ultimately, such characteristics in connection with complex problems such as e.g.

- the goal setting,
- the responsibility,
- the existing laws and
- the controllability,

set the decisive content framework for the describability of the systemic basics.

Therefore, it is advisable that such a system understanding also serves as a sensible basis for a general or generic SE thinking model of the future for dealing with complex problems (Mutius 2004; Pruckner 2005). This thinking in systems, i.e. the assignment of problems to definable systems, has been neglected. It must become a mandatory part of the to-be-developed Generic SE approach.

Most SE approaches start directly with problem analysis. The ***thinking in systems*** and the modeling based on it, i.e. the creation of a system image in the form of a model, is often not explicitly shown or partly neglected. This must be reintegrated into the to-be-developed Generic SE approach.

GSE must therefore be a ***general problem-solving approach***. The ***GSE philosophy*** is universally valid and forms the framework for problem solving. This is so broad that an adaptation for all types of specific problems is possible.

As the ***basis of the GSE approach***, the following are required

- system thinking and
- the basic principles of systematic thinking and action.

The ***focus of system thinking*** is the ***crystallization of the essential.*** The principle of system thinking follows the ***division of reality into two parts,*** i.e., the ***system*** as the essential object of consideration and the ***system environment.*** Consequently, the problem can initially be clearly assigned to an object of consideration and the problem-solving process can be started. If it should turn out during this process that the object of consideration, i.e., the defined system, was defined too narrowly, as was the case with the example "train slips on wet leaves", the system boundaries and thus the object of consideration must be expanded. For this example, among other things, the track system was included in the problem-solving process (see Sect. 3.1). This means that system thinking must be permanently included in the problem cycle. The same applies to the ***basic principles of systematic thinking and action.*** These principles, summarized from the evaluation of the literature (see Sect. 2.3), contribute to making both the thinking model and the procedural concept efficient. For this reason, they must also be the basis of the GSE approach.

Based on system thinking and the basic principles of systematic thinking and action, the following must be

- a standardized thinking model and
- a general, modular procedural concept

necessary ***components of the GSE***. Both form an inseparable unit.

With the help of system thinking, the system to be considered can be separated from its environment and made transparent and simplified via a thinking model, i.e., an image. Only when the complexity is recognized can it be analyzed, changed, or controlled according to clear rules. This requires a modularized, universal, but individually configurable GSE procedural concept. Knowledge gained through the implementation of the GSE procedural concept during the problem-solving cycle must always be recorded, i.e., they lead to the refinement of the GSE thinking model. In order for the GSE procedural concept to be efficiently controlled, the phases of project management must be synergistically linked with the approach concept.

Regardless of the complexity level of the problem to be solved and regardless of whether it involves technical, human, natural, social, or other systems, the systemic approach has proven to be a universal approach to problem solving. Consequently, the thinking model must be based on a general system consideration, which at the same time provides sufficient transparency regarding the interaction between the system and its environment. It should also link common model approaches from the black box to dynamic modeling with reality.

Current development trends underline that initially the characteristic of the system via the black box model, i.e., the description of the interactions of the system with its environment, proves to be sufficient to roughly characterize and delimit the problem and to derive essential goals. The existence of initial solution variants consequently leads to opening the black box model of the system and creating hierarchies from subsystems or system components according to possible change potentials, the system structure, or the degree of abstraction—in principle depending on the known or recordable data for system description. These different perspectives, in turn, may form the basis for a systematic delimitation of the problem-solving space and thus shorten the temporal dimension of problem solving.

Especially against the background that in complex systems the properties of their components do not fully explain the system properties, the thinking models between the black box model, the structured and the dynamic model should not be distinguished so strictly in the first place. This is all the more true as the different types of thinking models can be traced back to a common understanding of the system and merely highlight differentiated aspects of system consideration, such as the interactions with the environment, existing change potentials, the system structure, or a certain degree of abstraction. In addition, complex structures in practice usually have fluid boundaries. Therefore, the thinking models of the future are characterized by the fact that they effectively connect or combine the different thinking model approaches depending on the problem to be solved, so to speak, as a ***"general thinking model"*** (metamodel), instead of distinguishing between them. This also includes describing system states via their possibilities or their possibility fields in order to achieve a comparably accurate system description despite existing uncertainties. Consequently, a procedure must be developed for creating the thinking model that can be adapted to the specific problem. This leads to the generic further development or refinement of the GSE thinking model, i.e., its ***genesis.*** The ***thinking model of the GSE approach to be developed*** must not only allow the synergistic coupling of the model types (Black-Box-Model, structured and dynamic model), but also stand as a description, explanation, forecast, design-,and optimization model ***in interaction with the procedural concept.***

The GSE approach to be designed must distinguish itself from traditional SE approaches by sensibly integrating worthwhile modules from the multitude of differentiated approaches into a uniform, universally valid problem-solving approach based on the SE approach.

The initially drafted ***general procedural concept*** of the GSE approach is characterized by a kind of problem- and object-independent universality, which secures the problem and object specificity by means of a corresponding method and procedure database. It consists of the four modules, i.e., the goal formation-, the analysis and the design as well as the project management module. The GSE procedural concept ensures an iterative, planable, documentable, and traceable sequence of phase-specific, systematizable methods and procedures with these modules. The goal formation, the analysis, and the design module are summarized as phases of system design to establish a reference to the classic SE approaches of BAHILL, LINDEMANN, and WEILKINS (Bahill and Gissing 1998; Lindemann 2016; Weilkiens 2019).

The following specific demands are made on them:

- The goal derivation is based on the determination of requirements with the possibility of a permanent goal precision through the updating of goals depending on the progress of knowledge.
- The analysis includes a systematic scanning of the problem-solving space using constant accesses from the analysis results, which are suitable both for their standardized documentation and for ensuring the reality of the analysis data.
- The design is based on generic principles that implement a quick and systematic approach to solution ideas and the development of design variants.

The temporally logical coupling of the goal formation, analysis, and design module is carried out via the project management module. There, the planning, implementation, and control of the system design process take place. Therefore, this is initially referred to as an arrangement of the phases in the procedural concept of the draft of the Generic SE approach.

Consequently, the first draft of the ***Generic SE approach,*** shown in Fig. 3.8, has the following two components:

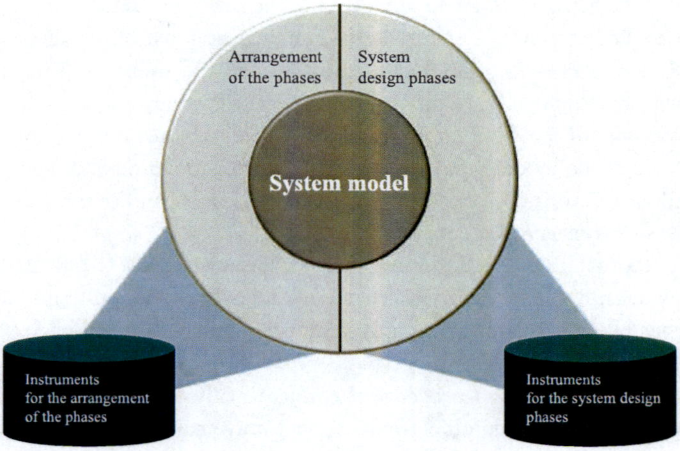

Fig. 3.8 The first GSE approach. (Based on Sitte and Winzer P. 2004)

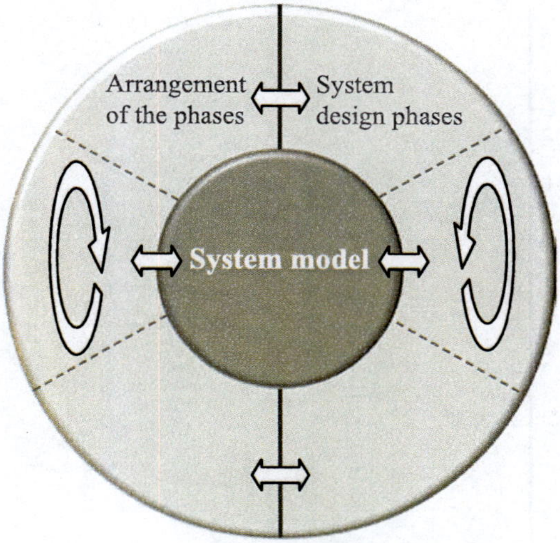

Fig. 3.9 Interactions of the components of the first GSE approach. (Based on Sitte and Winzer P. 2004)

1. the ***system model,*** i.e., the GSE thinking model and
2. the GSE procedural concept consisting of the ***phases of system design (goal formation, analysis and design module)*** and the ***arrangement of the system design phases*** via the ***project management module.***

In addition to the general problem-solving approach, there is a demand to use ***problem-specific or subject-specific methods and procedures in problem solving***. Undoubtedly, for example, the analysis of a microprocessor requires other methodsand procedures than the analysis of a turbine in a power plant. To meet this demand for problem specification, the GSE approach has the option of using appropriate problem-specific method databases as part of system design. Proposals already exist in the literature, such as for the design process, which systematically assigns "construction-specific" methods to the individual phases of system design and their arrangement (Ehrlenspiel and Meerkamm 2017; Franke 2005; Lindemann 2016). However, when and how to integrate which methods and procedures must be investigated in the further refinement of the draft of the GSE approach.

Due to the genesis of the GSE approach, there is inevitably an exchange of information between the individual phases of system design, i.e., between the goal formation-, the analysis-, and the design module, at least via the GSE thinking model, which is controlled and documented via the project management module. While the analysis results together with the goals form the basis of the design, all system design phases simultaneously contribute to describing or further improving the complex system itself. All these roughly outlined interactions are logically reflected in the possible ***generic arrangements of the system design phases*** (see Fig. 3.9).

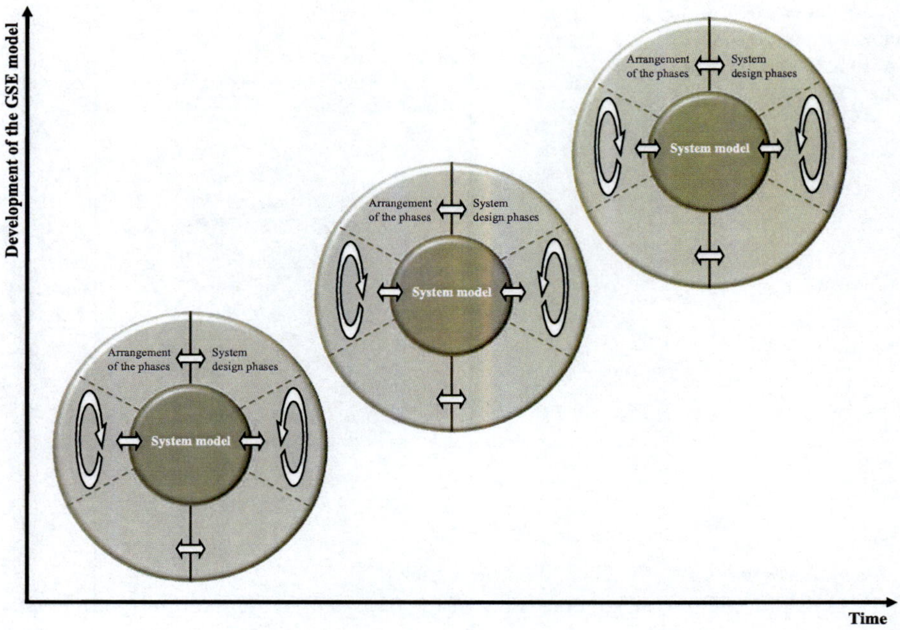

Fig. 3.10 Continuous improvement of the system over its life cycle using the first GSE approach. (Based on Sitte and Winzer P. 2004)

The application of the GSE approach in solving complex problems provides a series of information and data during the problem-solving cycle. Here, it is important to collect them systematically per phase in such a way that they are accessible to the other phases over the entire problem-solving process and guarantee traceability. The different possibilities of data collection, systematization, and structuring must still be investigated. In principle, however, the GSE thinking model should enable this. In addition, this first draft of the GSE approach is suitable for multiple applications to the same system in order to continuously improve it within its life cycle (see Fig. 3.10).

As a basic conclusion, it appears that the derived first draft for a GSE approach is based on system thinking and the basic principles of systematic thinking and action and has two components, i.e., the GSE thinking model and GSE approach concept. While the GSE thinking model is not yet further specified, the object-independent, i.e., also universally applicable GSE procedural concept consists of two parts, the phases of system design and the arrangement of these phases. Despite all universality, the accesses of the individual components of the GSE procedural concept to specific methods and procedures allow the necessary modifications of different problem situations. Thus, the GSE approach described here meets the essential demands worked out for a universal problem-solving approach.

What can the first draft of the GSE approach already achieve at present and what needs to be further specified in this concept?

It was found that systematic thinking was and is used in many disciplines. Thus, transdisciplinarity is implied in systematic thinking. If interdisciplinary development teams use systematic thinking, they must, for example, develop a common model when developing a new asynchronous machine that can be used by computer scientists, electrical engineers, and mechanical engineers alike. These teams must understand the product to be developed as a product system, be able to represent it in its interaction, but also describe the individual components, functions and their interactions in such a way that every team member understands it in the same way, regardless of which discipline they originally come from. However, how the system in its dynamics is to be described in a standardized way using a ***GSE system model is to be described***, which elements are to be characterized and how the relationships can be represented or attributed while simultaneously observing the basic principle of minimal models and the basic principle of minimizing interfaces, must be examined in more detail in Chap. 3. As a result, the ***GSE thinking model*** can emerge, which has to fulfill description, explanation, prediction, design, and optimization functions.

The demand for transdisciplinarity is already met by the proposed GSE procedural concept through two aspects. On the one hand, the modules analysis, goal settingand design are the subject in every discipline, as proven in Sect. 3.3, on the other hand, phases of project management are known across disciplines. They are used to manage and steer projects of any kind efficiently and agilely. Basically, modules in the first draft of the GSE approach concept were designed in such a way that a transdisciplinary solution to problems of any kind is fundamentally enabled. However, this general approach should not obscure the fact that specific questions require specific methods, procedures, and (IT) tools to be able to work on them. How these specific methods, procedures, and (IT) tools can be selected, coupled, and applied in the respective universal modules so that a transdisciplinary team succeeds in developing the industry-specific problem solution must be explained in more detail in Chap. 5. As a result, the first draft of the GSE procedural concept is further specified through the targeted selection and coupling of methods.

Furthermore, the life cycle of the system must also be captured and depicted via the procedural concept (Züst 2004; Lindemann 2016; Arnaut et al. 2016; Benno et al. 2018; Bender and Gericke 2021; Dumitrescu et al. 2021). This poses special requirements for both the GSE procedural concept and the GSE thinking model, because the procedural concept and the specification of the thinking model, or the corrections in the thinking model, must be traceable and transparently designed for the global market over the system life cycle. This is a particular challenge in specifying the GSE approach. The new dimensions of complexity management (see Chap. 1) require the ***principles of systematic thinking and action*** to be fundamentally observed and consistently implemented.

Table 3.4 provides an overview of where and how these principles could be applied in the first draft of the GSE approach.

Table 3.4 Overview of the application of the principles of systematic thinking and action in the first draft of the GSE approach

Basic principles	Thinking model	Procedural concept
Basic principle of systems thinking	X	
Basic principle from the whole to the detail	X	X
Basic principle of recurrent reflection	X	X
Basic principle of structuring	X	
Basic principle from the abstract to the concrete	X	X
Basic principle of minimal models	X	X
Basic principle of comprehensibility	X	X
Basic principle of application multiple layers	X	X
Basic principle of neutrality	X	X
Basic principle of multiple usability	X	X
Basic principle of standardization	X	X
Basic principle of information encapsulation	X	X
Basic principle of discursive procedure	X	X
Basic principle of thinking in alternatives	X	X
Basic principle of changing modalities	X	X
Basic principle of problem decomposition	X	
Basic principle of minimizing interfaces	X	X
Basic principle of subtraction of systems or system elements	X	
Basic principle of multiplication of systems or system elements	X	
Basic principle of division or decomposition of systems or system elements.	X	

It becomes clear that all principles can be used in both the GSE thinking model and the GSE procedural concept. This is a prerequisite for an efficient problem-solving cycle.

In summary, it can be estimated that through the comparative analysis of various thinking and procedure models that have emerged in the past and present in SE, it was possible to develop a first draft for the GSE approach. It is a general problem-solving approach. Consequently, the problem, for example an unfulfilled requirement, must first be assigned to the system (system thinking) and subsequently to the system model. The first draft of the GSE approach serves to solve overall and partial problems, which it implements into the system model in the form of partial or overall solutions and evaluates with its help. The problem-solving process in the first GSE approach requires that all modules of the GSE procedural concept use a common system model, i.e. the GSE thinking model, and appropriately apply the principles of systematic thinking and action.

These building blocks of the first GSE approach, i.e. the GSE thinking model and the GSE approach concept, need to be further developed.

With regard to the GSE thinking model, the following questions, among others, need to be clarified in Chap. 4:

- What views does the thinking model need?
- How are the interactions between the views to be defined captured?
- How are the views and their relationships to be represented and made transparent?
- How are changes illustrated by the thinking model?
- What procedure is required for creating and maintaining the thinking model?

Only when these questions have been clarified can the interactions between the GSE thinking model and the GSE prodecural concept be examined in detail and appropriate solution concepts be developed for this.

The further development of the GSE procedure concept requires answers to the following issues in Chap. 5:

- What synergies or interactions exist between its modules?
- Which methods, procedures, and (IT tools) can be used when and how in which module of the procedural concept?
- How can the methods, procedures, and (IT tools) be coupled?
- How can the results of the method, procedure, and (IT) tool application be incorporated into the thinking model and what consequences does this have for the thinking model?

As a result, a second draft of the GSE approach is possible, the testing of which on selected examples is presented in Chap. 6.

References

3Sat (2010): Gleiten statt Blockieren – Laub macht Zugbremsen im Herbst zu schaffen. Online verfügbar unter http://www.3sat.de/page/%3Fsource=/nano/technik/150250/index.html, zuletzt aktualisiert am 24.09.2012.

acatech (2021): Industrie 4.0 – Forschung für die Gestaltung der Zukunft. Impulsbericht des Forschungsbeirats der Plattform Industrie 4.0. Hg. v. Forschungsbeirat der Plattform Industrie 4.0 und acatech – Deutsche Akademie der Technikwissenschaften.

Arlt, Gregor (1999): Systemansatz eines produkt- und ablauforientierten Qualitätsmanagements durch Integration der Systemtechnik. Univ., Diss.-Duisburg, 1999. Als Ms. gedr. Düsseldorf: VDI-Verl. (Fortschritt-Berichte VDIReihe 16, Technik und Wirtschaft 109).

Arnaut, Bruno M.; Ferrari, Denise B.; Oliveira e Souza, Marcelo Lopes de (2016): A requirements engineering and management process in concept phase of complex systems. In: 2016 IEEE International Symposium on Systems Engineering (ISSE). 2016 IEEE International Symposium on Systems Engineering (ISSE). Edinburgh, United Kingdom, 03.10.2016–05.10.2016: IEEE, pp. 1–6.

Atkins, Peter W.; Paula, Julio de (2006): Physikalische Chemie. Set aus Lehrbuch und Arbeitsbuch. 4., vollständig überarb. Weinheim: Wiley-VCH.

Bahill, A. Terry; Gissing, Bruce (1998): Re-evaluating systems engineering concepts using systems thinking. In: *IEEE Transactions on Systems, Man, and Cybernetics, Part C (Applications and Reviews)* 28 (4), S. 516–527.

Bender, Beate; Gericke, Kilian (2021): Pahl/Beitz Konstruktionslehre. Methoden und Anwendung erfolgreicher Produktentwicklung. 9th ed. 2021. Berlin, Heidelberg: Springer Berlin Heidelberg; Imprint: Springer Vieweg.

Benno, Stützel; Borchart, Laura; Illa, Thomas; Gerling, Christoph (2018): Systems Engineering in Deutschland. Die deutsche Unternehmenslandschaft im Vergleich. Studie. Frankfurt am Main: Dialogistiker GmbH.

Bernhard, R.; Jahn, B. U. (2009): Modellgetriebene Entwicklung von serviceorientierten Architekturen. In: H. R. Hansen, D. Karagiannis und H.-G. Fill (Eds.): Business Services: Konzepte, Technologien, Anwendungen. Tagungsband der 9. Internationalen Tagung Wirtschaftsinformatik. 2 Bände. Wien (1), pp. 89–98.

Boy, Jacques; Dudek, Christian; Kuschel, Sabine; Wagner, Hardy R (2003): Projektmanagement. Grundlagen, Methoden und Techniken, Zusammenhänge. 11th ed., 49.–54. Tsd. Offenbach: GABAL. Online verfügbar unter http://www.gbv.de/dms/hebis-darmstadt/toc/125968515.pdf.

Braunholz, Helge (2006): Informationsflüsse im Generic-Management-System. In: Petra Winzer (Ed.): Generic-Management und Möglichkeiten der Stakeholderintegration. Aachen: Shaker (Berichte zum Generic-Management, 1/2006).

Burghardt, Manfred (2002): Projektmanagement. Leitfaden für die Planung, Überwachung und Steuerung von Entwicklungsprojekten. 6., wesentlich überarb. und erw. Aufl. Erlangen: Publicis Corp. Publ. Online verfügbar unter http://www.gbv.de/dms/hebis-darmstadt/toc/105958395.pdf.

Dörner, Dietrich (2007): Die Logik des Misslingens. Strategisches Denken in komplexen Situationen. 6th ed. Reinbek bei Hamburg: Rowohlt-Taschenbuch-Verl.

Dumitrescu, Roman; Riedel, Oliver; Gausemeier, Jürgen; Albers, Albert; Stark, Rainer (2021): Engineering in Deutschland – Status quo in Wirtschaft und Wissenschaft. Ein Beitrag zum Advanced Systems Engineering. Padaborn. Online verfügbar unter www.advanced-systems-engineering.de, zuletzt geprüft am 06.05.2021.

Ebert, Christof (2019): Systematisches Requirements Engineering. Anforderungen ermitteln, dokumentieren, analysieren und verwalten. 6., überarbeitete und erweiterte Auflage. Heidelberg: dpunkt.verlag.

Ehrlenspiel, Klaus; Meerkamm, Harald (2017): Integrierte Produktentwicklung. Denkabläufe, Methodeneinsatz, Zusammenarbeit. 6., vollständig überarbeitete und erweiterte Auflage. München, Wien: Hanser. Online verfügbar unter http://www.hanser-fachbuch.de/buch/Integrierte+Produktentwicklung/9783446440890.

FQS (2016): Leitfaden zur Nutzung virtueller Realität in der Produktentwicklung. 1st ed. Frankfurt am Main: Hanser (FQS-83-07).

Franke, Hans-Joachim (2005): Kooperationsorientiertes Innovationsmanagement. Ergebnisse des BMBF-Verbundprojektes GINA, „Ganzheitliche Innovationsprozesse in modularen Unternehmensnetzwerken“. Berlin: Logos-Verl.

Gausemeier, Jürgen; Gaukstern, Tobias; Tschirner, Christian (2013): Systems Engineering Management Based on a Discipline-Spanning System Model. In: Conference on Systems Engineering Research (CSER 13) (Ed.): Proceedings, 19.–22. März 2013: Elsevier B.V.

Haberfellner, Reinhard; Daenzer, Walter F. (2002): Systems engineering. Methodik und Praxis. 11., durchges. Aufl. Zürich: Verl. Industrielle Organisation.

Haberfellner, Reinhard; Weck, Olivier de; Fricke, Ernst; Vössner, Siegfried (2012): Systems Engineering: Grundlagen und Anwendung. 12. In: *Auflage. Zürich: Orell Füssli*, 978-3280040683.

Haberfellner, Reinhard; Weck, Olivier de; Fricke, Ernst; Vössner, Siegfried (2018): Systems Engineering Test. Grundlagen und Anwendung. 14. überarbeitete Auflage. Zürich: Orell Füssli Verlag.

Habermann, Frank (2013): Hybrides Projektmanagement – agile und klassische Vorgehensmodelle im Zusammenspiel. In: *HMD Praxis der Wirtschaftsinformatik* 50 (5), pp. 93–102. https://doi.org/10.1007/BF03340857.

Häuslein, Andreas (2004): Systemanalyse. Grundlagen, Techniken, Notierungen. Berlin: VDE-Verl. Online verfügbar unter http://www.gbv.de/dms/hebis-darmstadt/toc/113453388.pdf.

Heinke, Jonas; Mistler, Marian (2019): Agiles und modellbasiertes Projektmanagement in der Produkt- und Dienstleistungsentwicklung. In: Nadine Schlüter und Markus Reiche (Ed.): Herausforderungen im Umgang mit Anforderungen in Zeiten des industriellen Wandels. 1. Auflage. Düren: Shaker (Berichte zum Generic-Management, 2019,1), pp. 1–26.

Heinrich, Harald (2015): Systemisches Projektmanagement. Grundlagen, Umsetzung, Erfolgskriterien. München: Hanser.

Hoffmann, Karsten; Mörike, Michael (2008): IT-Projektmanagement im Wandel. Heidelberg: Dpunkt (HMD – Praxis der Wirtschaftsinformatik, 45.2008, 260).

Hoppe, St. (2014): Standardisierte horizontale und vertikale Kommunikation. Status und Ausblick. In: Thomas Bauernhansl (Ed.): Industrie 4.0 in Produktion, Automatisierung und Logistik. Anwendung, Technologien und Migration. Wiesbaden: Springer Vieweg, pp. 325–343.

IEEE Computer Society (IEEE) (2005): IEEE-STD-1220-2005 – IEEE standard for application and management of the systems engineering process.

Imboden, Dieter M.; Koch, Sabine (2005): Systemanalyse. Einführung in die mathematische Modellierung natürlicher Systeme ; mit 8 Tabellen. 2. korrigierter Nachdr. der 1. Aufl. aus 2003. Berlin: Springer.

Jeckle, Mario (2000): Konzepte der Metamodellierung – zum Begriff Metamodell. In: *Softwaretechnik Trends* Bd. 20, Heft 2 (2), zuletzt geprüft am 20.03.2015.

Jenkins, Gwilym Meirion; Youle, Philip V. (1971): Systems Engineering. A unifying approach in industry and society. London: Watts (The new thinker's Library).

Jumar, U.; Schnieder, E.; Diedrich, C. (Eds.) (2010): Entwurf komplexer Automatisierungssysteme. Beschreibungsmittel, Methoden, Werkzeuge und Anwendungen ; 11. Fachtagung mit Tutorium, 25. bis 27. Mai 2010 in Magdeburg, Denkfabrik im Wissenschaftshafen. EKA <11, 2010, Magdeburg>. Magdeburg: Ifak.

Kaufmann, Th. (2014): Die horizontale Integration der Wertschöpfungskette in der Halbleiterindustrie – Chancen und Herausforderungen. In: Thomas Bauernhansl (Ed.): Industrie 4.0 in Produktion, Automatisierung und Logistik. Anwendung, Technologien und Migration. Wiesbaden: Springer Vieweg, pp. 359–369.

Krcmar, Helmut (2015): Informationsmanagement. 6., überarbeitete Auflage. Berlin, Heidelberg: Springer Gabler.

Kuster, Jürg; Bachmann, Christian; Huber, Eugen; Hubmann, Mike; Lippmann, Robert; Schneider, Emil et al. (2019): Handbuch Projektmanagement. Agil – klassisch – hybrid. 4., vollständig überarbeitete und erweiterte Auflage. Berlin: Springer Gabler.

Lehner, Johannes M. (2001): Praxisorientiertes Projektmanagement. Grundlagenwissen an Fallbeispielen illustriert. 1st ed. Wiesbaden: Gabler.

Lindemann, U. (2005): Methodische Entwicklung technischer Produkte. Methoden flexibel und situationsgerecht anwenden. Berlin: Springer.

Lindemann, Udo (Ed.) (2016): Handbuch Produktentwicklung. München: Carl Hanser Verlag.

Luhmann, N. (1980): Komplexität. Enzyklopädie der Betriebswirtschaftslehre. Stuttgart: Poeschel.

Mamrot, M.; Marchlewitz, S.; Nicklas, J.-P.; Riekhof, F.; Schlüter, N.; Seider, G.; Winzer, P. (2012): Begriffe im Kontext des Generic Systems Engineering – Ansatzes. In: Petra Winzer (Ed.): Generic Systems Engineering als Basis für die Weiterentwicklung des WGMK-Modells. Aachen: Shaker (Berichte zum Generic-Management, 2012,2), pp. 21–30.

Mamrot, Michel (2014): Entwicklung eines Ansatzes zur modellbasierten Felddatenrückführung in die Produktentwicklung. 1st ed. Herzogenrath: Shaker (Berichte zum Generic-Management, 2014, 1).

Maucher, Irene; Paul, Hansjürgen; Rudlof, Christiane (2002): Bericht EMISA-AG: Modellierung in Soziotechnischen Systemen. In: Jörg Desel (Ed.): Prozessorientierte Methoden und Werkzeuge für die Entwicklung von Informationssystemen – Promise 2002. Bonn: Ges. für Informatik (GI-Edition. Proceedings, Vol. 21), pp. 128–137.

Mistler, Marian (2021): Entwicklung eines Vorgehenskonzeptes zum modellbasierten agilen Anforderungsmanagement (Requirements Engineering und Requirements Management) für Organisationen – REMOt. Dissertation. Wuppertal: Bergische Universität Wuppertal.

Mutius, Bernhard von (2004): Die andere Intelligenz. Wie wir morgen denken werden; ein Almanach neuer Denkansätze aus Wissenschaft, Gesellschaft und Kultur. Stuttgart: Klett-Cotta.

Parnell, Gregory S.; Driscoll, Patrick J.; Henderson, Dale L. (2008): Decision Making in Systems Engineering and Management. 1st ed. New York, NY: Wiley, J (Wiley series in systems engineering and management, 1).

Pohl, Klaus (2016): Requirements Engineering. Fundamentals, principles, and techniques. Berlin, Heidelberg: Springer-Verlag.

Preußig, Jörg (2015): Agiles Projektmanagement. Scrum, User Stories, Timeboxing Co. 1st ed. s.l.: Haufe Verlag (Haufe TaschenGuide, v.270). Online verfügbar unter http://gbv.eblib.com/patron/FullRecord.aspx%3Fp%3D2009682.

Pruckner, Maria (2005): Die Komplexitäts-Falle. Wie sich die Komplexität auf den Menschen auswirkt: vom Informationsmangel bis zum Zusammenbruch. 1st ed. Norderstedt: Books on Demand.

Rink, Anton W. (2002): Entwicklung einer Methode für die systemtechnische Auslegung verteilter und sicherheitskritischer Führungsfunktionen für Fahrzeugantriebe. Dissertation. Bergische Universität Wuppertal, Wuppertal.

Rupp, Chris (2021): Requirements-Engineering und -Management. Das Handbuch für Anforderungen in jeder Situation. 7., aktualisierte und erweiterte Auflage. München: Hanser.

Sage, Andrew P.; Rouse, William B. (Eds.) (2009): Handbook of systems engineering and management. 2. ed. Hoboken, NJ: Wiley (Wiley series in systems engineering and management).

Schnieder, Eckehard (1999): Methoden der Automatisierung. Beschreibungsmittel, Modellkonzepte und Werkzeuge für Automatisierungssysteme; mit 56 Tabellen. Braunschweig: Vieweg (Studium Technik).

Schnieder, Eckehard; Schnieder, Lars (2013): Verkehrssicherheit. Maße und Modelle, Methoden und Maßnahmen für den Straßen- und Schienenverkehr. Berlin, Heidelberg: Springer Vieweg (VDI-Buch).

Schuh, Günther; Anderl, Reiner; Dumitrescu, Roman; Krüger, Antonio; Hompel, Michael ten (2020): Industrie 4.0 Maturity Index. Die digitale Transformation von Unternehmen gestalten. Update 2020: acatech STUDIE.

Seghezzi, Hans Dieter (2007): Konzepte-Modelle-Systeme. In: Walter Masing, Tilo Pfeifer und Robert Schmitt (Hg.): Masing: Handbuch Qualitätsmanagement. 5, Bd. 5. 5th ed. München.

Sell, Robert (1989): Angewandtes Problemlösungsverhalten. Denken u. Handeln in komplexen Zusammenhängen. 2., korr. Aufl. Berlin: Springer.

Simon, Fritz B. (2006): Einführung in Systemtheorie und Konstruktivismus. 1. Aufl. Heidelberg: Carl-Auer-Verl. (Compact).

Sitte, J.; Winzer, P. (2011): Systemmodellierung im Fokus von Generic Systems Engineering. In: Gesellschaft für Systems Engineering e. V. (Ed.), Tag des Systems Engineering.

Sitte, J.; Winzer P. (2004): Systems Engineering: Old ideas, new potential. In: IEEE- SMC (Ed.): IEEE- SMC Konferenz. Oktober 2004. Den Haag. Netherland, pp. 51–61.

Tabeling, Peter (2002): Ein Metamodell zur architekturorientierten Beschreibung komplexer Systeme die Praxis, Arbeitstagung der GI, 25.–27. März 2002 in Tutzing, Proceedings. In: Martin Glinz und Günther Müller-Luschnat (Eds.): Modellierung 2002, Modellierung in der Praxis – Modellierung für die Praxis, Arbeitstagung der GI, 25.-27. März 2002 in Tutzing, Proceedings: GI (LNI, 12), pp. 51–61. Online verfügbar unter http://subs.emis.de/LNI/Proceedings/Proceedings12/article227.html.

Viertl, Reinhard; Yeganeh, Shohreh Mirzaei (2013): Mathematische Modelle für die Ungewissheit. In: Sabina Jeschke, Eva-Maria Jakobs und Alicia Dröge (Eds.): Exploring Uncertainty. Ungewissheit und Unsicherheit im interdisziplinären Diskurs. [S.l.]: Gabler, pp. 271–280.

Weilkiens, T. (2007): Systems engineering with SysML. Modeling, analysis, design. Amsterdam: Morgan Kaufmann OMG Press/Elsevier.

Weilkiens, Tim (2019): Systems Engineering mit SysML/UML. Anforderungen, Analyse, Architektur. 3., überarb. und aktualisierte Aufl. Heidelberg: dpunkt.verlag.

Weinberg, Gerald M. (2001): An introduction to general systems thinking. Gerald M. Weinberg. Silver anniversary ed. New York: Dorset House. Online verfügbar unter http://www.gbv.de/dms/goettingen/32222120X.pdf.

Welge, Matthias; Martinez, Nicolas; Steblou, Katarina; Friedrich, Christian (2015): Einsatz agiler Projektmanagement Methoden zur Erfüllung von Automotive SPICE Anforderungen. Erreichung von MAN.3 Projektmanagement Level 2 unter Anwendung des Scrum Frameworks. In: Maik S. Maurer, Jutta Abulawi und Sven-Olaf Schulze (Eds.): Tag des Systems Engineering. Bremen, 12.–14. November 2014 ; [TdSE]. München: Hanser, pp. 155–164.

Westkämper, Engelbert (1991): Integrationspfad Qualität. Berlin: Springer [u. a.] (CIM-Fachmann).

Winzer, Petra (1997): Chancen zur umfassenden Unternehmensgestaltung. Methodischer Ansatz zur qualitäts-, human- und ökologieorientierten Gestaltung von Arbeits- und Fabriksystemen. Techn. Univ., Habil.-Schr.-Berlin, 1996. Frankfurt am Main: Lang (Europäische HochschulschriftenReihe 5, Volks- und Betriebswirtschaft, Bd. 2189).

Winzer, Petra (2015): Generic System Description and Problem Solving in Systems Engineering. In: *IEEE Systems Journal* 11 (4), pp. 2052–2061. https://doi.org/10.1109/JSYST.2015.2428811.

Züst, Rainer (2004): Einstieg ins Systems Engineering. Optimale, nachhaltige Lösungen entwickeln und umsetzen. 3., überarb. Aufl. Zürich: Verl. Industrielle Organisation.

Züst, Rainer; Schregenberger, Johann W. (2003): Systems engineering. A methodology for designing sustainable solutions in the fields of engineering and management; a short summary. Zürich: Verl. Eco-Performance.

System Modeling in the GSE Approach

4

The demand for a metamodel, which teams, consisting of different disciplines, can use for problem-solving, crystallized as a result from the discussions in Chap. 3. However, how this can look and be used by the various specialists, there are different opinions. As part of integrated product development, a metamodel was created that is intended to connect reference, implementation, and application models (see Fig. 4.1).

Another metamodel, shown in Fig. 4.2, which is intended to connect the models of the various disciplines, is proposed by PREGITZER et al. (Pregitzer et al. 2014).

The current metamodel approaches assume that the model design in product design is interdisciplinary. Subsequently, the involved specialists create their specific model designs, which are integrated into the metamodel in a third step (see Gausemeier and Plass 2014). How this can be done in detail, whether this can be used for all types of systems, and how the trace links as a connection between the models are created, are questions that are open.

Chapter 4 aims to find answers to these questions and to systematically implement the requirements for a metamodel, which were derived in Chap. 3. First, it needs to be clarified how many views a metamodel should have at a minimum and how the relations within and between the views can be described. (see Sect. 4.1 and 4.2).

The result is a generalistic (usable for all types of systems and transdisciplinary), generic (changeable over time) thinking model, which should form the basis for the GSE. The first step is to clarify how complex systems, especially complex technical and sociotechnical systems, must actually be represented in order to make the structure and the relations within the system as well as the interaction of the system with its environment transparent (see Sect. 4.3).

Based on the results in Sect. 4.3, possible sequences for the creation of technical and sociotechnical systems are presented in Sect. 4.4 and 4.5. It is highlighted how a consideration of the system under the consideration of the basic principle from abstract to

N. Schlüter, *Generic Systems Engineering*, https://doi.org/10.1007/978-3-662-67994-4_4

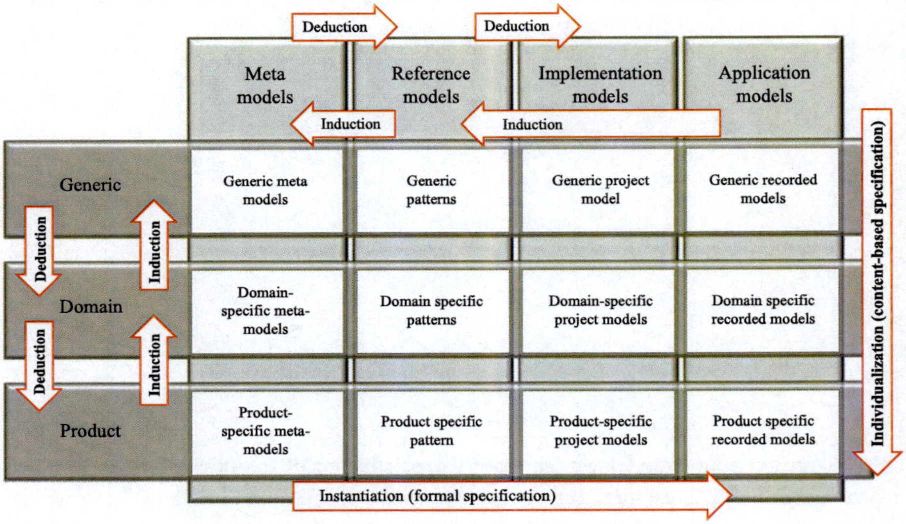

Fig. 4.1 The metamodel for integrated product development. (After Muschik 2011)

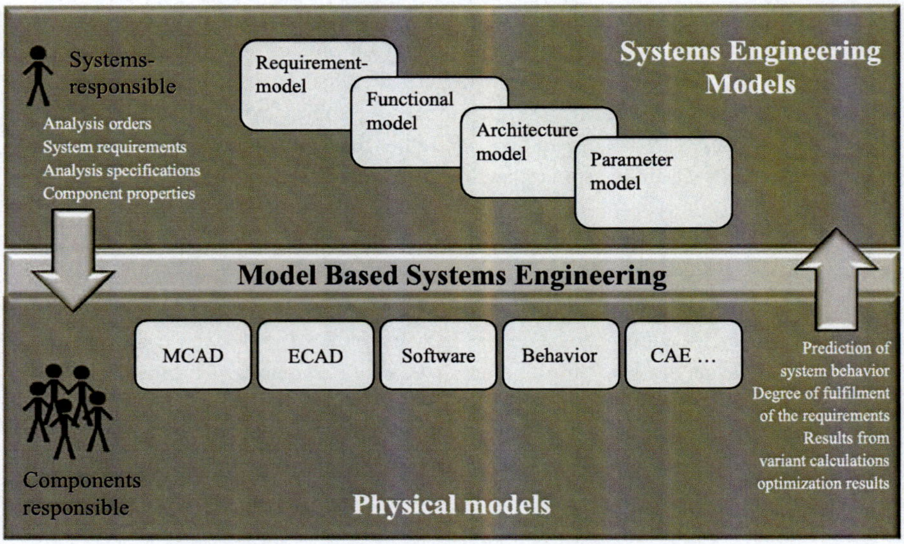

Fig. 4.2 Linking of system models and physical models in the disciplines. (After Pregitzer et al. 2014)

concrete can be modeled (see Sect. 4.4). It must be distinguished whether it is the design of a completely new system or the re-engineering of an existing system.

Finally, the advantages and disadvantages of the developed thinking model for the GSE should be discussed (see Sect. 4.6).

4.1 Derivation of the Views for System Modeling

Systems consist—so the essential consensus in the literature—of elements and their relations, a system boundary over which the interaction of the system with its environment, but also through the dynamics of the system itself can be described.

The thinking model of the GSE approach should be a generalistic representation of systems, such as the representation of the basic principle in Fig. 4.3. The disc corresponds to the image of the system with its environment, i.e., the GSE thinking model, and the cube to the views of the system image that are yet to be defined and standardized. An autonomous, self-learning robot to be developed could be abstractly represented in this way.

But how many and which views of the robot are needed to develop it together with the various disciplines?

Since there are a multitude of conceptual models in the various disciplines, these were classified according to the requirements for the GSE thinking model (see Sect. 3.2) into the following classes in order to find an answer to this question:

Fig. 4.3 Thinking model as a generalistic representation of systems without defined views

a) Conceptual models that represent the interaction of the system with the environment,
b) Conceptual models that characterize two views of the system,
c) Conceptual models that represent three views of the system,
d) Conceptual models that describe systems variably,
e) Conceptual models that are a result of intuitive thinking, and
f) Conceptual models that reflect systems using attributes.

a) CONCEPTUAL MODELS THAT REPRESENT THE INTERACTION OF THE SYSTEM WITH THE ENVIRONMENT

In the context of requirements engineering, which is a sub-discipline of system engineering, the focus is on modeling a partial aspect, i.e., the interaction of the system with its environment (see also Sect. 2.4.2). By focusing on the interaction between the requirements and the system, the requirements management implements the basic principle of minimal models of systematic thinking and action in an excellent way. A standardized, interdisciplinary definition of the system and its environment does not take place. The system is considered in the context of requirements management via a black-box model. The system environment is often generally represented as a surrounding system. A breakdown of the system environment into different systems, whose interrelationships with the considered system are visualized via a block diagram, can also be a possible form of representation of the interaction of the system with its environment (see Simon 2009). Since a variety of requirements can arise from the product idea to its realization or even in the usage, maintenance, and servicing phase of the product, clustering of the requirements and the representation of the change between system and environment over time are necessary. The current conceptual models of requirements management do not yet allow this (Mamrot et al. 2012).

b) CONCEPTUAL MODELS THAT CHARACTERIZE TWO VIEWS OF THE SYSTEM

However, the basic idea of requirements engineering mutated over time. Thus, currently in the literature, a variety of different approaches to requirements engineering can be found (see Danner 1996; Davis et al. 2007; Partsch 2010; Hallerstede et al. 2012; Pohl 2016; Pohl and Rupp 2015; Rupp 2021; Ebert 2019). While some of these requirements engineering approaches focus on identifying the sources of requirements (stakeholders) and their demands on a system, their weighting and prioritization, other approaches, as shown in Fig. 4.4, focus on describing the interactions between requirements and components of the system (Kanie 2009).

Consequently, the requirements engineering approaches are limited to two types of modeling in their conceptual model, i.e., considering the requirements via the system-environment interaction (black-box model) or representing the relations between requirements and components, or requirements and functions (hierarchical model with two or more system levels). If the requirements are transformed into functions, this is also referred to as functional system modeling or FAS method (Lamm and Weilkiens 2014;

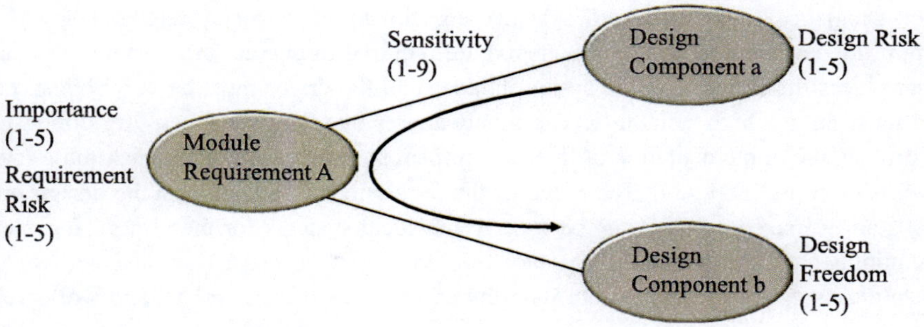

Fig. 4.4 Schema for describing interactions between A and K. (After Kanie 2009)

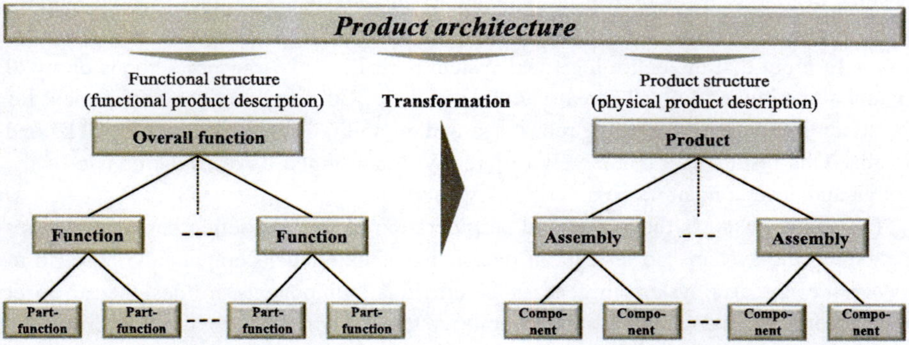

Fig. 4.5 Transformation of the function structure into a product structure. (After Rudolf 2013)

Grundel et al. 2014). The resulting functional system architecture defines the system functionality in the solution space without specifying concrete components.

The mentioned model types of requirements engineering are often used independently of each other and each corresponds to a tendency in requirements management. A sequential application, i.e., a coupling of the two tendencies, would be more sensible. Since in a first step the conceptual model via the black-box approach represents the interaction between the system and its environment, which corresponds to the first approximation of the implementation of the basic principle from the whole to the detail, the interactions between the requirements and the components or the functions can be considered as a hierarchical model with two levels (the system and the component or function level) in the second step of system modeling (see Fig. 4.5). When reflecting the requirement and component view in the conceptual model of requirements engineering, it is subsequently necessary to question to what extent the components and their degree of requirement fulfillment contribute to the requirement conformity of the overall system. The necessity to answer these questions, which requires the sequential application of the two model types, results from the following roughly outlined example.

In a logistical system (overall system), e.g., in the conveyor system of Düsseldorf Airport, thousands of drives (components) are required to operate this system. The customer places the demand on the system builder that the drives must be very robust, i.e., they must have a high reliability. The manufacturer of drives will therefore dimension the drive in the implementation of this requirement so that it can move a maximum load, drive many rolls, and so on. The result of the requirement realization in the design process is an oversized drive. It can be used in logistical systems for the airport, for opencast mining, for the glass industry, etc. Consequently, the drive was designed for all possible uses of a logistical system with the aim that the drive is robust and works reliably.

This same drive could be dimensioned quite differently if the overall system, i.e., the logistical system, was first designed to meet requirements (i.e., robust drive for logistical systems regardless of their field of application) and consequently then specific requirements for their special use at Düsseldorf Airport, i.e., also for the drive, were derived. So it may turn out that the entire logistical system basically only transports goods up to 20 kg and a maximum of 5 rolls are driven by one drive. Consequently, the requirement for the component "drive" regarding robustness and reliability is now specified for 20 kg and 5 rolls. This results in a completely differently dimensioned drive in the course of the design and development process.

The example shows that a minimal consideration of the interaction between requirements and the system can lead to an oversizing of individual components, but also to an ***over-design*** of a system in general. If individual components are designed to meet requirements independently of the overall system (second tendency in requirements engineering), this can lead to problems when integrating the designed components (drive) into the overall system (logistics system). If the design of the drive implements the demand for energy and resource conservation, the linear drive may be preferred as a solution alternative to various drive concepts. However, this linear drive results in a completely differently designed logistics system.

The complex issue presented can be simplified using the puzzle analogy. Basically, assembling individual puzzle pieces results in a complete puzzle picture. If individual puzzle pieces are changed and then an attempt is made to reintegrate them, gaps may appear in the puzzle picture or parts may not be inserted. Consequently, changing a single puzzle piece can result in the entire puzzle picture being incomplete. The same happens with the drive and the logistics system. If the drive is changed without considering the overall system, it may not provide the desired functionality in its interaction with the logistics system. Staying with the puzzle analogy, a second issue should be clarified. If a part of one puzzle picture is inserted into another puzzle picture, both puzzle pictures are faulty in their entirety. Picture A has a gap and picture B has a wrong puzzle piece. Transferred to the drive of the logistics system, this means that the drive was precisely dimensioned for the logistics system, i.e., it has a system reference exactly to this logistics system. However, if this drive is used in another system, e.g., a conveyor bridge, it may not provide the desired function.

The description of perspectives in requirements engineering is also very differentiated. Two further tendencies can be identified. One tendency only names the requirements or the components, while the other tendency describes the requirements and the system or the components through various forms of attribution. Standardization of the description only takes place within the framework of software support. However, this is not uniform. Opinions on the structuring of requirements in requirements engineering also diverge (see Schlund 2011). The same applies to the structuring of components within a system. The structural images range from classic hierarchical to cluster to chaotic structures. One thing becomes clear in all tendencies of requirements engineering: The description of the system's interaction with its environment can be represented in the model via the requirement view (representation of the interaction of the requirements with the system) and the component view (representation of the interactions of components with requirements). The basic principle of neutrality of systematic thinking and action can be implemented in these two views if the respective views are defined in a standardized way. A coupling of the black-box model and the hierarchical model is recommended in order to implement the basic principle from the whole to the detail or the basic principle of discursive approach of systematic thinking and action in modeling. In principle, however, models of systems can be created via the requirement and component view. These are incomplete because the functions are not made transparent and the changes in the system over its life cycle are not recognizable.

c) CONCEPTUAL MODELS THAT REPRESENT THREE VIEWS OF THE SYSTEM

The classic construction methodology—leading representatives are mentioned here as representatives for others, e.g. (Ehrlenspiel and Meerkamm 2017; Bender and Gericke 2021; Lindemann 2016), represent the view that in extension of the above-mentioned requirements engineering approaches, systems are to be depicted via a combined requirement, component and function view (see Lindemann 2016; Bender and Gericke 2021). The language of the designer is the drawing. It is used to represent the components. If a design is checked for its requirement conformity, the drawing is the basis for this. The functions are implicitly derived from the drawing, i.e. they cannot be explicitly represented at the moment. The shapes of the components allow conclusions to be drawn about their functions. Functions are thus always built into components or systems. Although the product system is captured in the construction methodology in the views of requirements and components as well as functions, the view of functions cannot be transparently identified (Bender and Gericke 2021). Since some of the requirements are functional requirements, a direct correlation between functional requirements and functions, or between functions and components, cannot be made transparent by this approach.

Currently, some software developers of CAD systems are starting to partially represent function views (see CATIA[1] V6). Functions are partly only verifiable in the usage phase of the system. The following example briefly illustrates this issue. A miniature robot, which is to be used for playing soccer, must be able to orient itself on a limited playing field (requirement). Consequently, the robot to be developed must have an orientation function (translation of the requirement "robot must orient itself" into the function "orientation function"). Orientation functions can be realized in various ways using sensors, such as ultrasonic and infrared sensors. However, it only becomes clear in the usage phase whether the robot can actually orient itself, i.e. whether the sensors in their combination jointly realize the orientation function in the desired manner. The user cannot see which sub-functions are necessary to implement the desired orientation function of the robot. The lack of transparency of functions can also become problematic in the re-design of product systems—as the following example roughly outlines.

A re-design of a pantograph was necessary. In this context, the designers asked themselves which of the existing components of the pantograph fulfill which functions. This question could not be answered completely for all components, because some functions were not recognizable even in the usage or maintenance phase. For this reason, it was no longer possible to understand why the pantograph was built exactly from these and not from other, cheaper components. This significantly complicated its re-design (Winzer 2012). To avoid such situations, it would be useful if the components were also depicted and described in their function view.

Current trends in systems engineering generate a product requirement model (requirement view) from the requirement analysis. From this, a solution-neutral function architecture (function view) is derived as far as possible. This in turn is the basis for the system structure (system design, component view). Thus, the function is transformed into a product architecture.

The model-based safety analysis according to the PROFUND approach proceeds analogously. As a result, three models are created that need to be connected (Slovak 2006). A continuous securing of the traceability process between the three views or models is demanded (see Auricht et al. 2014).

When modeling the system over these three views, no recommendation is given as to which types of models can be used when and how. The principles of systematic thinking and action should in principle be observed in modeling (Ehrlenspiel and Meerkamm 2017; Lindemann 2016; Bender and Gericke 2021). There are no standardized specifications for the system views. The same applies to the networking between the system views, i.e., the requirement, function, and component view.

---

[1] CATIA *(Computer Aided Three-Dimensional Interactive Application)* is a 3D CAD/CAM software for integrated product design by the French company Dassault Systèmes.

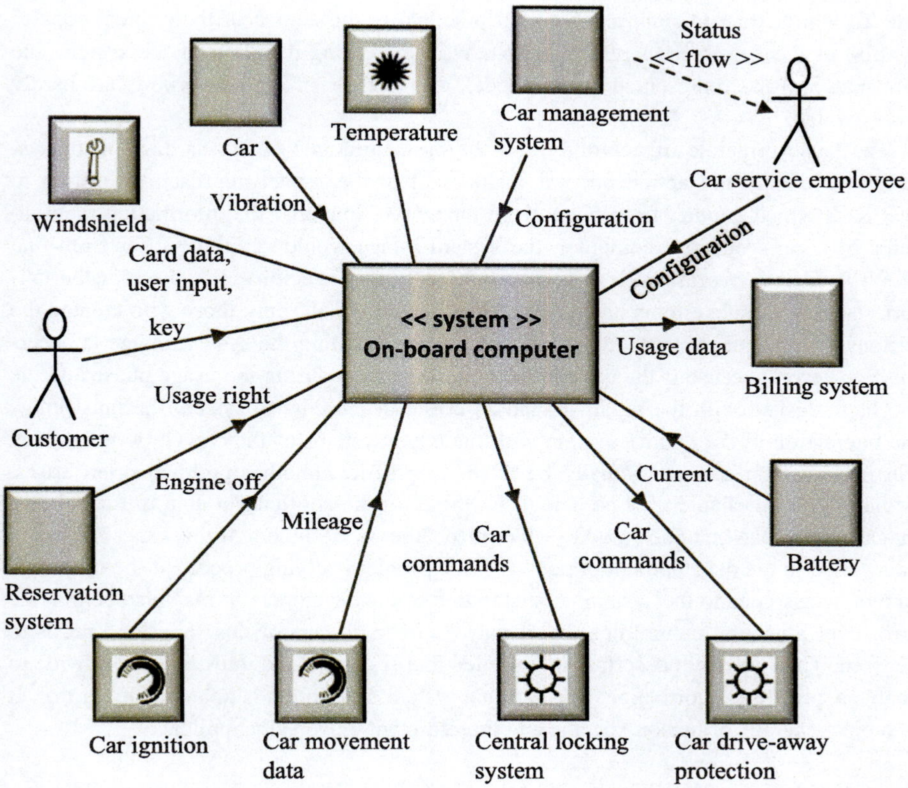

Fig. 4.6 System description via the information flow. (After Weilkiens 2007)

d) CONCEPTUAL MODELS THAT DESCRIBE SYSTEMS VARIABLY

A completely different type of description of systems is proposed by WEILKIENS, as illustrated in Fig. 4.6.

According to his view, systems can be described exclusively via their information flow. This is supported and extended by GAUSEMEIER, who emphasizes that mechatronic systems can be described via the information, energy, and material flow.

While GAUSEMEIER describes the mechatronic system via subsystems that fulfill a certain function and can consist of various components, the subsystems in WEILKIENS are freely definable (Weilkiens 2019).

This applies to the overall system "car", the subsystem "battery", the customer, or the temperature (see Fig. 4.6). The advantage of the approach is that the system is first represented as a black-box model (car's onboard computer). This type of complexity reduction initially allows exclusive concentration on the interaction of the overall system with its environment. On this basis, the modeling and the first solution draft are also carried out. To what extent this approach is still practical in the course of the gradual concretization of the design draft remains open. With increasing detailing of the system, the information flows represented by the model, due to their strong networking, are hardly recognizable.

The basic principle of neutrality and the basic principle of standardization of systematic thinking and action are not maintained in the model approach according to WEIKLIENS. If another team were given the task of mapping the information relationships of a car's onboard computer, the system image would certainly differ from that of WEILKIENS' representation, as there are no further specifications, e.g., for the uniform naming of subsystems or system elements. As a result, only those who created the various system images can understand them. Understanding the system image is necessary, however, because if the car comes to the workshop during the usage phase and the mechanic tasked with the repair diagnoses errors in the control system, he must know the interaction of the control system with the other systems of the car. Only in this way can the error for this car actually be fixed. In practice, another problem often arises. Although the mechanic may be able to fix the error, he is often not able to formulate a diagnosis that can be understood by other mechanics or designers. If this exact error is to be avoided in the development of new cars, the problem-solving process of the mechanic cannot be assigned to the system understanding of the developers. It may also be that the error is of a general nature, i.e., that it may occur in all cars of this series and needs to be fixed. Then a system description is needed that is understood uniformly worldwide to enable a quick error correction. The variable system description is not suitable for this. It is subject-specific or person-specific and therefore not universally applicable.

e) CONCEPTUAL MODELS THAT ARE A RESULT OF INTUITIVE THINKING
A completely different type of system description as a result of intuitive thinking is presented by PAHL/BEITZ.

Using the example of the bearing, an excerpt of a semantic network is shown in Fig. 4.7. The designers sketch freely their ideas that they associate with the concept of the bearing. Facts and relations are consciously analyzed, varied and newly combined, checked or discarded. The terminology, as well as the logic of the semantic network, can only be retraced by the participants. This can be counteracted if the terms are standardized and defined across disciplines. This approach makes intuitive thinking transparent, but does not generate a comprehensible structure of the product system "bearing". Requirements, such as mobility, are compared with different system types, such as the plain bearing, the roller bearing or the ball bearing, as well as various components like roller, cylinder, housing etc. If a structured and standardized system image is created from the semantic network, which observes the basic principle of neutrality and the basic principle of minimizing interfaces of systematic thinking and action, this way of creating a system image is very useful in the early phases of idea generation, because relationships between various system views are already illustrated via the semantic network. At

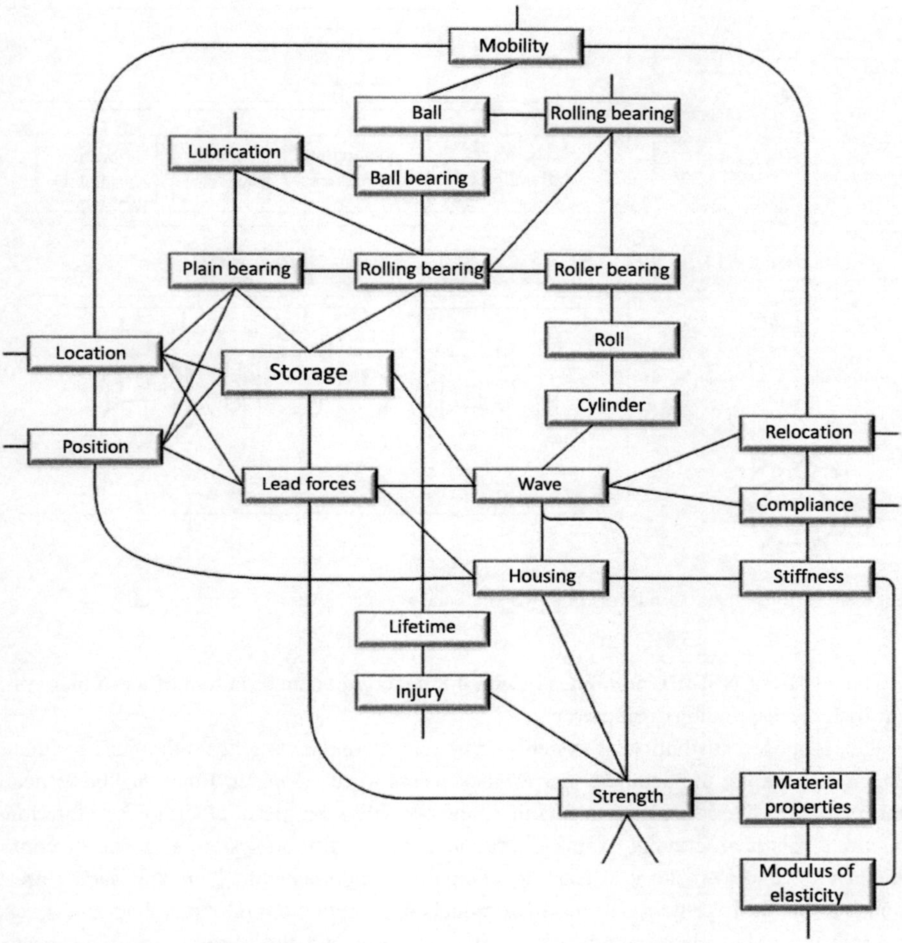

Fig. 4.7 Excerpt of a semantic network for storage. (After Bender and Gericke 2021)

the same time, in this approach, the black-box model "bearing" is coupled with the hierarchical model (roller bearing, rolling body, cylinder etc.).

Thus, the interaction between the system and its environment can be considered, i.e., what impact does the desired mobility have on the bearing as a whole and what effects does it have on the components, i.e., the rolling body or the cylinder.

f) CONCEPTUAL MODELS THAT REFLECT SYSTEMS USING ATTRIBUTES

Another approach is the representation of conceptual models through attributes. It assumes that problems can be described in an abstract form and then solved using a model.

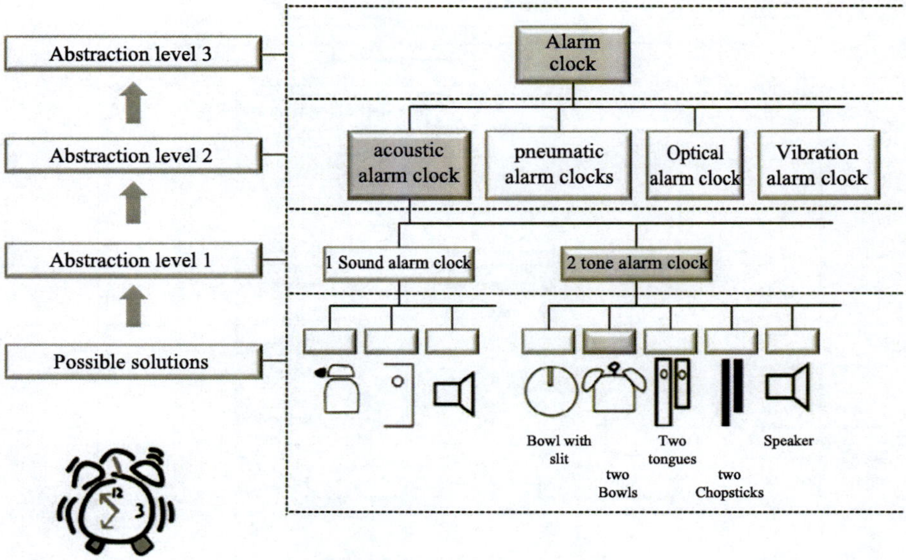

Fig. 4.8 Various abstraction levels of a two-shell alarm clock. (After Lindemann 2005)

Thus, EHRLENSPIEL combines in Fig. 4.8 the component structure of a product system with corresponding attributes.

This proposed attribution of systems or system components reflects the built-in function. For example, the acoustic alarm clock refers to the acoustic function, the optical alarm clock to the optical alarm function, and the vibration alarm clock to the vibration function. This is an attempt to link the function view with the system element, or component view and to make it at least conceptually comprehensible. This approach corresponds to the basic principle of minimal models of systematic thinking and action. It has few interfaces, but also requires standardization to ensure the observance of the basic principle of neutrality in the future.

In summary, the following table illustrates how the individual system models meet selected requirements that are placed on a general thinking model.

In the majority of the conceptual model approaches presented in Table 4.1,

- Requirements,
- Components and
- Functions

are depicted as views of the system, but are not defined in a standardized way. Consequently, these views must necessarily be incorporated into the GSE thinking model and standardized. Furthermore, it must be checked whether there is a need for further views.

Table 4.1 Requirement-related comparison of the various representation possibilities of conceptual models

Requirements	Display options					
	A) System/Environment	B) System with two views	C) System with three views	D) Variable system description	E) Intuitive system description	F) Attributive system description
Describe interaction between the system and the environment	applicable	partially applicable	partially applicable	partially applicable	partially applicable	partially applicable
Representation of the elements of the system	not applicable	applicable	applicable	not applicable	partially applicable	partially applicable
Representation of the Relations between the system elements	not applicable	partially applicable	partially applicable	partially applicable	partially applicable	partially applicable
Generalized and not subject-specific description of the system	partially applicable	partially applicable	partially applicable	not applicable	partially applicable	partially applicable
Coupling of black box with structured and dynamic model	partially applicable	partially applicable	partially applicable	not applicable	partially applicable	partially applicable
Basic principle of minimizing interfaces	partially applicable	partially applicable	partially applicable	partially applicable	partially applicable	partially applicable
Basic principle of minimal models	partially applicable	partially applicable	partially applicable	partially applicable	partially applicable	partially applicable
Basic principle of neutrality	partially applicable	partially applicable	partially applicable	not applicable	partially applicable	partially applicable
Basic principle of standardization	not applicable	partially applicable	partially applicable	not applicable	not applicable	not applicable
Basic principle from whole to detail	partially applicable	partially applicable	partially applicable	not applicable	partially applicable	partially applicable

Legend: not applicable; partially applicable; applicable

Through the ***requirement view***, the interaction of the system with the environment as well as the problem-oriented approach in the thinking model can also be depicted, assuming that the problem is equivalent to an unfulfilled requirement. When a problem arises, the requirement view must therefore be checked first.

As a result, two main cases can occur: Either a requirement was not met or a requirement was not yet recognized, i.e., new requirements must be defined from the problem. Considerations of the life cycle of systems are missing in these models. Their necessity can best be explained by the construction of a single-family house. Regardless of whether this house is built in prefabricated construction or traditionally monolithic, a considerable amount of time passes from its planning to its realization (approx. 1 to 2 years). During this time, it is not uncommon for the builder to change his original requirements or add new ones. For example, the requirements for the installation of electrical connections may change or additional water extraction points may be needed. These changed requirements must now be implemented in the house project and during the construction of the house. In such cases, changed requirements can often be easily implemented if the builder takes over the additional costs. However, it is different if certain aspects that were overlooked in product development only become apparent in the course of the production of the product system. The heat development of a gearbox on the test bench cannot be avoided afterwards by an additional fan. The solution to this problem requires a complete re-design of the gearbox, although both components were developed and implemented separately according to the requirements set. These examples show that in the process of product creation, its realization, use and reprocessing, new requirements arise and components and functions can change. Furthermore, it can often only be proven in the test and/or usage phase of a product by the processes taking place there how the components actually fulfill the functions. Consequently, ***the processes must also be represented as a view in the system image,*** i.e., as a **process view,** must be represented. In addition to this insight, the question arises whether the system views are sufficient if, in addition to the technical system, the socio-technical system is also considered? This means, how can people be included in the system? Because as in the described example, the person "builder" has requirements for the house project. Another person is responsible for implementing this requirement. Accordingly, for example, new requirements also influence a person involved in the construction of the house. This can in turn influence the process of house construction. Therefore, a **person view** is also absolutely necessary for the representation of socio-technical systems. Consequently, the requirement, component, process, person and function view must be defined in a standardized way.

But how can systems be described in a standardized way through the various views? This question will be illuminated in the next chapter with the presentation of the GSE thinking model.

4.2 Derivation of the Description Possibilities of the Interrelationships in and Between the Views in System Modeling

The required transparency of the GSE thinking model includes two subsequent conditions:

1. The GSE thinking model must be understood by interdisciplinary teams and be traceable over the product life cycle. As a result of Sect. 4.1, it was worked out that the

GSE thinking model must have at least five views. These are the requirement, component, function, process, and person views.
2. The description of these views of the GSE thinking model must first refer to the views themselves, i.e., the elements and the relations between the elements in the respective view must be described. But also between the views, the relations are to be characterized.

Furthermore, it is evident from Sect. 4.1 that the designations in and between the views must be described in a standardized way. It now needs to be clarified what description possibilities exist for the views and between the views. The forms of description must be applicable to all views that still need to be defined for the GSE thinking model.

The following are some ***description possibilities,*** that could possibly be used for the GSE thinking model:

a) Tree structures
b) Matrices
c) Clusters
d) Graphs
e) Petri nets
f) Flow diagrams

The listed forms of description are described in the following by way of example in relation to the GSE thinking model. This means that even if specific views are focused on, such as the requirement view, **the principle of modeling can be transferred to all other system views.** In addition, it is discussed why the selected description possibilities are significant for the modeling of systems.

a) *Tree structures*
The classic product development derives requirements, functions and components without necessarily establishing a transparent connection between these views (domains) (Ehrlenspiel and Meerkamm 2017; Bender and Gericke 2021). By establishing the connection between the functions and components, modular structures of product systems can be derived (Rudolf 2013). On this basis, a typification of the product system architecture in the form of **tree structures** is possible, as shown in Fig. 4.9.

The main advantages of tree structures are that they offer a hierarchical structure. This means they can break down a system into coarse components, but also go into more detail (Lindemann 2016; Bielefeld 2020). As a result, such tree structures offer the possibility of forming modules at defined levels (Hornby 2007; Suh 1998). This fact is also outlined in Fig. 4.9. The advantage of forming modules is that this structure can generate agility in system modeling (Mistler 2021, b). However, it also carries the risk that changes at various levels of a tree structure may require the entire modeling architecture

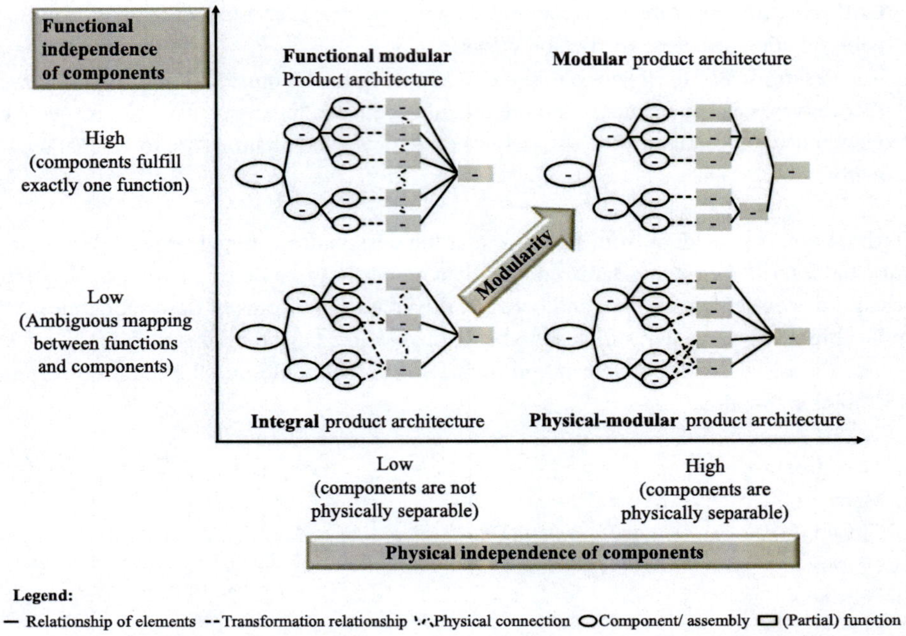

Fig. 4.9 Types of product system architectures. (After Feldhusen et al. 2013)

to be changed. This can in turn also have the effect that the modeling effort is very high or the logic of the tree structure is no longer comprehensible after these changes.

b) *Matrices* (unrated, rated and *attributed*)
In the following, an unrated matrix-based modeling is described using the example of the requirement view.

Before requirements can be described in detail, they must be captured and structured. Stakeholders are the sources of requirements. Their number is increasing, consequently the variety of requirements is also growing. By comparing the requirements in relation to the stakeholders, as shown in Table 4.2, identical, similar or contradictory requirements can be determined and as a result a precise requirement list can be created. This is referred to as a ***stakeholder-oriented approach to requirement structuring***.

Furthermore, the requirements could be subdivided into "must-have requirements" and "can-have requirements". The "must-have requirements" represent a group of requirements that must be complied with by law or that have been contractually agreed with the customer. The result is a ***tree structure of requirements*** (see also Fig. 4.9). ***Other structuring approaches for requirements*** have been extensively compared by CROSTACK and HOLZMÜLLER (Jockisch and Holzmüller 2009).

Table 4.2 The stakeholder-oriented approach to requirement structuring. (after Lex 2004)

Organizational			Technical-economic					REQUIREMENT STRUCTURE	
Auxiliary means	Personal	Time	Standards/Laws/Patents/Guarantee	Cost receivables	Interface: Regarding people/society/environment	Interface: Regarding Tech. Implementation	ONLY-technical	GENERAL REQUIREMENTS	
					+			Easy handling	CUSTOMER
	+			+				low maintenance	CUSTOMER
				+				reliable	CUSTOMER
		+						Long service life	CUSTOMER
				+				optimal price/performance ratio	CUSTOMER
					+			hazard-free	CUSTOMER
					+			energy-saving	CUSTOMER
						+		Small, light	CUSTOMER
							+	functionally expandable	CUSTOMER
							+	Remote maintenance	CUSTOMER
					+			EMC unproblematic	LAWS, STANDARDS
					+			environmentally compatible	LAWS, STANDARDS
					+			reprocessable	LAWS, STANDARDS
					+			low radiation	LAWS, STANDARDS
					+			Hazard-free for humans and animals	LAWS, STANDARDS
					+			low noise	LAWS, STANDARDS
			+					CE conformity	LAWS, STANDARDS
			+					VDS Recognition	LAWS, STANDARDS
						+	+	modular	FINAL PRODUCER
						+		flexible	FINAL PRODUCER
						+		multifunctional	FINAL PRODUCER
		+						High reliability	FINAL PRODUCER
+	+							Easy mounting	FINAL PRODUCER
				+				serviceintensive	FINAL PRODUCER
						+		manufacturing that can be automated	FINAL PRODUCER
					+			low-noise	FINAL PRODUCER
					+			Product safety	FINAL PRODUCER
	+							Direct Sales	FINAL PRODUCER
								Market success of the product	EMPLOYEES
								Ease (simple) of manufacture	EMPLOYEES
				+	+			Low weight	EMPLOYEES
	+							Good working conditions	EMPLOYEES
	+							Flexible working time	EMPLOYEES
	+							Personal responsibility, self-determination	EMPLOYEES

Legend: „+" = relationship between requirement and requirement structure element exists

Table 4.3 Prioritizing requirements via a requirement-requirement matrix. after (Sitte and Winzer 2006)

Requirements / Requirements	Self-protection	Low external electomag	Power fail-safe	Functionality	Interface tolerance	Self-diagnostics	Serviceability	Vibration tolerance	Economy	sum
self-protection	-	1	3	3	3	3	2	1	3	19
low external electomag	1	-	0	0	1	0	1	0	3	6
power fail-safe	3	0	-	3	3	3	1	0	3	16
functionality	3	0	3	-	3	3	3	1	3	19
interference tolerance	3	1	3	3	-	3	1	2	3	19
self-diagnostics	3	0	3	3	3	-	2	1	3	18
serviceability	2	1	1	3	1	2	-	0	3	13
vibration tolerance	1	0	0	1	2	1	0	-	3	8
economy	3	3	3	3	3	3	3	3	-	24
Legend: 0 less important, 1 equally important, 2 more important, 3 clearly more important										

Furthermore, with the Quality-Function-Deployment (QFD), the relations between the requirements of customers can be described matrix-based (Herrmann and Fritz 2016). This idea of pairwise comparison of requirements, however independent of the stakeholders, leads to a prioritized requirement list. Through the pairwise comparison of the requirements, the importance of each individual requirement is determined when compared with other requirements (Pohl 2016) (see Table 4.3). Subsequently, the priority per requirement (highest sum number) can be determined via the row sum.

Furthermore, ***the relations between the requirements*** can also be questioned via a matrix for each requirement, i.e., whether it is related to other requirements and if so, how. These relations can be marked in color, for example (e.g., red = strong dependency, blue = dependency to be considered) or as in Table 4.4 rated with numbers from -3 to +3 ***(rated matrix)***. Here, -3 means: “This requirement strongly negatively affects the following requirement”, or +3: “This requirement strongly promotes the respective other requirement”.

Referring to the example of the logistics system (see Table 4.4), this means that the requirement “Securing the reliability of the logistics system” is strongly interrelated with the requirement “Securing the reliability of the drive”. Since thousands of drives are installed in a logistics system, such as the one at Düsseldorf Airport, their reliability affects the logistics system as a whole, i.e., if several drives fail, the logistics system may come to a standstill.

Thus, there is a very strong correlation between these two requirements, namely the “reliability of the logistics system” and the “reliability assurance of the drive”. This was

Table 4.4 Excerpt from a requirement comparison for a logistics system

Requirements / Requirements	Ensuring the reliability of the logistic facility	Ensuring the reliability of the drive	Create low cost logistic facility
Ensuring the reliability of the logistic facility	-	3	-1
Ensuring the reliability of the drive	3	-	-1
Create low cost logistic facility	-1	-1	-
Legend: negative influence -3 -2 -1 0 1 2 3 positive influence			

rated with "3". If the requirement for "reliability of the logistics system" is compared with the requirement to create a cost-effective logistics system, high reliability assurance will certainly result in increased costs. The dependency between these two requirements can, for example, be rated with -1. As a result of the relations matrix, on the one hand, ***requirement clusters*** can be created and, on the other hand, ***relationships of requirements*** can be transparently represented. Both can also be represented using graphs, or mind maps, as descriptive tools (Künne and Richard 2009; Schlund 2011; Winzer 2012; Mistler 2021; Nicklas 2016).

Furthermore, there is the possibility to attribute matrices individually. This is shown in Table 4.5 using the example of material flow and energy flow.

The energy flow in Table 4.5 goes from the voltage source (3.1) to the phase (1.1.1.1) to the copper plate (5.1). At the same time, energy is transferred from the phase to the subsystems of the primary part, i.e., tooth (1.1.2.1), slot (1.1.2.3) and body (1.1.2.3.). In this way, the energy flow can also be represented at lower hierarchy levels and possible causes for energy losses can be identified (Riekhof et al. 2012). From this, solutions can be developed in a targeted manner using the GSE approach, which is described in more detail in Chapter 4, to avoid heat development when using the linear drive in the logistics system.

c) *Clusters*

Within the framework of the ***DSM matrix (Design Structure Matrix) as another description option***, any views of the design can be represented, including components in their interaction (Wynn et al. 2010). If this restriction of the DSM matrix is made, it can be referred to as a ***component-component matrix***. Its principle is based on the representation of the interrelationships between the components. When creating this matrix, it is asked through pairwise comparison which component influences which other components in what way. In this DSM, structures such as circuits, hierarchies, clusters,

Table 4.5 Hierarchy and relations between components (Riekhof et al. 2012)

		1 Primary section	1.1 Primary section	1.1.1 Three-phase winding	1.1.1.1 Phase	1.1.2 Sheet Iron	1.1.2.1 Tooth	1.1.2.2 Groove	1.1.2.3 Body	2 Air Gap	3 Power Electronics	3.1 Voltage source	4 Basic mechanical structure	5 Secondary part	5.1 Copper Plate	5.2 Iron Reverse	5.2.1 Recesses for KLT	6 Small load carriers (KLT)	6.1 Copper plate receptacle/connector	7 Connecting element
1	Primary section																			
1.1	Primary section	B.v.																		
1.1.1	Three-phase winding		B.v.																	
1.1.1.1	Phase			B.v.			E	E	E	E					E	E				
1.1.2	Sheet iron		B.v.																	
1.1.2.1	Tooth					B.v.														
1.1.2.2	Groove					B.v.														
1.1.2.3	Body					B.v.														
2	Air gap																			
3	Power electronics																			
3.1	Voltage source				E						B.v.									
4	Basic mechanical structure								E											
5	Secondary part																			
5.1	Copper Plate													B.v.		E			E	E
5.2	Iron Reverse													B.v.	E					E
5.2.1	Recesses for KLT															B.v.			E	
6	Small load carriers (KLT)																			
6.1	Copper plate receptacle/connector														E		E	B.v.		
7	Connecting element														E	E				

Legend: **B.v.** = component of / **S** = material flow / **E** = energy flow

and bridges are already recognizable (Browning 2001). Fig. 4.10 provides an overview. A loop of components exists when the influence of a component affects itself through another component. Hierarchies represent influences from superior to subordinate components. On the other hand, clusters describe groups of components that have strong interrelationships. If clusters are connected via individual components, this is referred to as a bridge (see Fig. 4.10).

However, for very complex products, such as the logistics system, the component-component matrix can become very extensive, so it is advisable to structure components using the SoS approach or to hierarchize them using brainstorming.

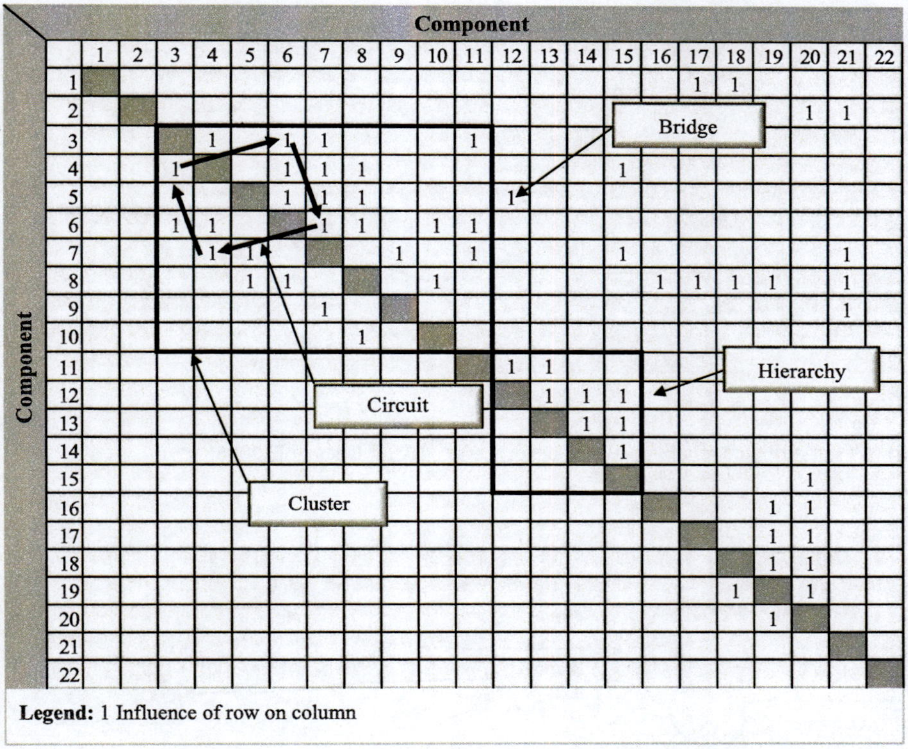

Fig. 4.10 Classification of structure types in components—component—matrices based on (Browning 2001)

d) *Graphs*

The representation of processes is discussed in the literature in a diverse and differentiated manner. Due to the large number of process modeling software, a large variety of variants for describing the object "process" is created. As a result of a comprehensive comparison, DANILOVIC and BROWNING (Danilovic and Browning 2007) tried to standardize the attribution of the process. For the modeling of processes in the GSE thinking model, a ***simple designation of the processes and their short characteristics via the input and output information*** is initially sufficient. Subsequently, ***process hierarchies can be mapped via the tree structure***. The representation of the networking of processes can, as SCHNIEDER and SCHNIEDER (Schnieder and Schnieder 2013) summarize, be done via simple ***block diagrams***, ***event-oriented process chains,*** via ***graphs*** or via ***Petri nets*** (Scheer and Cocchi 2006; Sendler 2009; Eigner and Stelzer 2009).

An example of the graphical representation of the temporal logical sequence of processes is provided by Fig. 4.11.

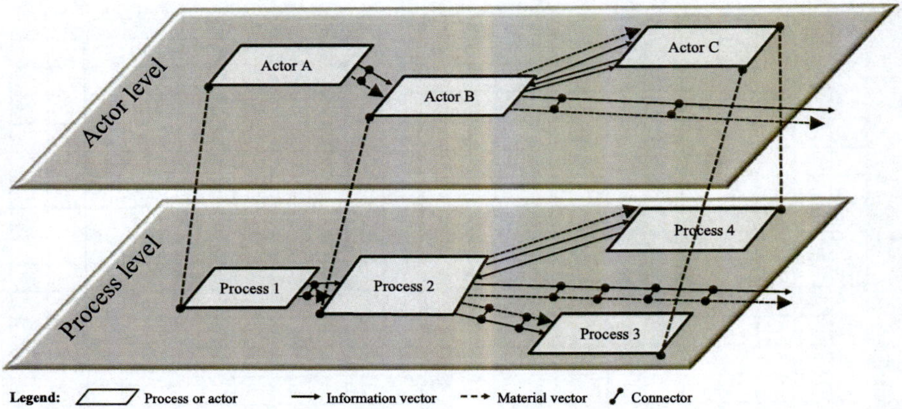

Fig. 4.11 Attributed relation between the processes. (After Braunholz 2006)

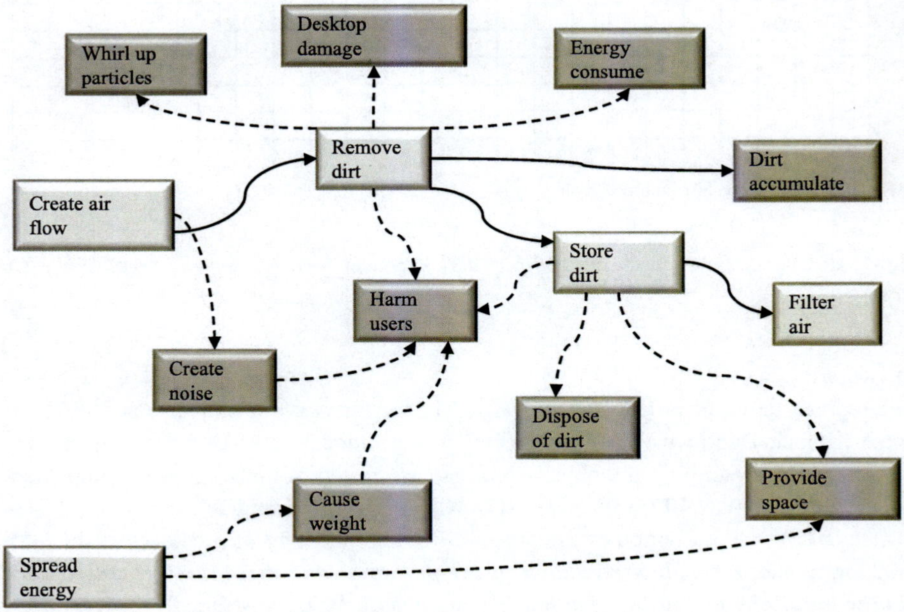

Fig. 4.12 Relation-oriented function model of a table vacuum cleaner with useful functions (white text fields) and harmful functions (black text fields). (After Lindemann 2005)

In addition to such temporal logical representations in the sequence of processes, functions can also be represented with so-called ***function networks*** as shown in Fig. 4.12.

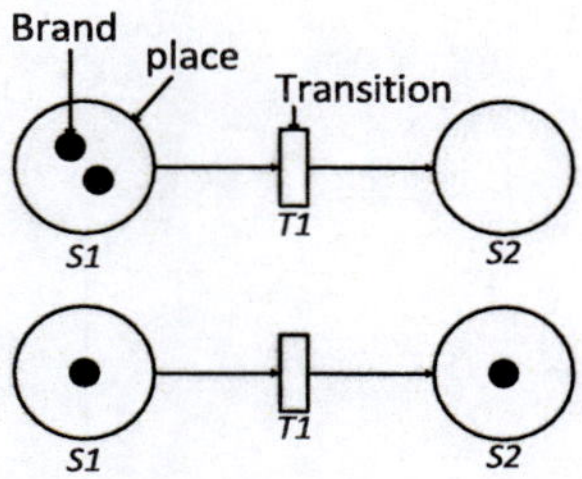

Fig. 4.13 Occupying a place with a token by activating a transition. (After Huber et al. 2008)

Within the framework of function networks, as in the aforementioned figure, a distinction can be made between useful and harmful functions. The dirt storage function of the vacuum cleaner from Fig. 4.12 is represented as a positive, i.e., useful function, whereas the disposal function is represented as a disturbing, harmful function. Through the additional ***attribution of the function*** as useful and disturbing or harmful, a kind of attribution of the function is already represented within the network. This representation encourages the developer to immediately compensate for the harmful functions by adding further useful functions. This in turn means that the developer must already think in terms of solution alternatives. The same applies to the classification of functions into basic, auxiliary, special, adaptation and order-related functions (Rudolf 2013).

e) *Petri nets*

Petri nets are finite graphs that represent various state elements and their transitions. They are used similarly to UML and EPCs, for example, for modeling processes or for RE for system designs (Desel and Reisig 2014; Partsch 1991). They have the basic property of describing systems and processes dynamically (see Fig. 4.13).

This is shown by Fig. 4.13, in which Petri nets fundamentally consist of *places* (represented by circles) and *transitions* (represented by lines/rectangles). The different elements are connected by arrows *(edges)* and can be occupied with so-called *tokens/marks*, which are moved from place to place using the transitions (Partsch 1991).

Fig. 4.14 represents this through a variety of different elements. Thus, processes in systems can be described at different times and states, i.e., for example, how a product moves through a system (Ochel 2017; Reisig 2010). A Petri net can consist of a variety of sequences, contents, and actions, but at its core, it is always about the basic connection of places and transitions and the transport of tokens (Reisig 2010).

Unlike similar representation options, such as flowcharts, Petri nets can describe processes as well as sub-processes in detail and dynamically. They also offer the advantage that they can be analyzed mathematically and thus allow the derivation of, for example, time and application technical sizes (Desel and Reisig 2014). Processes and problems can also be easily visualized and read. Likewise, concurrency can be detailed and

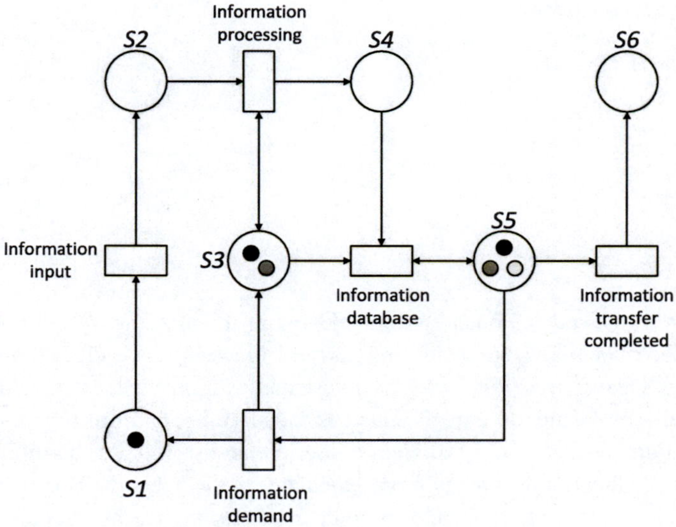

Fig. 4.14 Abstract Petri net for information transmission. (After Huber et al. 2008)

traceably depicted (Hering 1989). However, a disadvantage of Petri nets is that improvements and problem solutions cannot be directly derived from the model (Desel and Reisig 2014). The mapping method can also only be applied to a certain degree of complexity and does not offer the possibility to represent detailed information about the individual components (Hering 1989; Lill 2021).

f) *Flow diagrams*

The dependencies of the components can be further detailed through the representation of material, energy, and information flows and can also be represented via matrices (attributed matrices). These can simultaneously be depicted as flow diagrams, as illustrated by a linear drive in Fig. 4.15. A classification of the components according to their spatial arrangement is also possible. All forms of viewing the dependency between the components are necessary, elaborate, and difficult to present clearly or to process programmatically.

Another example of a ***flow diagram*** is illustrated in Fig. 4.16 for a soccer robot.

In summary, there are many ways to model system elements using various representation methods. The examples shown are intended to apply to all system elements of the GSE thinking model and can be logically combined. The following chapter explains what the GSE thinking model can look like based on these findings.

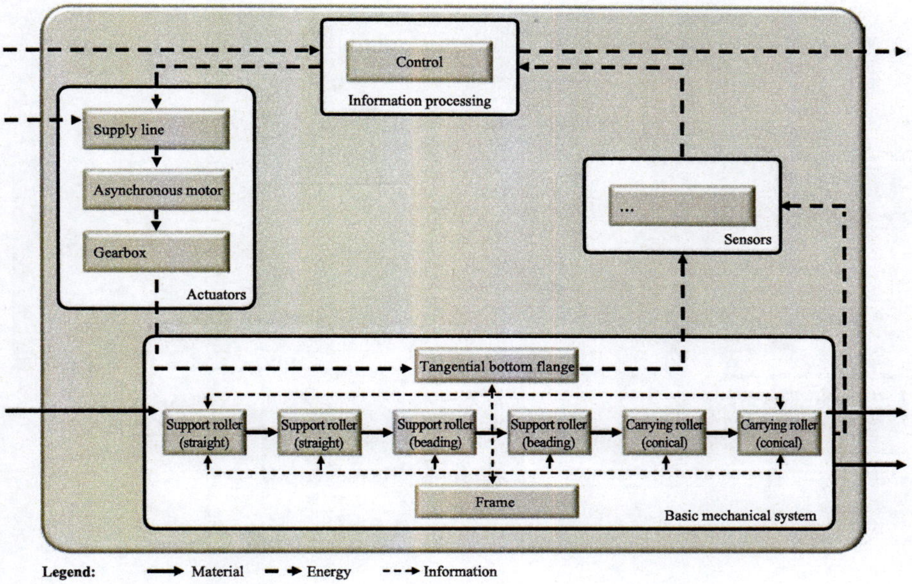

Fig. 4.15 Possibility of attributing relations between the components (work result in (SFB 696 2010))

4.3 General Description of Systems with the Metamodel (e-) DeCoDe

Basically, the question arises from sections 4.1 and 4.2 as to how systems can be described generally or generically. This means, which perspectives are essential to be able to describe any type of system uniformly? The diversity of how system mapping is viewed by renowned authors (see Sage and Rouse 2009; Lindemann 2016; Lindemann et al. 2009; Haberfellner et al. 2019; Ehrlenspiel and Meerkamm 2017; Kaiser 2014; Dumitrescu et al. 2021; Gausemeier et al. 2014; Gausemeier et al. 2012; Weilkiens 2019; Pohl 2016; Pohl and Rupp 2015; Rupp 2021; Bender and Gericke 2021; Farid and Suh 2016; Ebert 2019) shows why this is such a great challenge. It is not in line with the basic idea of SE that there is such a diverse understanding of how systems are described. This carries, among other things, the risk that systems are developed under different conditions and thus there is a lack of comparability between the systems.

An example of the range of different perceptions of system modeling can be a comparative consideration of the approaches of HABERFELLER et al., GAUSEMEIER et al. and WEILKINS (Haberfellner et al. 2019; Gausemeier et al. 2014; Weilkiens 2019). While HABERFELLNER et al. describe systems in general with system elements, GAUSEMEIER et al. and WEILKIENS specify the perspectives on a system.

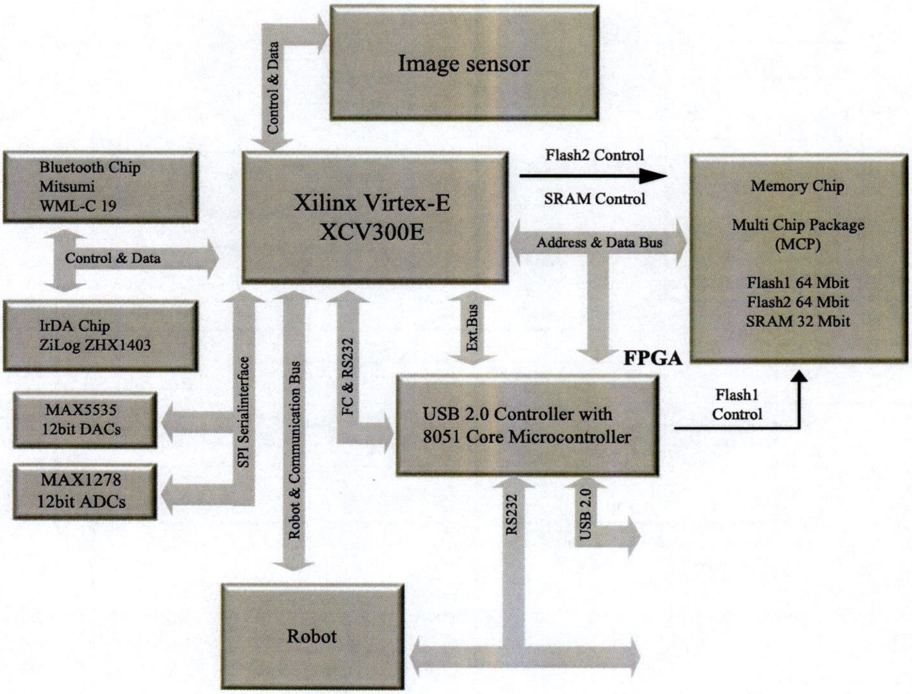

Fig. 4.16 Representation of the relation between the components via a flow diagram (according to Sitte and Winzer 2007)

For this, GAUSEMEIER et al. use the CONSENS approach (CONceptual design Specification technique for ENgineering of complex Systems) and WEILKIENS, among others, the SysML (Systems Modeling Language). GAUSEMEIER et al. try to keep the system modeling very open with the CONSENS approach by dividing the view of a system basically into two views: environment elements and system elements. WEILKIENS, on the other hand, uses SysML to provide a much more detailed view of the system, specifying the views of a system with a variety of applicable diagrams, such as the requirement diagram, internal block diagram, block definition diagram, assurance diagram, package diagram, activity diagram, use case diagram, sequence diagram, and state diagram. All approaches have their justification in a certain way. Because it makes sense to be able to describe systems very roughly. But it is also important to specify this. However, the questions remain open as to how the various approaches with the implicitly different views of a system can be brought together and what is the minimum number of views of a system. Because on the one hand, it can be determined that the more views of a system are given, the more complex the system modelling becomes. On the other hand, too few views of a system can be too vague, so that the viewer can imagine little about the depicted system. These findings result from a large number of research works from the GSE (see Heinrichsmeyer 2020; Heinrichsmeyer et al. 2020a; Schlueter et al. 2019, 2018; Mistler 2021; Mistler et al. 2021b; Mistler et al. 2019; Heinke and Mistler 2019; Bielefeld 2020; Bielefeld et al. 2016; Schlueter 2016; Beyerer and Winzer 2018; Sitte and Winzer 2005; Sitte and Winzer 2007, 2011b; Mistler et al. 2021a; Nicklas 2016; Mamrot 2014; Winzer 2015; Sitte and Winzer 2011a; Marchlewitz et al. 2015; Nicklas et al. 2016; Ott 2009; Mamrot et al. 2014; Sitte and Winzer 2006). The research works reflect, just like the conclusion from section 4.1, that the requirement, function, process, component, and person view are needed to generically model technical and sociotechnical systems. Thus, the GSE thinking model consists of these five interconnected views. The developed system modeling approach is the **Demand Compliant Design (DeCoDe)** for technical systems and the **enhanced Demand Compliant Design (e-DeCoDe)** for sociotechnical systems. The approaches see themselves as a superordinate metamodel to enable a uniform basis for interdisciplinary modeling (Mistler 2021; Winzer 2015).

In order for the DeCoDe or e-DeCoDe approach to be understood uniformly, it is fundamentally explained in the following. To achieve a basis for a common thinking model, the respective system views must first be defined. These are explained in Table 4.6.

The system views and their definitions presented in Table 4.6 are essential for system modeling. Because as MISTLER (Mistler 2021) shows, without this basis, terms and definitions from other modeling approaches or specific company language cannot be structured, systematically, and comprehensibly classified into a thinking model. This would make the ontological integration in system modeling unmanageable (Mistler 2021). However, in order for the system views to be logically connected, the semantics and syntax must be determined in a metamodel (Pohl 2016; Gausemeier et al. 2013a). This is outlined in Fig. 4.17.

Fig. 4.17 presents the syntax and semantics of the (e-)DeCoDe approach. At the same time, the difference between the DeCoDe and e-DeCoDe approach is visualized here. It becomes clear that the e-DeCoDe approach is basically based on the DeCoDe approach. This means that the functioning of both approaches is the same. The distinction of the approaches is historically grown. The DeCoDe approach by SITTE and WINZER was specifically developed for the modeling of technical systems (Sitte and Winzer 2005, 2006, 2007, 2011a). However, NICKLAS (Nicklas 2016) found that the DeCoDe approach had to be extended by the person view in order to be able to model sociotechnical systems. Thus, the e-DeCoDe approach was created and was further specified in MISTLER (Mistler 2021) by specifically working out the meaning of input and output. So it makes sense to distinguish the approaches in order to be able to linguistically delimit whether the modeling of a technical or sociotechnical system is being talked about.

However, it is not necessary to use each view of the approaches necessarily. The deeper thought behind the DeCoDe approach is that only the views from the metamodel that are necessary for the development of the respective system are used. Thus, the (e-) DeCoDe tools shown in Fig. 4.18 result from modeling with (e-)DeCoDe. These are

Table 4.6 Definition of the (e-)DeCoDe system views (see Mistler 2021; Nicklas 2016; Heinrichsmeyer 2020; Bielefeld 2020; Schlueter 2016; Mamrot 2014)

<table>
<tr><th rowspan="7">e-DeCoDe approach</th><th rowspan="5">DeCoDe approach</th><th>System view</th><th>Definition</th><th>Example</th></tr>
<tr><td>Requirements</td><td>Requirements are stakeholder needs or expectations of a system that are specified, usually assumed, or mandatory.</td><td>The "campsite" system is subject to the requirement that the shower must be warm.</td></tr>
<tr><td>Functions</td><td>Functions describe the purpose or task that a system has to fulfill. They thus give a target direction to the conversion of inputs into outputs of a system. Thereby functions make a description of the "what" possible. That means, what a system or parts of it are to realize.</td><td>The campsite must offer a "shower function" with hot water.</td></tr>
<tr><td>Processes</td><td>Processes describe how the inputs of a system are converted into outputs, i.e. the "how". The built-in functionality of the system is realized via the process, i.e. within processes, functions are implemented in technical systems through the use of components. If the integration of persons into processes takes place, the latter are often also called work or business processes (process of a socio-technical system).</td><td>In the "shower function", the "shower process" converts cold water into hot water.</td></tr>
<tr><td>Components</td><td>Components are physical or logical, individual or combined parts of a system.</td><td>In order to ensure the "shower function" with hot water, the component "instantaneous water heater" needed.</td></tr>
<tr><td></td><td>People</td><td>People describe people. They use and realize components as well as processes and provide input and output for the performance of services. Thus, they realize functions, which in turn fulfill requirements.</td><td>For the implementation of the "Shower function" is the responsibility of the person "Caretaker".</td></tr>
<tr><td></td><td>Input / Output</td><td>Inputs (inputs) and outputs (outputs) are understood as matter, information and energy.</td><td>For the implementation of the "Shower function" is used in the "Shower process" the input "cold water" transformed into the output "warm water".</td></tr>
</table>

deliberately based on matrices in order to create a continuous link between the views. This is essential to show the complexity of systems in context. This is illustrated by a large number of research works in which system development is based on (e-)DeCoDe (see Mamrot 2014; Nicklas 2016; Schlueter 2016; Bielefeld 2020; Heinrichsmeyer 2020; Mistler 2021).

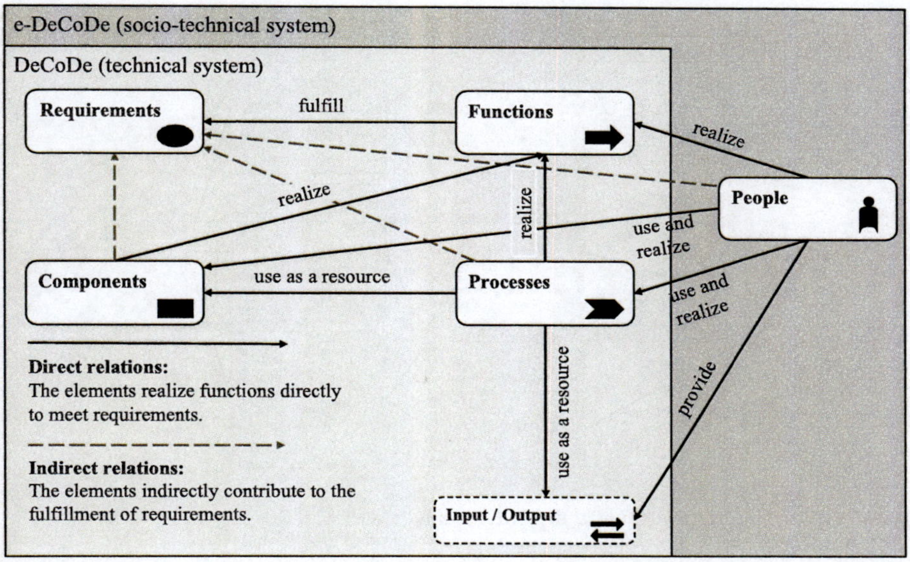

Fig. 4.17 (e-)DeCoDe modeling. (After Mistler 2021; Ott 2009; Nicklas 2016; Mistler et al. 2021b)

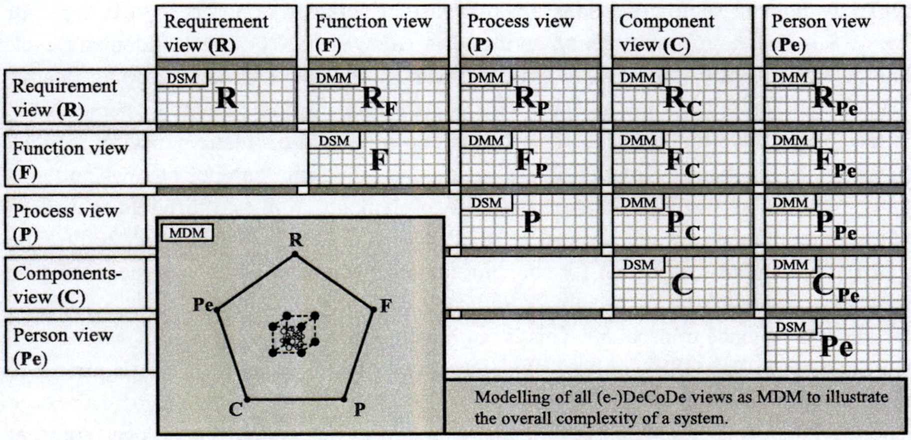

Fig. 4.18 (e-)DeCoDe basic schema. (After Mistler et al. 2021a)

It should be emphasized here that (e-)DeCoDe is fundamentally understood as a **philosophical approach** that aims to achieve an interdisciplinary cognitive understanding of system mapping (Mistler et al. 2021a).

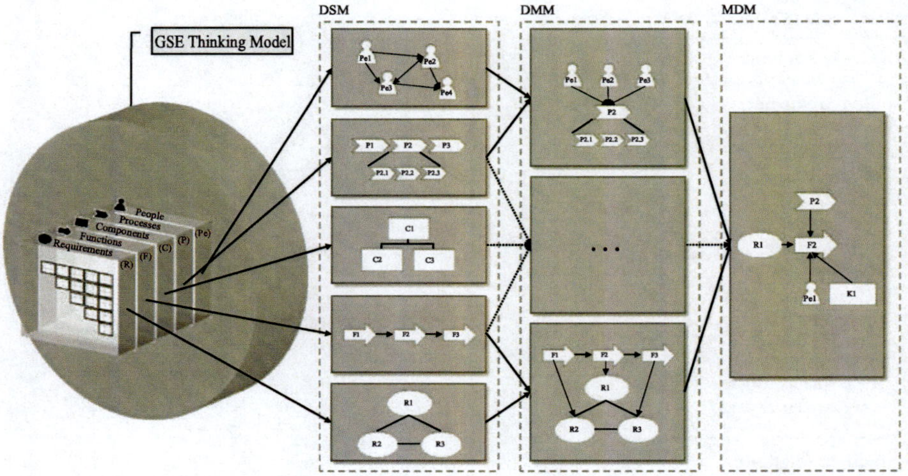

Fig. 4.19 Principle representation of the GSE thinking model with five views

Fig. 4.18 illustrates that matrices serve as tools for describing the e-DeCoDe views and their interactions with each other (Winzer 2015). A consistent link is particularly important according to EHRLENSPIEL and MEERKAMM (Ehrlenspiel and Meerkamm 2017) so that the same conditions always exist in collaboration. The matrix logic in Fig. 4.18 is based on graph theory, as found in LINDEMANN et al. (Lindemann et al. 2009). Consequently, the Design Structure Matrix (DSM), Domain Mapping Matrix (DMM) and Multiple Domain Matrix (MDM) are implicit components or perspectives on the (e-)DeCoDe tools. These perspectives aim to generate situation-specific models. The DSM forms a square matrix, i.e., a matrix with the same number of rows and columns. In addition, the DMM can represent two elements from two different views. To represent more than two opposing views, the MDM is used (Lindemann et al. 2009). The MDM is particularly important for showing the overall complexity of a system (Mistler 2021; Mistler et al. 2021a). As already indicated in Fig. 4.18 with the MDM, the matrices can also be made transparent. This is outlined in the summary Fig. 4.19.

In summary, the GSE thinking model consists of five system views: requirement, function, process, component, and person view. These views are intended to create a generic system image across disciplinary boundaries. The system image can represent a cognitive construct in the form of a thinking model, but can also be visualized transparently. Depending on the questions about a system, different system views are to be selected individually. An overview of the different matrix-based system views and their implicit questions is summarized in Table 4.7. This also offers possible results for the respective questions.

Table 4.7 Contents of the (e-)DeCoDe matrices (Mistler 2021, p. 50)

Matrix	Question	Results
R (Requirements vs. Requirements)	Which requirements influence each other?	Prioritizing, systematizing, but also partially eliminating requirements
$\mathbf{R_F}$ (Requirements vs. Functions)	Which functions influence which requirements (and vice versa)?	Representation and determination of the functions required to meet the respective requirements and their effects on other requirements
$\mathbf{R_P}$ (Requirements vs. Processes)	Which requirements influence which processes (and vice versa)?	Derivation of new requirements from processes or fulfillment of requirements by processes (for example, by representing input and output)
$\mathbf{R_C}$ (Requirements vs. Components)	Which components influence which requirements (and vice versa)?	Derivation of new requirements from components as well as representation of the impact of existing components on requirement fulfillment
$\mathbf{R_{Pe}}$ (Requirements vs. People)	Which requirements influence which people (and vice versa)?	Derivation of responsibilities that fulfill requirements or contribute to the fulfillment of requirements
$\mathbf{S_F}$ (Functions vs. Functions)	Which functions influence each other?	Determination of relationships at the function level (for example, possible target conflicts)
$\mathbf{S_{F, P}}$ (Processes vs. Functions)	Which processes influence which functions (and vice versa)?	Determination of additional functions or processes or input and output as well as target conflicts between functions and processes, for example, through input and output
$\mathbf{S_{F, C}}$ (Functions vs. Components)	Which functions influence which components (and vice versa)?	Determination of necessary components for function fulfillment or target conflicts between functions and components
$\mathbf{S_{F, Pe}}$ (Functions vs. People)	Which functions influence which people (and vice versa)?	Representation of responsibilities through the role of people in connection with service provision
$\mathbf{S_P}$ (Processes vs. Processes)	Which processes influence each other?	Determination of relationships at the process level (for example, possible target conflicts in the conversion of input into output)
$\mathbf{S_{P, C}}$ (Processes vs. Components)	Which processes influence which components (and vice versa)?	Determination of necessary components for process execution or target conflicts between processes and components
$\mathbf{S_{P, Pe}}$ (Processes vs. People)	Which processes influence which people (and vice versa)?	Representation of responsibilities and input and output regarding process realization in connection with service provision
$\mathbf{S_C}$ (Components vs. Components)	Which components influence each other?	Determination of possible effects of the use or change of certain components

(continued)

Table 4.7 (continued)

Matrix	Question	Results
$S_{C, Pe}$ (Components vs. People)	Which components influence which people (and vice versa)?	Representation of responsibilities of components in connection with service provision
S_{Pe} (People vs. People)	Which people influence each other?	Determination of relationships in the organizational structure of a company in connection with service provision

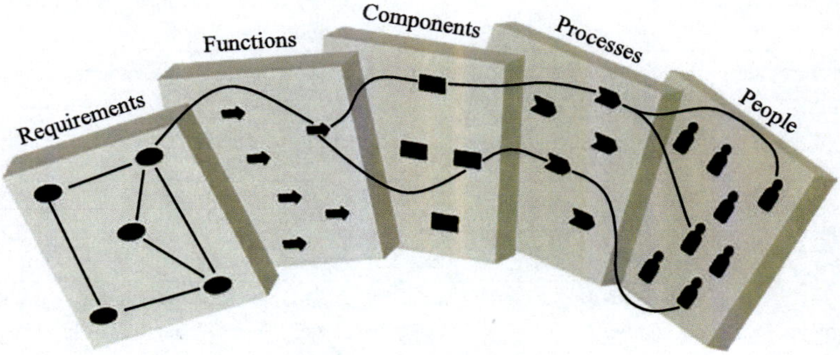

Fig. 4.20 The principle of networking the five views in the GSE thinking model

In summary, all e-DeCoDe views must be consistently linked and modeled with each other (see Fig. 4.20).

As various works show, the system model changes over the temporal course of a project (Mamrot 2014; Nicklas 2016). This means that the system model in all five views at the time t_0, t_n, $t_{n+1,}$ as Fig. 3.21 represents, must be archived. The changes over time are possible by overlaying the images t_0, t_n, t_{n+1}. This idea of the image of the temporal change of the system is illustrated in Fig. 4.21.

The deviations of the images of the system from time t_0 to time t_{n+1} thus correspond to the changes in the system. It is assumed that the designations of the views and the fixed formulations for the respective requirements, functions, components, processes, and people ideally should not be changed.

To implement the e-DeCoDe approach, IT tools are needed, otherwise the complexity of very extensive system models in projects cannot be managed. For this purpose, a software analysis was carried out, the results of which are summarized in Table 4.8. The software analysis specifically considered the e-DeCoDe approach to modeling organizations. Since the approach does not differ in the basic scheme from the DeCoDe approach, the results can also be transferred to this approach for modeling technical systems. The consideration of the software could be narrowed down from the previous findings on GSE to the following: MS Excel, LOOMEO, Quam, Cameo and iQUAVIS (see

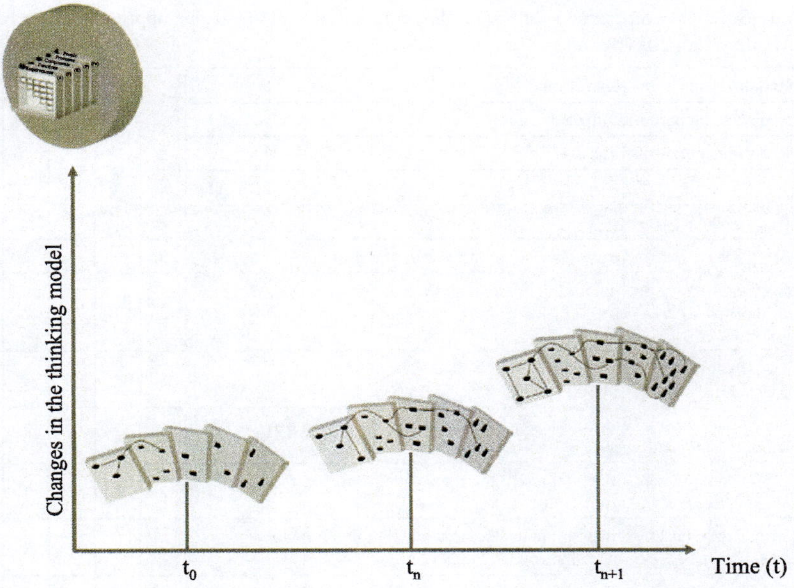

Fig. 4.21 The principle of the temporal change of the GSE thinking model

Schlueter et al. 2018; Bielefeld 2020; Heinrichsmeyer 2020; Mistler 2021; Mistler et al. 2021b; Mistler 2020b).

The following describes the IT tools listed in Table 4.8 and explains why they were selected for evaluation. Finally, a conclusion on the evaluation is given. It should be noted that the following content is mainly quoted from the research work of MISTLER (Mistler 2020b; Mistler et al. 2021b; Mistler 2021). Therefore, the formulation of the content is very closely based on these works.

MS Excel was developed by Microsoft for tabular visualization and editing (Microsoft 2021). Due to its easy application and low cost, it is often used in companies for SE (Marques-Lucena et al. 2015; Schuhmann et al. 2010). Moreover, due to its tabular structure, it is suitable for implementing the basic e-DeCoDe tools (Heinrichsmeyer et al. 2020b; Heinrichsmeyer 2020). As Table 4.8 shows, when using Excel, it can be observed that the e-DeCoDe views can be made transparent through the free design of matrices. At the same time, it is possible to attribute the e-DeCoDe views independently (R3). However, these cannot be made visible graph-based, as no special functions are integrated for this purpose (R2). As a consequence, such inflexible matrix formation with missing visualization leads to a difficult representation of the temporal logical sequence of, for example, processes and functions, since pure matrix formation is no longer manageable from a certain model size (R4). These findings are also supported by SCHUHMANN et al. (Schuhmann et al. 2010), who highlight MS Excel as a globally popular IT tool due to its

Table 4.8 Software comparison for the implementation of the e-DeCoDe approach. (after Mistler 2020a; Mistler et al. 2021b)

3	Requirement completely fulfilled					
2	Requirement partially fulfilled					
1	Requirement not fulfilled					
	IT Tools	MS Excel	LOOMEO	Quam	Cameo	iQUAVIS
No.	Requirements (R)					
R1	The IT tool must make the e-DeCoDe views transparent.	3	3	2	2	3
R2	The IT tool must take into account the interrelationships between all the Visualize e-DeCoDe elements both matrix-based and graph-based.	2	3	2	2	3
R3	The IT tool must be able to attribute the e-DeCoDe elements and their interrelationships.	3	3	3	1	3
R4	The IT tool must be able to arrange functions and processes in a temporally logical sequence.	2	2	3	3	3
R5	The IT tool must be able to represent the information flow between the e-DeCoDe elements.	2	2	2	1	3
R6	The IT tool must be able to graphically and textually model the inputs and outputs of e-DeCoDe elements .	2	2	2	1	2
R7	The IT tool must be able to systematically focus on the e-DeCoDe elements and their interactions in order to concentrate on the essentials.	1	3	2	1	3
R8	The IT tool must enable a continuous data flow through all e-DeCoDe views.	1	3	3	2	3
R9	The IT tool must enable platform-based project management for interdisciplinary and cross-location exchange and collaboration.	1	1	3	3	3
R10	The IT tool must be able to save system states in order to track phases of project management.	2	2	1	3	3
R11	The IT tool must be able to make the project results transparent in the company through sustainable and R11 platform-based organizational management	1	1	3	2	2
Total valuation		**20**	**25**	**26**	**21**	**31**

intuitive and easy use of freely configurable tables and worksheets. They also show that it can realize extended functionalities through programming with Visual Basic for Applications (VBA). This has already contributed to enabling the model-based processing of complaints via the e-DeCoDe approach (Heinrichsmeyer et al. 2020b; Heinrichsmeyer 2020). However, SCHUHMANN et al. also point out the need for graph-based representations for more accurate analyses. For example, system elements such as requirements or components cannot be adequately represented at deeper levels with MS Excel. This circumstance is also reflected in the representation of the information flow through input and output. Accordingly, a graphical visualization is needed to be able to trace the information flow between the elements (Winzer and Braunholz 2000; Braunholz 2006)

(R5 and R6). In addition, MS Excel lacks pre-configured filter and structuring functions to reduce the complexity of the e-DeCoDe views. In this context, SCHUHMANN et al. highlight the advantage of the extensive filter options of MS Excel. However, MS Excel does not offer a function for isolating system elements with a focus function like LOOMEO (Schlueter et al. 2018; Bielefeld et al. 2018) or iQUAVIS (Mistler 2021; Mistler et al. 2021a) (R7). In addition, MS Excel does not offer a consistent data flow without considering other software tools (R8). This means that MS Excel is not suitable for reliably supporting distributed work (R9). Although data or cell contents can be shared with other team members via cell references, neither the data consistency nor the existence of the referenced files can be guaranteed (Schuhmann et al. 2010) (R10). Finally, it should be noted that MS Excel is not automatically platform-based and is not suitable for organizational management. It lacks a visualization to sustainably implement a process-oriented organizational model in a company, such as for integrated management systems (Mistler et al. 2019; Kamenický et al. 2014) (R11); (cf. Mistler 2020b; Mistler et al. 2021b; Mistler 2021).

LOOMEO was explicitly developed for complexity management (REDPOINT. TESEON 2021). In this consideration, the version LOOMEO 2.9 from 2015 is focused. LOOMEO has been used in a variety of research projects due to its free design based on matrices and supports the e-DeCoDe basic scheme through the use of matrices as well as graph-based representations of system elements (Bielefeld 2020; Lindemann et al. 2009; Schlueter et al. 2018; Luft et al. 2014; Maurer and Braun 2008; Mirson et al. 2011; Shimomura and Kimita 2013). As Table 4.8 shows, it is possible with LOOMEO to create and make transparent the e-DeCoDe views both matrix-based and graph-based. In addition, it offers the possibility to freely attribute the interactions between the system elements (Lindemann et al. 2009; Schlueter et al. 2018; Bielefeld et al. 2018; Luft et al. 2014; Maurer and Braun 2008) (R1, R2 and R3). When modeling, for example, processes and functions with LOOMEO, it is noticeable that it is very difficult to arrange the functions and processes in a time-logical order (R4). Furthermore, LOOMEO can represent the information flow by considering attributions between the elements (Lindemann et al. 2009). However, the complex graph-theoretical visualization and the lack of functionality in the textual modeling make it difficult to represent the information flow from a certain number of system elements (R5 and R6). To address this problem, LOOMEO offers a focus function that allows you to focus specifically on the interactions between system elements. This function has already been tested and can be transferred to the e-DeCoDe approach (Schlueter et al. 2018; Bielefeld et al. 2018) (R7). Furthermore, when modeling with e-DeCoDe, it should be noted that changes to the e-DeCoDe views are transferred to all other views. Thus, a continuous data flow is ensured (R8). However, LOOMEO is not designed for platform-based project management and has the same problems in versioning as described by SCHUHMANN et al. also for MS Excel (R9 and R10). Although LOOMEO offers a variety of visualization options, it is unsuitable as a sustainable, transparent, and platform-based software tool for organizational management (R11). As some research works show (Lindemann et al. 2009; Schlueter et al. 2018; Bielefeld et al. 2018; Luft et al. 2014; Maurer and Braun 2008; Schmidt III et al. 2011), it is more for the development of complex product systems than for the sustainable process-oriented management of organizations (cf. Mistler 2020b; Mistler et al. 2021b; Mistler 2021).

Quam, unlike MS Excel and LOOMEO, is an IT tool based on the Business Process Modeling Notation (BPMN). It connects MS Share-Point with MS Visio and enables a matrix-based approach for process management in organizational development (Join GmbH 2021). The first approach of Quam was developed by GRAUP (Graup 2005) for Gerneric Management and has since been used within the GSE for organizational management (Mistler 2021; Mistler et al. 2021a; Kamenický et al. 2014). As Table 4.8 shows, Quam can only partially make the e-DeCoDe views transparent, as Quam can only compare two views at the same time. Consequently, a clear multi-domain matrix is missing (R1). However, since Quam follows a process-oriented approach, it mainly focuses on the process organization. This means that, for example, only the processes and functions are represented as system elements in one view (R2). In addition, Quam can still realize the consistent linking of the views with independent attributes among each other (R3). The connection to other system elements, such as the person view, is made via the attributes using BPMN. This approach allows Quam to visualize the temporal sequence of processes and functions and, for example, to link them with the person view or organizational structure (R4). However, the representation of the information flow in the used flow diagrams is partly neglected and can only be further understood by additional modeling of input and output elements in matrices or tables (R5 and R6). With the use of matrices and tables, Quam also offers functionalities to give different perspectives on the system (R7). If system elements or attributes are changed, it shows that Quam, through the use of MS Share-Point, enables a continuous data flow (R8). In addition, the integration of a project management application (CPM) in MS SharePoint with Quam enables platform-wide project management (R9). The use of MS SharePoint as a platform also ensures a permanent versioning of system states. However, there is no explicit function that realizes the backup and traceability of system states for specific phases in project management of organizational development (R10). Therefore, Quam is not, like LOOMEO, Cameo or iQUAVIS, explicitly suitable for enabling system analysis and design in projects. Rather, Quam is designed to implement process-oriented organizational management in companies sustainably and transparently via MS SharePoint (R11). Quam has been used in research only rarely so far. However, it has proven itself in the context of the GSE for organizational management and the research work carried out so far can support the described findings (Mistler et al. 2019; Kamenický et al. 2014; Mistler 2021). The fundamental difference between Quam, MS Excel, and LOOMEO is that it is based on BPMN and embedded platform-based in MS SharePoint. BPMN characterizes the modeling of business processes and is therefore often preferred to other modeling languages in this context (Weilkiens 2019) (cf. Mistler 2020b; Mistler et al. 2021b; Mistler 2021).

Cameo is a comprehensive software tool for modeling complex systems. The Cameo Systems Modeler is based on the Systems Modeling Language (SysML) (3D 2021). SysML is often used for the development of technical systems. Consequently, it does not include the process environment that is considered in organizational development. Therefore, it can be supplemented by BPMN (Weilkiens 2019; 3D 2021). SysML is popular for modeling various systems and was also used in the GSE (Nicklas et al. 2016; Winzer 2015). The use of SysML in combination with BPMN has not yet been considered in the context of the GSE. Potentially, this opens up new possibilities for organizational development. In addition, it could be possible to realize the e-DeCoDe basic scheme graphically and based on matrices (Salehi et al. 2018; Pavalkis et al. 2011). As Table 4.8 shows, when modeling with the Cameo Systems Modeler, the e-DeCoDe views can only be partially made visible. The reason for this is the predefined views of SysML, which need to be adapted to the e-DeCoDe basic scheme. In addition, the linking of the SysML and BPMN views with each other is inconsistent, as are the views within SysML (R1, R2 and R3). The advantage of the additional use of BPMN is that Cameo enables process modeling for organizations (Pavalkis et al. 2011) (R4). Due to the consistency problems, however, it is unclear how the information flow between all e-DeCoDe views can be represented with Cameo (R5 and R6). Furthermore, the filter functions of Cameo were considered. Although Cameo has several filter functions, their suitability for the e-DeCoDe approach can be classified as impractical, as the views cannot be fully represented. In addition, no comparable focus function was found, as offered by LOOMEO or iQUAVIS (R7). The consistency problems of the SysML views in the Cameo Systems Modeler were also identified by SALEHI et al. (Salehi et al. 2018). It is to be criticized that there is a lack of a concrete description of how the predefined views or models of SysML can be merged. This circumstance complicates the adaptation of other approaches (Salehi et al. 2018; Espinoza et al. 2009). In addition, missing descriptions and inconsistency problems of BPMN models are also addressed by other authors (Pavalkis et al. 2011; Zensen and Kuster 2018). Therefore, it is not surprising that the adaptation of the e-DeCoDe approach into a SysML- and BPMN-supported software tool is challenging. Consequently, a continuous data flow between all e-DeCoDe views could not be achieved with Cameo (R8). Furthermore, Cameo is in principle suitable for ensuring platform-based project management across locations and for documenting and tracing system intermediate states for project management phases (Salehi et al. 2018; Taha et al. 2018) (R9 and R10). However, it does not serve, like Quam, as a transparent and interactive platform for organizational management that is accessible to every employee of a company. Rather, it is used, like LOOMEO or iQUAVIS, for project management for the analysis and design of model-based systems in interdisciplinary teams (R11); (cf. Mistler 2020b; Mistler et al. 2021b; Mistler 2021).

iQUAVIS is used for system modeling and project management and is based on the CONceptual design Specification technique for the Engineering of complex Systems (CONSENS) approach. CONSENS primarily targets the modeling of technical systems (Two Pillars 2021). Although the approach has already been discussed in NICKLAS and

SCHLUETER et al. (Nicklas 2016; Schlueter et al. 2018) for the modeling of systems within GSE, its technological implementation with iQUAVIS has not yet been analyzed. However, it should also be considered, as it can be shown that the CONSENS approach can also support matrix-based approaches and is therefore potentially suitable for the realization of the e-DeCoDe basic scheme. Furthermore, the goal of CONSENS is to enable simple, holistic, and interdisciplinary system modeling (Gausemeier et al. 2013b; Gausemeier et al. 2014; Kaiser 2014). Therefore, the implementation of the e-DeCoDe approach with iQUAVIS is also being investigated. As Table 4.8 shows, iQUAVIS allows the creation of all e-DeCoDe views that can be represented with matrices and graphs (R1 and R2) with an individual system model design. In addition, it has various attribution options for system elements and their interaction (R3). iQUAVIS also offers the possibility to create various diagrams and worksheets that are consistently linked with the created e-DeCoDe views. Thus, the temporal structuring of, for example, processes and functions as well as the textual and graphical representation of the information flow through input and output can be realized with the help of diagrams and worksheets (R4 and R5). However, it is to be criticized that no function combines the graphical representation with the tabular one in one window, as is possible, for example, with Quam (R6). It can also be noted that iQUAVIS has extensive preset filter and focus functions. It should be noted that the filter functions for the e-DeCoDe approach can be manually extended and that the focus functions offer several possibilities. For example, not only direct dependencies between system elements can be focused on, but also, for example, indirect dependencies (R7). However, it can be noted that iQUAVIS is based on the metamodel of the CONSENS approach, which consists of only a few views and notations. This basic systematics allows a consistent and logical linking of views that can be represented both matrix-based and graph-based (Mistler 2021; Gausemeier et al. 2014). Since iQUAVIS allows the functionality of the CONSENS approach and is comparable to the e-DeCoDe basic scheme, the implementation of e-DeCoDe with a consistent data flow between all views was very easily achieved (R8). In addition, the integration of project management tools and working in a cloud in iQUAVIS is noteworthy. Thus, it can realize platform-based, cross-location project management and has integrated functions to ensure system states and tracking during the project (R9 and R10). However, for sustainable and transparent organizational management in companies, iQUAVIS is unsuitable, just like MS Excel, LOOMEO and Cameo, as it is designed for the model-based analysis and design of systems in interdisciplinary teams and not for organizational management (Mistler 2021). This challenge is also partially addressed with the CONSENS approach (Gausemeier et al. 2013b) (R11); (see Mistler 2020b; Mistler et al. 2021b; Mistler 2021).

This chapter has detailed the basics of modeling with the e-DeCoDe approach. It also explained which IT tools are suitable for implementing the e-DeCoDe approach. The following chapters will illustrate how modeling with the e-DeCoDe approach can be specifically realized.

4.4 Possible Sequence of Steps for Creating the GSE Thinking Model for Technical Systems with DeCoDe

While the previous chapter described how the views of the system can be transparently represented, it will now be illustrated how a representation of a system can be structured and systematically created step by step—while simultaneously implementing the requirements for system modeling. Only the ***procedure for creating a GSE thinking model*** is characterized. How a system analysis or system design using the model is to be carried out will be described in Chap. 5.

The procedure for creating a GSE thinking model is understood as the temporally logical, step-by-step description of creating the views of the system model and their links. The result of Sect. 4.3 determined that systems can essentially be represented in a standardized way in a minimal manner using the requirement, component, function, and process view, with the help of DeCoDe tools. Their logical and temporal combination in creating a system image is problem-oriented and characterizes the DeCoDe workflow.

The four-stage model of system design recommends first a requirement analysis, then a function analysis, and finally an architecture analysis to ultimately provide proof of the system's requirement fulfillment (Scheithauer 2014). This sequence of steps can also be realized with the DeCoDe tools. As a result of applying the four-stage model of system design with the DeCoDe tools, the requirement view can be created from the requirement analysis, the function view can be created from the function analysis, and the component view can be created from the architecture view. The proof of the system's requirement correctness is provided by the networking of the three aforementioned views. This linear approach of the four-stage model is hardly found in practice in system design. Here, other approaches, such as Smart Engineering, are in demand (Anderl 2012). In general, a metamodel is also needed in the Industry 4.0 era so that transdisciplinary teams can work in a targeted, problem-solving-oriented manner. The use of DeCoDe tools for this is a possible solution approach. This raises a number of questions that still need to be answered, such as:

- With which perspective is the description of the product system started?
- Is it possible to describe all four views simultaneously?
- How detailed should the description of the views be?
- When are the links within the respective view transparently represented and evaluated?
- When, where, and how are the respective views linked with each other?
- How detailed should the perspectives be described?

These and many other questions arise in system modeling, assuming that the ***DeCoDe tool is used to create a system image*** of a technical system in a problem-oriented manner.

The following will examine three different examples of how and in what order the individual DeCoDe tools were used to create a system image of a technical system. The goal is to develop a generalized sequence of steps (DeCoDe workflow) for creating a modified GSE thinking model.

Example 1:_The requirement-based design of a drive of the roller conveyor

As part of the special research area 696 "Logistics on Demand", it was the task, within the framework of a subproject, to optimize the drive (Künne and Richard 2009). To guarantee the failure safety of logistical systems, they are often oversized. This also applies to the drives. They must be robust and reliable. If a drive fails, the entire logistical system fails. Since drives are installed in a large number in a logistical system, they pose a high risk to the functionality of the logistical system. To avoid this, drives are also designed to be oversized. The task of an interdisciplinary team, consisting of logisticians, mechanical engineers, electrical engineers, and quality engineers, was to optimize the logistical system in interaction with the drive in such a way that efficient, robust operation becomes possible. This team set itself the goal of creating a common image of the system of the logistical system using DeCoDe. ◀

How was the creation of the image of the "Roller Conveyor" system approached?

The requirement view for the "Drive" system was initially not described, as it was an existing logistical system in the form of a roller conveyor and thus a re-engineering of the drive (asynchronous machine). The only requirement was the optimization of the drive. Consequently, the creation of the other system views, i.e., the component, function, and process view, could be started step by step, as explained below.

1. Step: Description of the component view.

In the logistical system, the drive was initially considered as a black-box model (see Fig. 4.22) and its interactions with other system components were recorded in a matrix. Before this matrix could be created, the components had to be recorded and precisely defined together in workshops, which was a lengthy process. In addition, the components had to be hierarchized, as shown in Table 4.9.

This hierarchization allows precisely locating which components of the logistical system interact with the drive at which level. The representation of the interactions between the elements was done in the so-called component-component matrix of Table 4.9. Initially, it was only estimated whether an interaction exists. Subsequently, the interaction of the components could be roughly represented via the energy, information, and material flow. By consistently applying the basic principles of systematic thinking and acting, especially the basic principle of minimal models, only the elements that interact with the drive were represented in the component-component matrix. Through this problem focus, i.e., the optimization of the drive, the complexity of the logistical system could be drastically reduced.

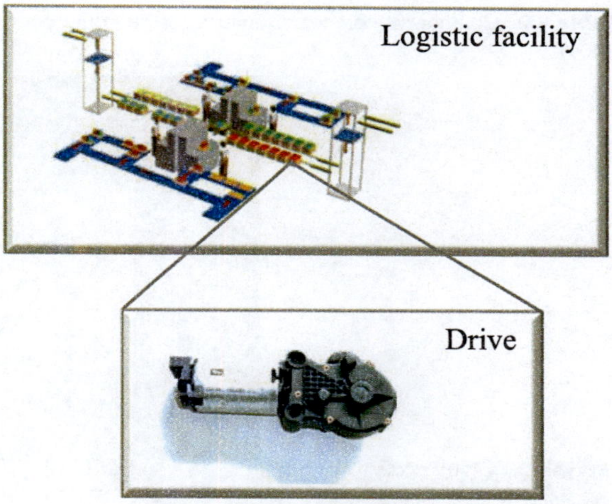

Fig. 4.22 The relationship of the logistical system and the drive via the black-box model approach (based on Jockisch and Holzmüller 2009)

2. Step: Description of the function view in relation to the components

In the following step, the function structure was worked out. In the context of a moderated workshop, it was determined which components fulfill which functions. This resulted in the function-component matrix of Table 4.10.

Thus, the interaction between components and functions can initially only be made transparent as a dependency. With reference to the example, i.e., the optimization of the drive of the logistical system, it was shown that the slip between the belt and the roller was among other things a cause of the losses of drive performances. But also the starting and stopping of the roller conveyor or the emergency shutdown of the logistical system under load, which results from the realization of the protection function, required further considerations. For this reason, the team defined the usage processes exactly in a subsequent step.

3. Step: Creation of the process structure

The discussion in the group resulted in the rough structure of the processes shown in Fig. 4.23.

The start-up and shut-down processes as well as the emergency shutdown process showed enormous effects on the drive. Subsequently, it had to be checked whether this resulted in changes of functions and components, especially of the drive, but possibly also of parts of the logistical system.

4. Step: Design review through the comparative consideration of all four views of the system model

The comparative consideration of the requirement, component, function and process view of the "Drive" system (ASM–Asynchronous machine) via the main matrix of the

Table 4.9 Component-component matrix for the roller conveyor. (after Jockisch and Holzmüller 2009)

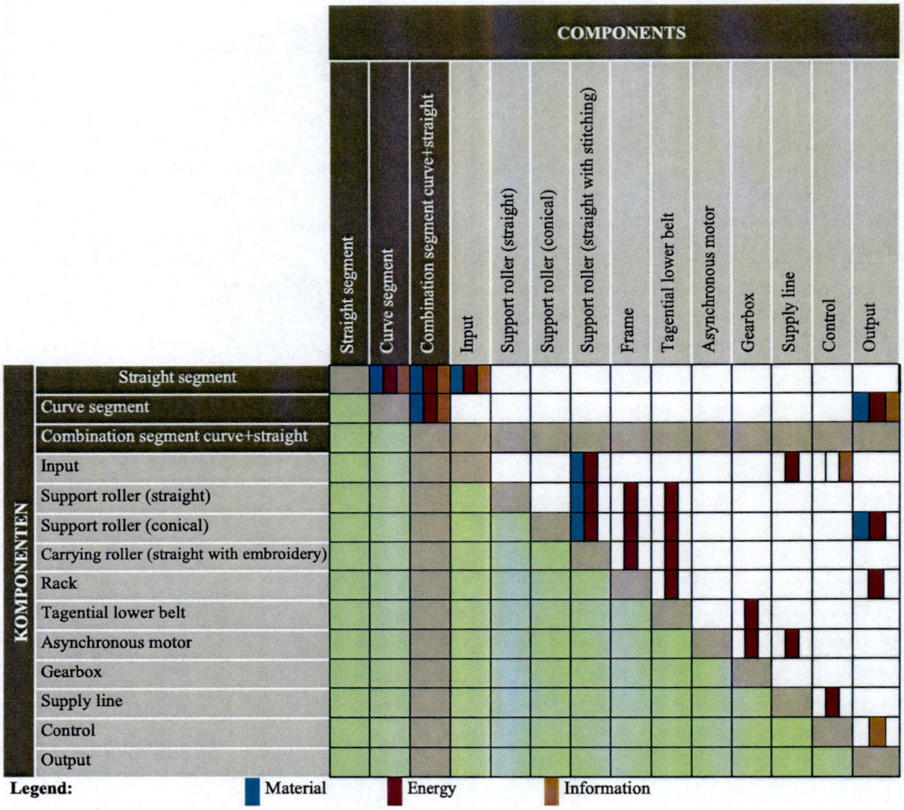

DeCoDe tools (see Fig. 4.24) were the basis for the decision whether a redesign of the drive is necessary.

The requirements "Ensure smooth running", "Promote regardless of direction" and "Emergency shutdown possible" could only be implemented to a limited extent. This could be demonstrated in the functions "Torque ripple", "Generate torque" and "Change direction of movement". All three functions can be attributed to the asynchronous machine. Thus, the system must be changed to meet the requirements. As a result, the main matrix was discussed step by step and changes to the drive were considered, as the following graph illustrates (see Fig. 4.25).

As a result of the design review, an adjustment of the rotor bars was made. Whether the change in the rotor bars of the asynchronous machine actually contributes to the increase in efficiency and thus to meeting the requirements cannot be proven using the

Table 4.10 Function-component matrix (after Jockisch and Holzmüller 2009)

	Convey	COMPONENTS										
		Input	Support roller (straight)	Support roller (conical)	Support roller (straight with stitching)	Frame	Tangential lower belt	Asynchronous motor	Gearbox	Supply line	Control	Output
Function	Convert electrical energy into mechanical energy								X			x
	Convert speed					X				X		
	Transfer speed					X			X	X		
	Convert torque					X				X		
	Torque transmitted					X			X	X		
	Convert rotation to translation		X	X	X	X						
	Convert translation to rotation		X	X	X	X						
	Transmit power		X	X	X	X	X	X	X	X		
	Protective function from and against external influences		X	X	X	X	X	X	X	X		
	Precise positioning of mutually immovable parts		X	X	x	X	X	X	X	X		
	Record information											
	Guide information								X		X	X
	Process information											
	Divert/guide				X	X	X	X				
	Guide transported goods (determine direction)		X	X	X		X	X				
	Dissipate temperature								X	X		
	Transmit electrical power								X			

DeCoDe tools. For this, appropriate simulation tools must be used in a targeted manner. This can only be done in conjunction with the GSE thinking model, which is revisited and illustrated in Sects. 5.3 and 6.1 for this example, as initially only the way of creating specific GSE thinking models should be explained by example.

This example illustrates that the drive optimization for a logistical system in the re-engineering of the drive initially requires the functions to be derived from the recording of the components and then the processes, especially the usage process, to be structured more precisely. From the process "Emergency shutdown", a sub-process of the usage process, new requirements emerged. They are the trigger for a new run-through of the system description of the drive. These new requirements led to new functions and

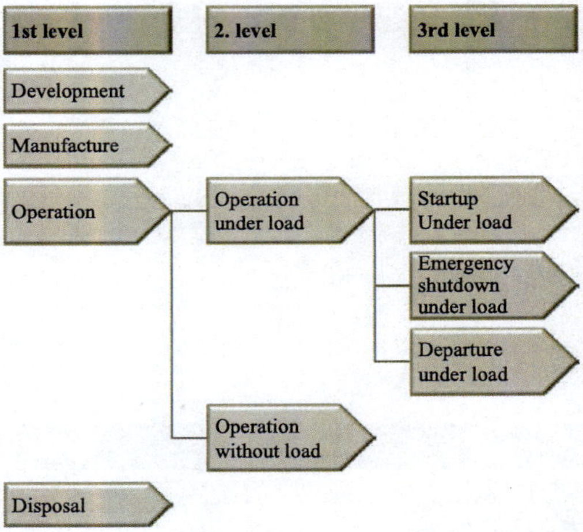

Fig. 4.23 An excerpt from the process hierarchy of the usage processes of the logistical system. (After Jockisch and Holzmüller 2009)

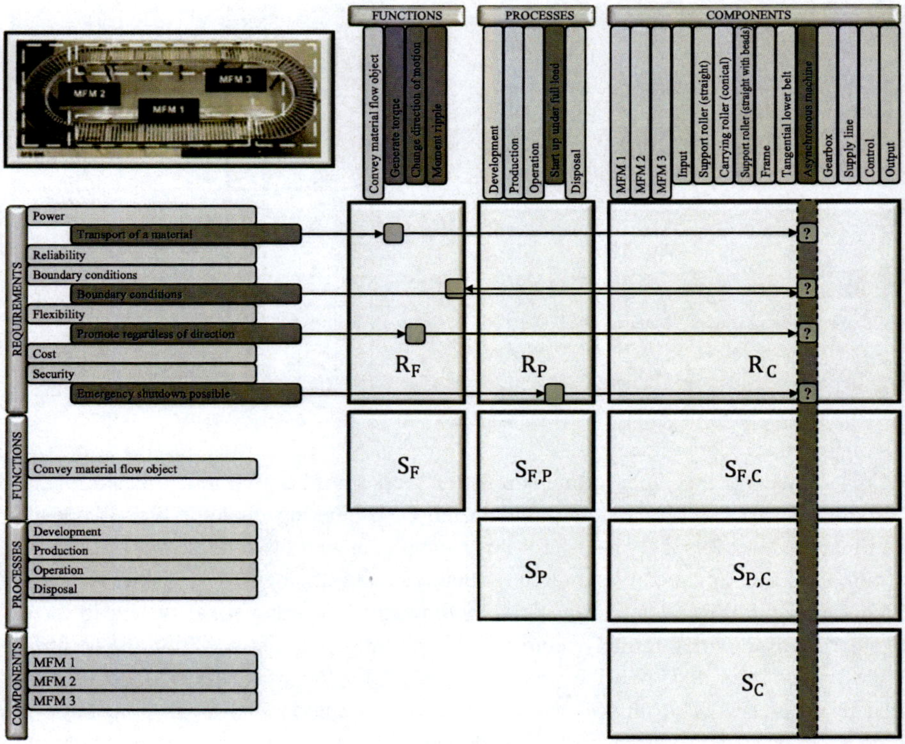

Fig. 4.24 A comparison of the four views of the "Drive" system using the DeCoDe tools. (After Jockisch and Holzmüller 2009)

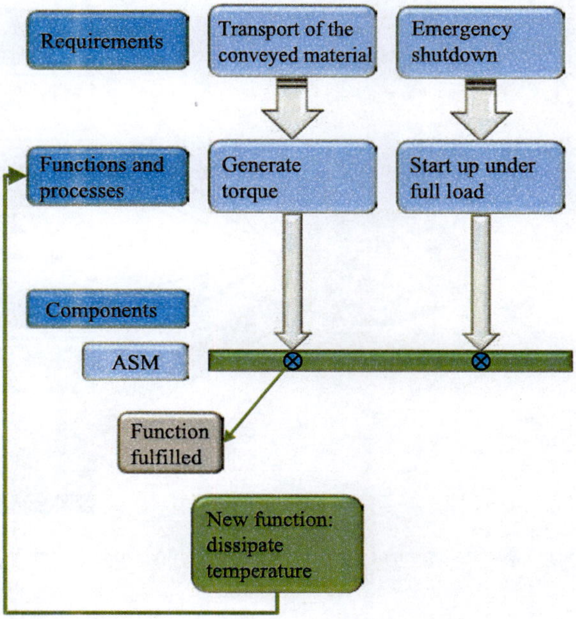

Fig. 4.25 Excerpt from the design review. (After Jockisch and Holzmüller 2009)

changed components, such as the rotor bars. Consequently, the creation of the system image itself already generates initial ideas for the re-design of the system.

Example 2: Creating an image of mechatronic systems for reliability consideration

Another approach to creating a GSE thinking model using the DeCoDe tools was realized in the PromeSys project (Winzer 2012).

The aim of this project was to predict or influence the reliability of mechatronic systems over the product life cycle at an early stage. Since this project was realized by a team of representatives from various industries, such as the automotive supplier industry, automobile construction and capital goods industry, it was a particular challenge to depict mechatronic systems over their product life cycle in a common understanding as a standardized model. In this case too, it was a matter of changing existing product systems, i.e. a re-engineering process for mechatronic systems. This sequence of steps is exemplified below using a pantograph, although it was used for all comparable mechatronic systems of the practice partners in this project. ◄

1. Step: Creating the component view

The teams initially started with the definition and creation of the component structure (see Fig. 4.26) in the form of a simple tree structure to create a uniform understanding of the product system.

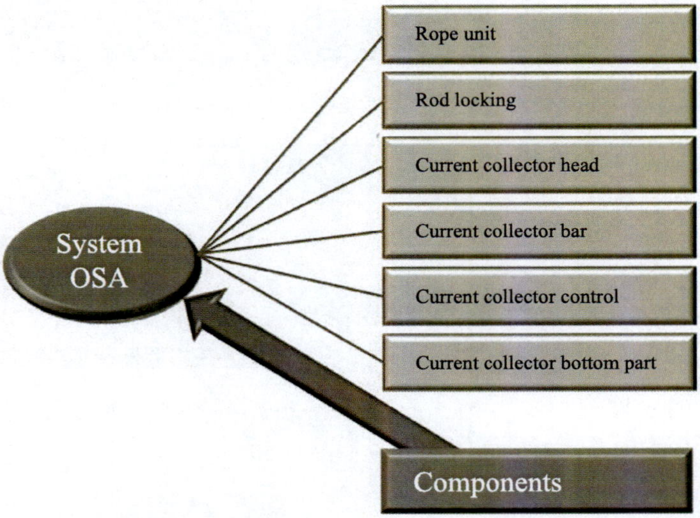

Fig. 4.26 Component structure of the pantograph. (After Winzer and Vossloh Kiepe Company 2008)

2. Step: Creating the process view

The reliability forecast of mechatronic systems requires a large amount of data, which can certainly arise in different processes at different companies, but must be comparable. This means that not only the component structure, but also the process structure of mechatronic systems must be standardized.

Thus, in several workshops, the process structure (see Fig. 4.27) was also defined as a simple tree structure, initially per company and then across companies.

If reliability forecasts of mechatronic systems are to be made, this can only be done at certain points—so-called key points—in the product life cycle of a mechatronic system. These could be fixed after the standardized process structure.

3. Step: Creation of the requirement view

After the component and process view were initially generated only via the designation of the elements, the requirement view of mechatronic systems was worked out. Requirements could in principle be derived from the specification. Their detailed representation is not possible here because confidentiality was agreed with the companies. To illustrate the principle, Fig. 4.28 provides an abstract representation of the requirement tree for the "Pantograph" system.

The requirement comparison for mechatronic systems across company boundaries showed clear similarities in the requirements. This supports the thesis that data for the degree of realization of the requirements by mechatronic systems can be collected in various comparable companies and used for the reliability forecast.

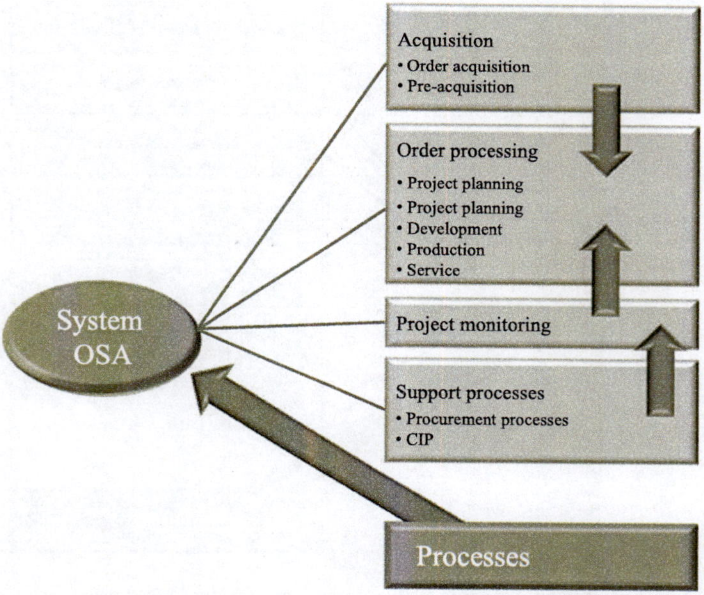

Fig. 4.27 Process structure for the pantograph (Winzer and Vossloh Kiepe Company 2008)

4. Step: Creating the function view
The generation of the function view proved to be the most difficult. Functions in already existing product systems are very difficult to describe, as they are partly built into the system (role—roll function) via the form of the components and are therefore not always transparent. This sparked the greatest discussions in the workshops. Nevertheless, the following function structure could be derived, as Fig. 4.29 illustrates for the pantograph.

This gave rise to the idea of fixing functions at the beginning of the product development process so that they are recognizable over the product life cycle and, if necessary, can be easily and verifiably changed.

5. Step: Problem-oriented networking of system views
After all four views of mechatronic systems were initially generated via their designation of the elements, i.e. the component view via the designation of the components, the process view via the designation of the processes, the requirement view via the designation of the requirements and the function view via the designation of the functions, the problem-oriented networking of the respective system views with each other now took place. This means: If a mechatronic system failed, it was described which requirement is no longer met and how these requirements are related to which functions, components and processes. The representation of these relationships was done via the individual matrices.

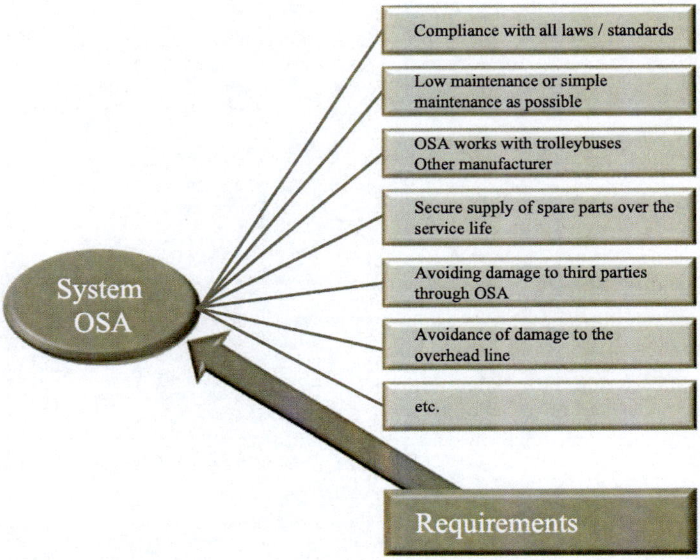

Fig. 4.28 Rough requirement structure of the pantograph (Winzer and Vossloh Kiepe Company 2008)

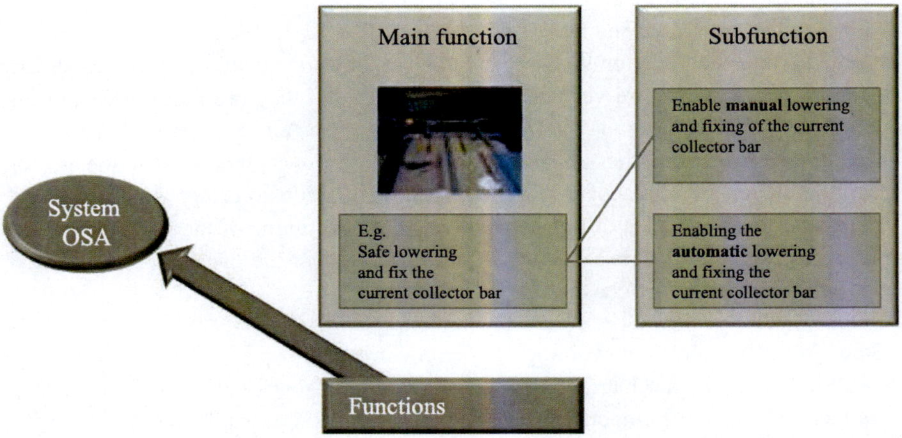

Fig. 4.29 Function structure for the pantograph (Winzer and Vossloh Kiepe Company 2008)

For this purpose, the PromeSys portal was developed and used as computational support (Winzer 2012). Fig. 4.30 illustrates the principle of networking the views using the example of the pantograph.

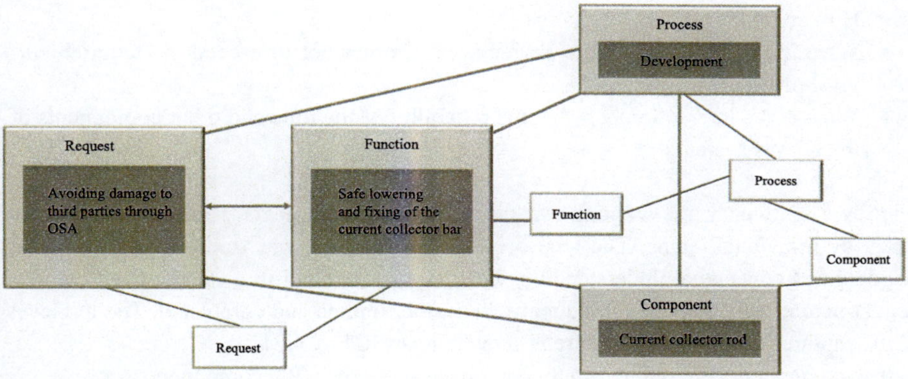

Fig. 4.30 Problem-oriented networking of the views, illustrated using the pantograph. (After Winzer and Vossloh Kiepe Company 2008)

In the following steps, which are not relevant in the phase of describing the procedure of the system image, targeted solution variants could be derived in interaction with the GSE approach concept. This is presented in Sect. 6.3.

In summary, the procedure for creating the system image within the framework of the PromeSys project consisted of the following steps:

1. the definition and description of the component view,
2. the structure definition and the description of the process view and their structure,
3. the definition and derivation of the requirement from the specifications,
4. the definition and description of the functions and
5. the problem-related networking of the individual views of the system image.

Example 3: Creating an image for a complex product system as a basis for risk analyses

The third example describes the application of the DeCoDe tools in the phases of product development for developing a GSE thinking model.

The task was to estimate the risks of the product design for a KitVes system, i.e. for a kite system[2], which converts wind energy on a ship into electricity. This cannot be done generally for the very large and complex overall system, but requires a detailed consideration of both the subsystems and their interaction. Consequently, the team had to answer the following questions, among others:

[2] KITVES—EU project in the 7th EU Framework Programme FKZ: SCS7-GA-2008–218.691, Airfoil-based solution for Vessel on-board energy production destined to traction and auxiliary services Proposal acronym KITVES, period: 01.10.2008 –30.09.2012.

- How reliable is the rope?
- When does the rope need to be replaced in order not to exceed the required survival probabilities?
- What is the interaction between the reliability of the rope and other components of the KitVes system? ◄

The rope, the winch, the warning system, and other components were developed separately by individual teams. The goal of several workshops was to create a cross-team, standardized conceptual understanding. This initially resulted in a standardized component structure, which could subsequently be further refined and completed. The top level of the captured component structure is shown in part in Fig. 4.31.

The component structure currently comprises about over 200 components.

2. Step: Derivation of requirements
In parallel, research was conducted on international laws, standards, and norms. The goal was to review, compare, and refine the requirements that are directly addressed to the safety and reliability of the KitVes system (Hartmann and Winzer 2011). These will not be presented in the following due to the confidentiality agreed upon in the project.

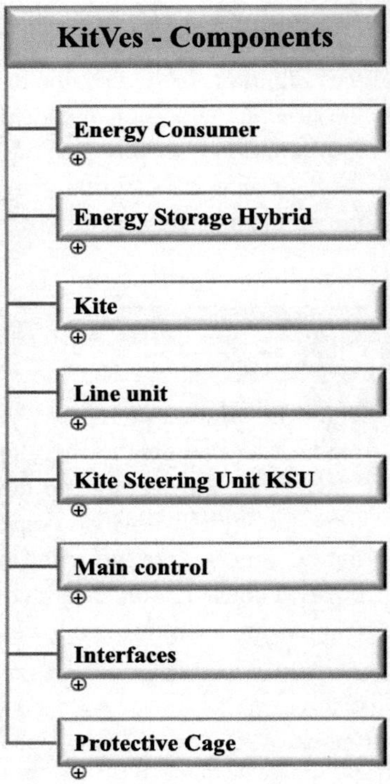

Fig. 4.31 KitVes component structure according to (Hartmann and Winzer 2011)

3. Step: Determination of the process structure
During the recording of the processes in the workshops, it became clear that some processes were missing, which are necessary for testing the reliability of the product system. Fig. 4.32 also shows a corresponding excerpt.

4. Step: Development of the functional structure
After the processes, the functions of the system were recorded as the last view of the KitVes system. These main functions are shown in Fig. 4.33. It only reflects the top level of the functional structure.

After the views of the KitVes system were recorded and jointly defined, the identification of core functions of the system could begin. The "lowering of the KitVes system" in emergencies as well as in normal situations was selected as the first core function by the partners. Only for this function was the views of the KitVes system networked in the next step.

5. Step: Networking of views via the main matrix of the DeCoDe tools
For the prioritized function, those components were linked that are needed for the fulfillment of the function. The same was done via the main matrix of the DeCoDe tools with the requirements and the corresponding processes in which this function was needed. In subsequent steps, the reliability of these components in terms of fulfilling the defined functions was checked using additional methods according to the specific task in the

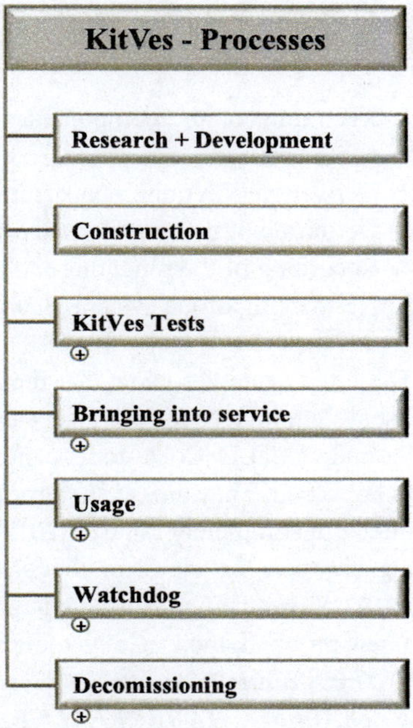

Fig. 4.32 KitVes process structure. (According to Hartmann and Winzer 2011)

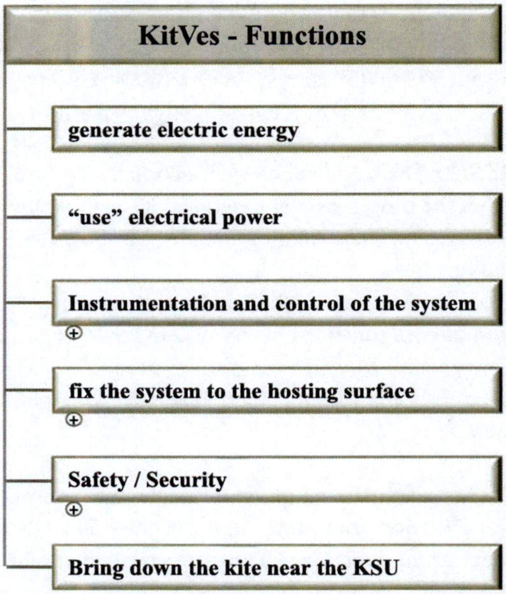

Fig. 4.33 KitVes functional structure. (According to Hartmann and Winzer 2011)

project. However, this is not part of the procedure for creating the system image, but the focus in the GSE approach concept. For this reason, this is described in Sect. 5.5.

In summary, the procedure for creating the system image in this project consisted of the following steps:

1. Recording of the components, definition of the components, hierarchization of the components,
2. Derivation of requirements from laws, norms, and standards,
3. Definition of the processes and derivation of the process structure,
4. Recording of the functions and their structure and
5. Networking of the system views for a core function.

The three examples show that the ***DeCoDe tools*** are suitable for systematically generating system images for each use case. At the same time, these examples also illustrate that the individual DeCoDe tools can be combined in any way—depending on the problem to be solved. They are, as the PromeSys project in particular shows, not necessarily to be filled out completely (Winzer 2012). As part of the PromeSys portal, a selective networking between the views of the system image was generated problem-oriented (Winzer 2012). Since the portal is able to store this knowledge and provide it again when solving a new problem, the DeCoDe tools can therefore be gradually supplemented or refined.

This gradually creates a more precise image of the respective technical system, which, unlike HABERFELLNER (see Haberfellner et al. 2019***), is not necessarily***

Table 4.11 Overview of step sequences for problem-oriented creation of GSE thinking models for technical systems

Steps	SFB 696	PromeSys	KitVes
1	Description of the component view	Creation of the component view	Development of the component structure
2	Description of the interactions between components and functions	Creation of the process view	Derivation of requirements from laws, norms, and standards
3	Recording of the process structure	Creation of the requirement view	Determination of the process structure
4	Design review and derivation of new requirements	Creation of the function view	Determination of the functional structure
5		Problem-oriented networking of system views	Networking of system views for the core function and design review

developed top-down, but problem-specifically. The DeCoDe tools thus enable a ***genesis of the image of the product system.*** It is not necessarily required to completely map the product system before it can be analyzed and designed. This is a decisive advantage of the DeCoDe tools, especially since a complete image of a system can never be created according to LUHMANN (Luhmann 1980). This is also not necessary, as only through the targeted reduction of the system image, i.e. the reduction of the complexity of the system via the image, can this be designed in a targeted manner. The genesis of the system, which is represented via the DeCoDe tools, is still presented in a punctual manner, but can, by linking the images of the system created at each point in time, trace the changes in the system over time (albeit laboriously).

The logically temporal sequence of the combination of the DeCoDe tools is described by the ***DeCoDe workflow***. Table 4.11 shows a comparative summary of how different methodological approaches were used in the individual application cases when creating the system image problem-oriented using the DeCoDe tools.

The following is intended to present an idealized sequence of steps by SITTE/WINZER for creating a GSE thinking model using the DeCoDe tools (Sitte and Winzer 2007) (see Fig. 4.34)

It is also demonstrated that the ***GSE thinking model*** created in this way ***can serve as a description, explanation, prognosis, design, and optimization model*** (see Sect. 3.2).

As a result of the use of the DeCoDe tools in product development, SITTE/WINZER come to another combination possibility of the matrices, as shown in Fig. 4.35.

The main matrix always serves to check the fulfillment of requirements (task of a ***forecast model***). This can be done in a ***first cycle*** by comparing requirements with functions. If it is determined that the functions meet the requirements, SITTE and WINZER recommend looking for components that can realize these functions (task of a ***design***

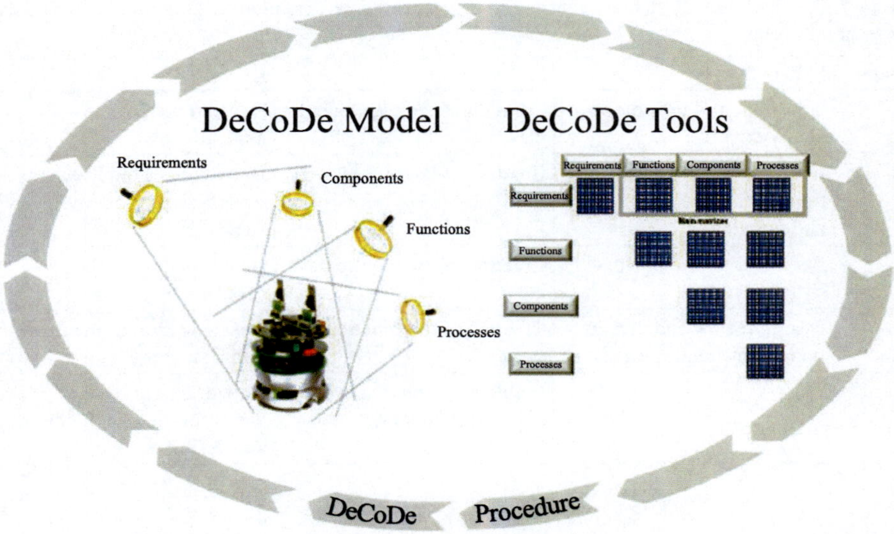

Fig. 4.34 Systematic creation of a system model using the DeCoDe tools via a DeCoDe workflow. (According to Sitte and Winzer 2011b)

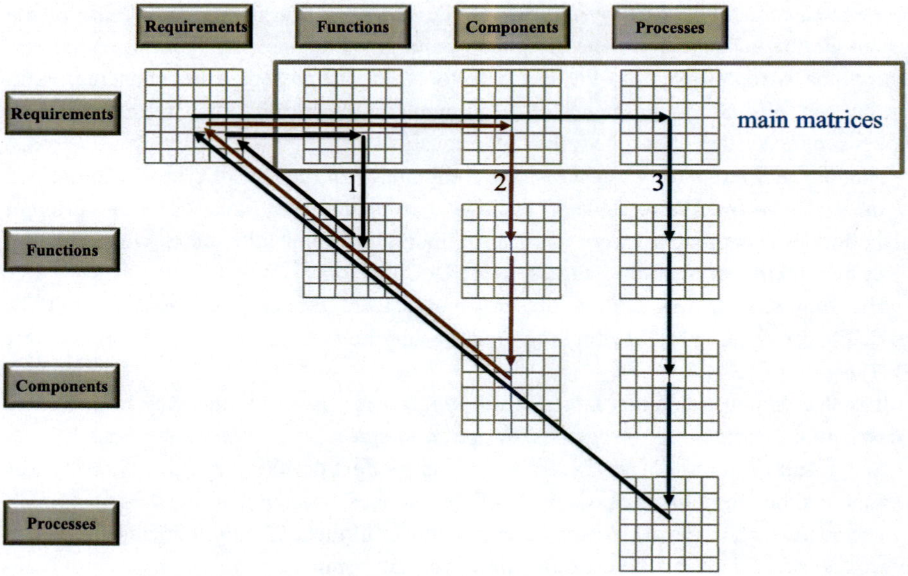

Fig. 4.35 Principal DeCoDe workflow in product development for creating a system model. (According to Sitte and Winzer 2011a)

model). For example, the robot's movement function can be implemented using rollers, legs, wheels, etc. How these components implement the functions in accordance with the requirements is the result of the ***second cycle*** (task of a ***description model***). The ***third cycle*** of Fig. 4.35 considers how the selected components can start, stop, turn on the spot, etc., in the process of "playing football", i.e., can perform the functions in accordance with the requirements (task of the ***explanation model***). The subordinate matrices consider the interactions of the elements in the respective views and between selected views in depth (task of an ***optimization model***). This can lead to new insights that need to be further investigated using the GSE approach. This will be explained in more detail in Chap. 5, specifically for the GSE approach, and illustrated in Chap. 6 using examples of the interaction between the GSE thinking model and the GSE approach.

In summary, it can be stated that with the help of the DeCoDe tools, a system image for technical systems can be systematically derived in the four views, i.e., the requirement, component, function, and process view, via the DeCoDe workflow. They thus serve to create the GSE thinking model, ensure its transparent representation, and its updating. ***Thus, the DeCoDe tools and the DeCoDe workflow are important tools for deriving the GSE thinking model.***

The next chapter will illustrate how sociotechnical systems can be modeled using the e-DeCoDe approach. This includes the addition of the person view.

4.5 Possible Sequence of Steps for Creating the GSE Thinking Model for Sociotechnical Systems with e-DeCoDe

In Sect. 4.4, an example was presented of what the modeling of technical systems with DeCoDe can look like and how the system modeling changes over the course of a problem-solving process. This chapter will now demonstrate the exemplary modeling of sociotechnical systems with e-DeCoDe.

The contents of this chapter are based on the recommended sequence of steps by MISTLER (Mistler et al. 2021a) for agile[3] development of organizational systems with the e-DeCoDe approach (see Fig. 4.36).

As can be seen from Fig. 4.36, the approach consists of four steps, which are connected with the e-DeCoDe thinking model. The temporal course of development is always documented via the states of system modeling (Mistler et al. 2021a). In an extended form, this procedure was also validated in MISTLER (Mistler 2021) in two different manufacturing companies. However, the validations are very extensive. Therefore,

[3] To ensure agility in modeling, a modular system architecture can be a possible solution. This means making the elements of systems reusable by determining modules at defined hierarchy levels. It must be selected which elements of the system should be constant and which elements should be designed to be variable (Mistler 2021; Hornby 2007; Schapiro and Henry 2012).

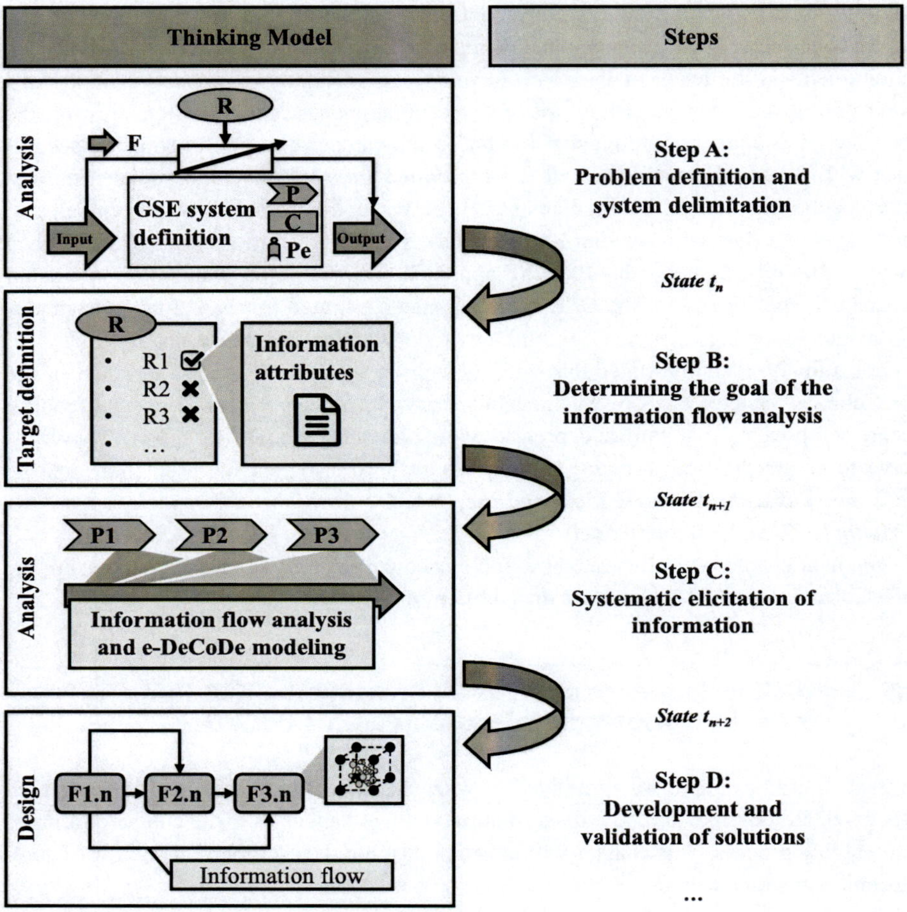

Fig. 4.36 Procedure for agile development of organizations (Mistler et al. 2021a)

the principle of agile system modeling will be explained below using a simplified application example.

Step A: Problem definition and system delimitation

In step A, a problem definition and system delimitation is carried out. It serves to create a uniform understanding of the system to be developed and the associated problem (Mistler et al. 2021a). The system definition is carried out in the GSE with the GSE system approach. This has already been tested and adapted to the development of organizational systems (Mistler 2021; Mistler et al. 2021b, 2019; Marchlewitz et al. 2015; Nicklas et al. 2014). Flip charts and prepared worksheets are suitable tools for visualizing the GSE

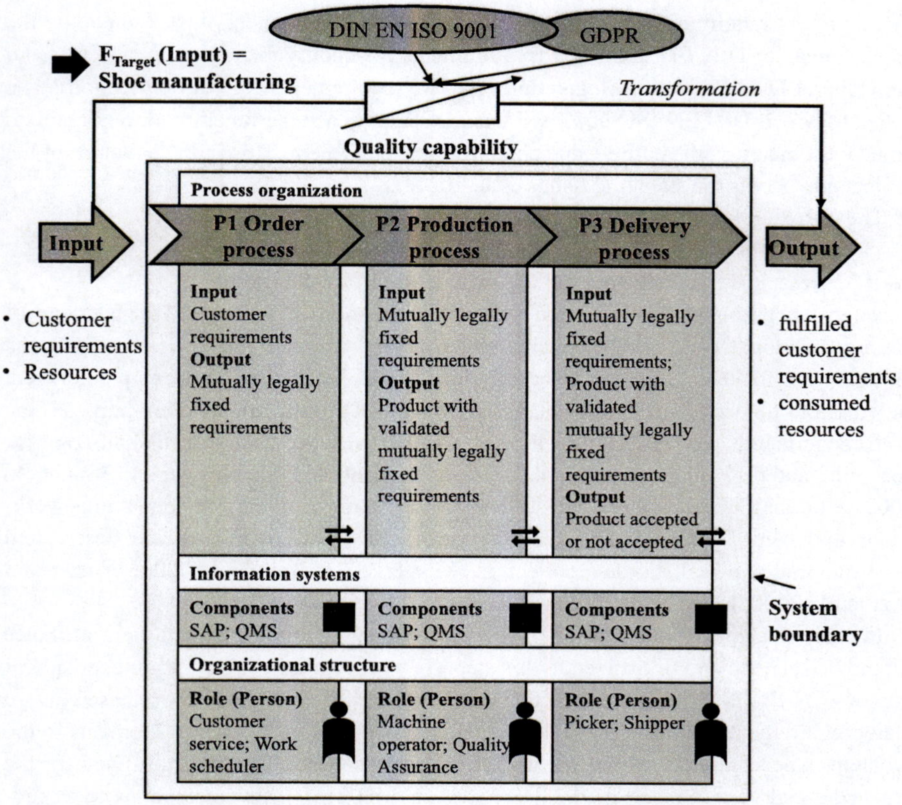

Fig. 4.37 GSE system approach for organizational systems (Mistler et al. 2021a)

system approach in a workshop (Mistler 2021; Mistler et al. 2021a). This facilitates the cognitive moderation of the e-DeCoDe approach.

Figure 4.37 shows the basic system delimitation with the GSE system approach.

The GSE system approach shown in Fig. 4.37 shows the function of the organizational system, the "shoe production". This is defined by the input "customer requirements" and "resources". As a result, the organizational system should produce "fulfilled customer requirements" and "consumed resources" with regard to the function "shoe production". Furthermore, the rough processes of the value chain are shown: "P1 offer process", "P2 production process" and "P3 shipping process". It should be emphasized that the processes are to be delimited from each other by the clear definition of input and output. Only then can components in the form of information systems and people with their roles be assigned to the processes. The interaction of the system elements serves to realize the function "shoe production". As Fig. 4.37 shows, the quality capability (degree of requirement fulfillment) of the function is influenced, for example, by the requirements of DIN EN ISO 9001 (requirements for quality management systems) and the General Data Protection Regulation (GDPR; requirements for data protection). The extent to which DIN EN ISO 9001 and the GDPR influence the function of the organizational system across the entire value chain forms the problem. This initial situation of the project now forms the state t_0 (Mistler et al. 2021a).

The next step B will specify the objective.

Step B: Description of the goal for the information flow analysis

Through the problem definition and system delimitation in step A, a rough image of the organizational system to be considered could be drawn. On this basis, in step B, the requirements for the system are to be prioritized. This is intended to develop a concrete and feasible objective for further analysis (Mistler et al. 2021a). For this purpose, the GSE recommends the **information flow analysis (IFLA)** as a suitable methodology for analyzing and designing organizational systems (Braunholz 2006; Winzer and Braunholz 2000; Sitte and Winzer 1999). The IFLA is particularly suitable for conducting workshops and using questionnaires and interviews. However, interviews are best suited for structured and realistic data collection (Mistler 2021; Braunholz 2006; Winzer and Braunholz 2000; Davis et al. 2006).

In order to achieve a targeted analysis in step C, the requirements and their attributes should be selected and prioritized in this step. As worked out in step A, a joint consideration of DIN EN ISO 9001 and the GDPR should be carried out. However, these contain many requirements on different topics. Therefore, a limitation of the requirements to the problem is necessary. Based on the limitation, attributes can then be determined for the organizational system (see Fig. 4.38) (Mistler et al. 2021a). This corresponds to a modeling within the requirements vs. requirements matrix.

As Fig. 4.38 shows, with regard to the requirement consideration for DIN EN ISO 9001, the traceability of personal data in documented information must be considered. In addition, with regard to the GDPR, it should be investigated in which documented information personal data of the customer is processed. The reason why this particular objective was discussed is due to the requirement of the GDPR, i.e., the customer must be shown in which processes personal data is processed, what it is used for, and whether it is necessary to use it. In addition, according to DIN EN ISO 9001, personal data is considered the property of the customer and must be marked, verified, protected, and secured accordingly. To determine in which information personal data is processed, the attributes “name”, “address”, “email”, “company name”, and “identification numbers” must be queried. For this purpose, the “inputs” and “outputs” must be queried for the processes along the value chain (“P1”, “P2”, and “P3”). Furthermore, it must be determined in which components or information systems (“software”, “machines”, and “equipment”) the information is processed and stored. In addition, the responsibility of persons must be discussed through their “role”. This means, for example, finding out to what extent people are responsible for processing personal data. The result of step B is

Fig. 4.38 Principle representation—selection of attributes for the IFLA (Mistler et al. 2021a)

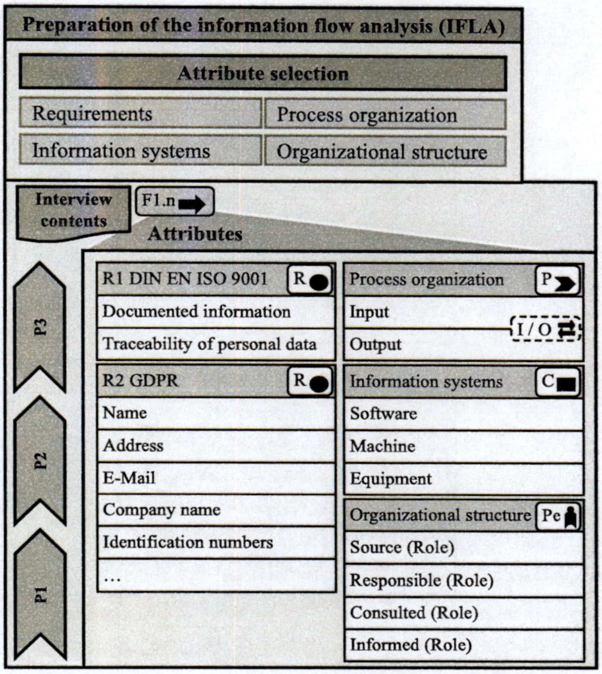

an interview guide with attributes that are to be collected in the IFLA in step C. This forms the state t_1 (Mistler et al. 2021a).

Step C: Systematic collection of information

In step C, the organizational system is analyzed with regard to the objective from step B. For this purpose, information is systematically collected with the information flow analysis (IFLA) (Mistler et al. 2021a). Suitable tools for using the IFLA already exist for collecting the process and structural organization (Braunholz 2006; Winzer and Braunholz 2000) and can be used together with the e-DeCoDe approach (Mistler 2021) (see Fig. 4.39).

Fig. 4.39 shows, on the one hand, the IFLA data sheet for the structural organization and, on the other hand, the IFLA data sheet for the process organization. These data sheets were adapted by MISTLER (Mistler 2021) based on the data sheets of WINZER and BRAUNHOLZ (Braunholz 2006; Winzer and Braunholz 2000) for the application of the e-DeCoDe approach to organizational system modeling. The number of interviews "N" to be conducted depends on how many people are to be interviewed in the processes "P1", "P2" and "P3". The IFLA data sheet for the structural organization is primarily used to determine whether the interviewee has been correctly classified in the value chain in step A. This allows deviations to be identified on the one hand, and on the

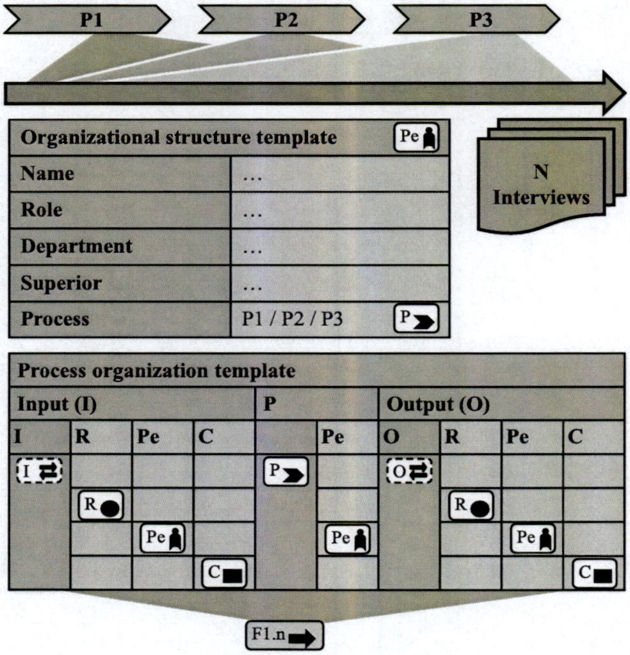

Fig. 4.39 Principle representation for the IFLA implementation (Mistler et al. 2021a)

other hand, it explains to the interviewee how his role is understood by the companies in the value chain. After querying the attributes with the IFLA structural organization, the processes are determined with the IFLA process organization data sheet. Only the processes for which the person to be interviewed is responsible are recorded. Based on this, the requirement, person, and component attributes selected in step B can then be specifically collected and assigned to the identified processes with the relevant inputs and outputs. This process concretization allows a derivation of the function for the respective processes. Because by querying the attributes of the processes delimited by input and output, it is possible to identify the complexity of a subsystem for the entire organizational system. This means it is clear, ***how the process is carried out***, ***who is responsible for carrying out the process*** and ***with what the process is implemented.*** Thus, it can be derived, ***what is done in the subsystem*** (Mistler 2021; Mistler et al. 2021a) (see Fig. 4.39).

Once all interviews "N" have been conducted in step C, it is necessary to document the interview results. This represents the state t_2. In the next step, the interview results are bundled and serve to design the organizational system model (Mistler et al. 2021a).

Step D: Development and validation of possible solutions
In step D, the interview results from step C are combined using e-DeCoDe modeling. This allows, on the one hand, to identify interfaces and redundancies and, on the other hand, to create a system model that shows the complexity of the organization across the entire value chain. Based on this, a solution space can be systematically defined and designed (Mistler et al. 2021a). What the basic system design looks like is outlined in Fig. 4.40.

As Fig. 4.40 shows, the interview results from step C are combined through e-DeCoDe modeling. To handle the enormous amount of system elements, a suitable IT tool is required. In this context, the software iQUAVIS has proven to be highly suitable in the course of an extensive software investigation (see Sect. 4.3). With regard to agile system modeling, the e-DeCoDe elements must be reusable by forming modules at a defined hierarchy level. For this, the most understandable level of the function view is selected for the organization. The function level is selected because only through the function view can the degree of requirement fulfillment of systems or subsystems be determined (see Sect. 4.3).

The merging of the system elements with e-DeCoDe in iQUAVIS is visualized in Fig. 4.41.

Fig. 4.41 represents the merging of the e-DeCoDe system elements at the most understandable level of the function view. To show the temporally logical sequence of the

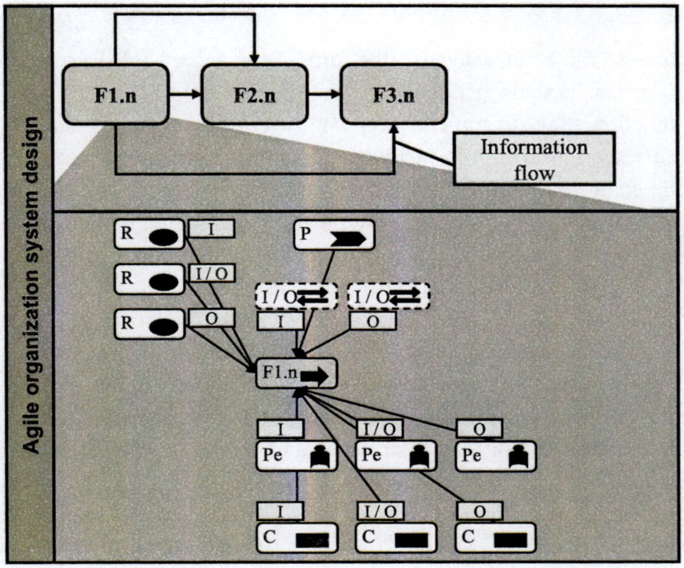

Fig. 4.40 Basic representation of organizational system design (Mistler et al. 2021a)

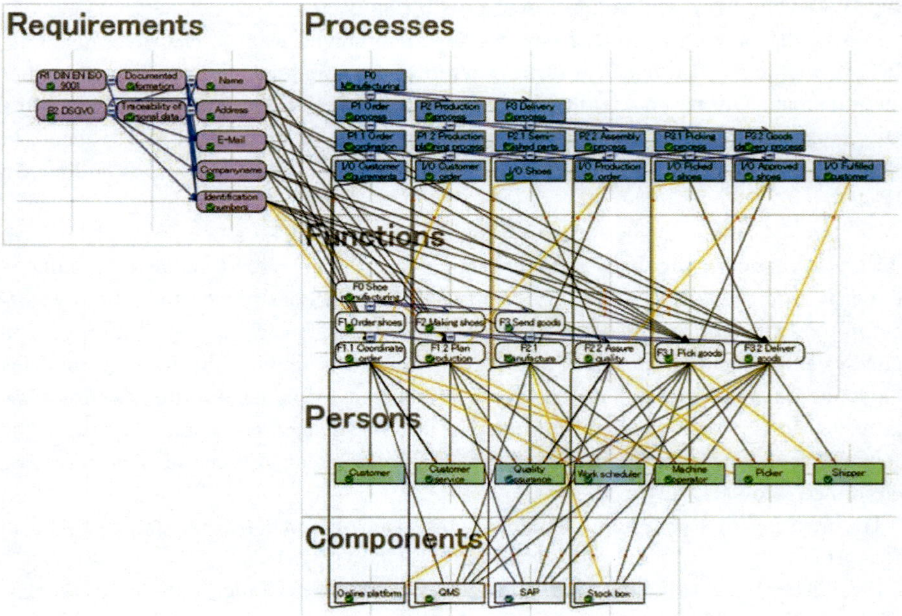

Fig. 4.41 Organizational system model in iQUAVIS (Mistler et al. 2021a)

function level, these can be visualized in the form of flow charts (cf. Farid and Suh 2016; Suh 1998, 2001). This is also feasible in iQUAVIS (see Fig. 3.42).

With the function flow diagram in Fig. 4.42, a uniform understanding of information flow can be created. This runs from the entry of customer requirements through the entire value chain to the fulfilled customer requirements. Each function represents a subsystem of the overall system. In order to be able to show the complexity of the subsystems, these can be focused in the organizational system model using integrated filter functions in iQUAVIS (Mistler et al. 2021a) (see Fig. 4.43).

Fig. 4.43 illustrates the complex relationships of the function "F1.1. Coordinate order". This can show, for example, which personal data the function processes. At the same time, it is traceable through the requirement structure which requirements this function affects. Furthermore, the flow of information is represented by modeling inputs and outputs. In addition, it can be visualized which roles are responsible for processing the information and why, and which information systems process the information. Using the modeling of the organizational system, specific solution suggestions can be derived and implemented successively (Mistler 2021; Mistler et al. 2021a). This represents the state t_3.

In summary, the approach was able to show, on the one hand, a procedure for agile organizational system modeling with e-DeCoDe and, on the other hand, how to ensure

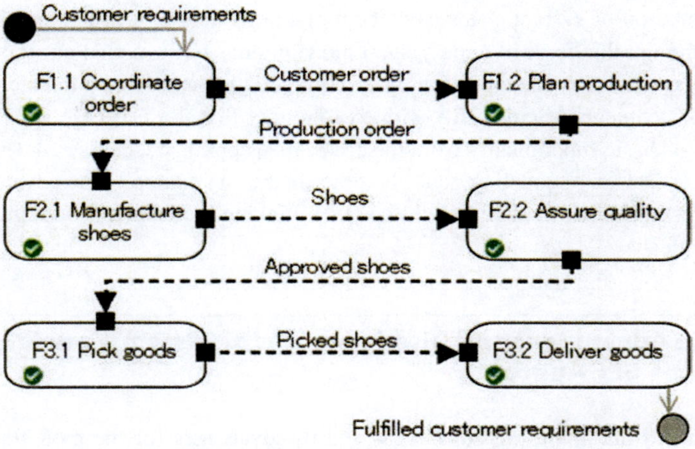

Fig. 4.42 Function flow diagram in iQUAVIS (Mistler et al. 2021a)

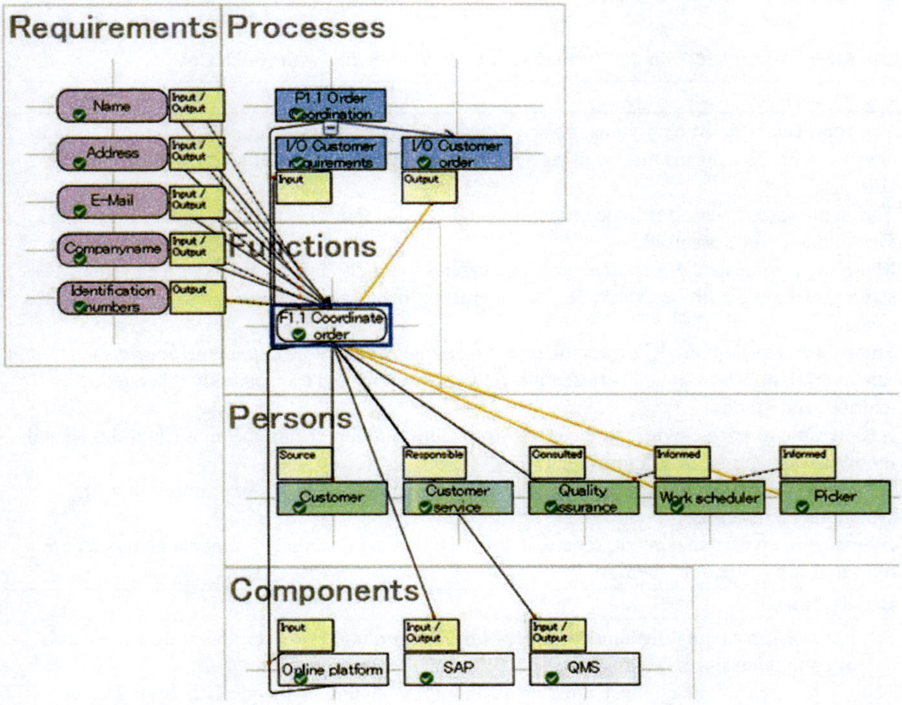

Fig. 4.43 Use of the filter and focus function in iQUAVIS (Mistler et al. 2021a)

the documentation of system states over the temporal course of a problem-solving process. Even though this form of agile system modeling has been carried out for organizations, it is suspected that this type of agile system modeling can also be transferred to the modeling of technical systems. This is due to the fact that the DeCoDe approach does not differ in its basic functionality from the e-DeCoDe approach (see Sect. 4.3).

After the e-DeCoDe modeling could be described as an example, the advantages and disadvantages of system modeling with the GSE thinking model are to be highlighted in the next chapter.

4.6 The Advantages and Disadvantages of System Modeling in the GSE Approach

It is to be noted that there are advantages and disadvantages for the modeling of both technical and sociotechnical systems with the GSE model. These result from the current state of science and technology.

Table 4.12 summarizes the advantages and disadvantages of system modeling of technical and sociotechnical systems with the GSE thinking model.

Table 4.12 Advantages and disadvantages of system modeling with (e-)DeCoDe

Advantages
• The approach is based on systems theory. Thus, the system is systematically delimited from its environment. This means that by using systems thinking, the object of study is clearly delineable • The standardized views form the corporate language and thus also the language of modeling • Few defined views are used • Modeling is done as the corporate language defines the individual elements of the system. The same applies to distributed work, i.e., the interdisciplinary language defines the elements of the system • The complex system modeling can be traced from the rough to the detail and vice versa • The interrelationships in and between the views are simply and transparently represented by matrices and graphs • It is possible to trace changes in the model over time and thus secure the architectural traceability of the system model relationships • Due to the structured, uniform, and logical approach in modeling, a higher reusability of the system model is ensured • By focusing on certain content, focus can be placed on the essential system elements that are relevant for a defined problem
• **Disadvantages**
• The initial effort of creating a model is very time-consuming. However, this is a general problem of comprehensive modeling • The interdisciplinary team must agree on a common modeling language. This is very time-consuming • Ideally, matrices of a maximum of 10 by 10 should be chosen, as the overview for the user is lost with too high matrix dimensioning

As a conclusion, Chapter 4 shows that system modeling requires a problem-solving-oriented approach to model building, the targeted application of problem-specific methods for system description, and the necessity of (IT) tools. This means, among other things, that coarse goals must be broken down into small goals and implemented. But even small goals can be formulated into coarse objectives. At the core, it must always be questioned what goals are being pursued in system development and whether these objectives are current. For example, for product development, it must be decided in which phases which objectives are pursued and which methods and tools are needed to solve specific problems. The same applies to the development of sociotechnical systems, such as organizations, as they also need to adapt to new objectives, for example, due to changing business models.

But what can a general approach in the GSE look like, which is applicable to any problem and at the same time provides problem-specific methods and (IT) tools? This is described in the next chapter.

References

3D (2021): 3D Design & Engineering Software. Online verfügbar unter https://www.3ds.com/, zuletzt aktualisiert am 01.09.2021, zuletzt geprüft am 02.09.2021.

Anderl, Reiner (2012): Smart Engineering. Interdisziplinäre Produktentstehung. Berlin: Springer Vieweg (acatech DISKUSSION).

Auricht et al (2014): Durchgehender Traceability-Prozess im Systems Engineering. In: Maik S. Maurer, Jutta Abulawi und Sven-Olaf Schulze (Hg.): Tag des Systems Engineering. Bremen, 12.–14. November 2014 ; [TdSE]. München: Hanser, S. 133–143.

Bender, Beate; Gericke, Kilian (2021): Pahl/Beitz Konstruktionslehre. Methoden und Anwendung erfolgreicher Produktentwicklung. 9th ed. 2021. Berlin, Heidelberg: Springer Berlin Heidelberg; Imprint: Springer Vieweg.

Beyerer, Jürgen; Winzer, Petra (Hg.) (2018): Beiträge zu einer Systemtheorie Sicherheit. Deutsche Akademie der Technikwissenschaften; Herbert Utz Verlag GmbH. München: Herbert Utz Verlag GmbH (acatech DISKUSSION). Online verfügbar unter http://web.archive.org/web/20181117000045/http://www.acatech.de/wp-content/uploads/2018/10/acatech_DISKUSSION_Systemtheorie_WEB.pdf.

Bielefeld, Ovidiu (2020): Entwicklung einer Methodik für eine modellbasierte und ganzheitliche Fehleranalyse. Dissertation. Wuppertal: Bergische Universität Wuppertal.

Bielefeld, Ovidiu; Dransfeld, Hendrik; Schlueter, Nadine; Yazdanmadad, Soroush; Winzer, Petra (2016): Modellbasierte Analyse komplexer Fehlerketten zur Erhöhung der Verlässlichkeit in der Produktentwicklung. In: Christian Tschirner: Tag des Systems Engineering. (Print-on-Demand). Hg. v. Sven-Olaf Schulze und Christian Muggeo. München: Hanser, Carl, S. 163–172.

Bielefeld, Ovidiu; Dransfeld, Hendrik; Schlüter, Nadine (2018): Development of a Procedure for Analysis of Failure Chains in Complex Mechatronic Systems to Improve Sustainability. 15th Global Conference on Sustainable Manufacturing. In: *Procedia Manufacturing* 21. DOI: https://doi.org/10.1016/j.promfg.2018.02.195.

Braunholz, Helge (2006): Werkzeugentwicklung für informationsflussorientierte Prozessmodelle. Aachen: Shaker Verlag GmbH.

Browning, T. R. (2001): Applying the design structure matrix to system decomposition and integration problems: a review and new directions. In: *Engineering Management, IEEE Transactions on* 48 (3), S. 292–306. DOI: https://doi.org/10.1109/17.946528.

Danilovic, Mike; Browning, Tyson R. (2007): Managing complex product development projects with design structure matrices and domain mapping matrices. In: *International journal of project management* 25 (3), S. 300–314.

Danner, S. (1996): Ganzheitliches Anforderungsmanagement für marktorientierte Entwicklungsprozesse. München, Wien: Hanser.

Davis, A.; Dieste, O.; Hickey, A.; Juristo, N.; Moreno, A. M. (2006): Effectiveness of Requirements Elicitation Techniques: Empirical Results Derived from a Systematic Review. In: 14th IEEE International Requirements Engineering Conference (RE'06). 14th IEEE International Requirements Engineering Conference. Minneapolis/St. Paul, MN, 11.09.2006–15.09.2006: IEEE, S. 179–188.

Davis, Alan M.; Hickey, Ann M.; Zweig, Ann S. (2007): Requirements Management in a Project Management Context. In: Peter W. G. Morris und Jeffrey K. Pinto (Hg.): The Wiley Guide to Project Technology, Supply Chain, and Procurement Management. Hoboken, N.J: John Wiley & Sons, Inc., S. 1–31.

Desel, Jörg; Reisig, Wolfgang (2014): Petrinetze. In: *Informatik Spektrum* 37 (3), S. 165–167. DOI: https://doi.org/10.1007/s00287-014-0789-1.

Dumitrescu, Roman; Riedel, Oliver; Gausemeier, Jürgen; Albers, Albert; Stark, Rainer (2021): Engineering in Deutschland—Status quo in Wirtschaft und Wissenschaft. Ein Beitrag zum Advanced Systems Engineering. Padaborn. Online verfügbar unter www.advanced-systems-engineering.de, zuletzt geprüft am 06.05.2021.

Ebert, Christof (2019): Systematisches Requirements Engineering. Anforderungen ermitteln, dokumentieren, analysieren und verwalten. 6., überarbeitete und erweiterte Auflage. Heidelberg: dpunkt.verlag.

Ehrlenspiel, Klaus; Meerkamm, Harald (2017): Integrierte Produktentwicklung. Denkabläufe, Methodeneinsatz, Zusammenarbeit. 6., vollständig überarbeitete und erweiterte Auflage. München, Wien: Hanser. Online verfügbar unter http://www.hanser-fachbuch.de/buch/Integrierte+Produktentwicklung/9783446440890.

Eigner, Martin; Stelzer, Ralph (2009): Product Lifecycle Management. Ein Leitfaden für Product Development und Life Cycle Management. 2. Aufl. Berlin u. a.: Springer.

Espinoza, Huascar; Cancila, Daniela; Selic, Bran; Gérard, Sébastien (2009): Challenges in Combining SysML and MARTE for Model-Based Design of Embedded Systems. In: Richard F. Paige, Alan Hartman und Arend Rensink (Hg.): Model driven architecture—foundations and applications. 5th European conference, ECMDA-FA 2009, Enschede, The Netherlands, June 23–26, 2009 ; proceedings. Berlin: Springer (Lecture notes in computer science, 5562), S. 98–113.

Farid, Amro M.; Suh, Nam P. (Hg.) (2016): Axiomatic design in large systems. Complex products, buildings and manufacturing systems. Cham: Springer International Publishing. Online verfügbar unter http://gbv.eblib.com/patron/FullRecord.aspx?p=4558514.

Feldhusen, Jörg; Grote, Karl-Heinrich; Göpfert, Jan; Tretow, Gerhard (2013): Technische Systeme. In: Pahl/Beitz Konstruktionslehre: Springer, S. 237–279.

Gausemeier, J.; Dumitrescu, R.; Steffen, D.; Czaja, A.; Wiederkehr, O.; Tschirner, C. (2013a): Systems Engineering in der industriellen Praxis. Universität Paderborn: Heinz Nixdorf Institut.

Gausemeier, Jürgen; Gaukstern, Tobias; Tschirner, Christian (2013b): Systems Engineering Management Based on a Discipline-Spanning System Model. In: Conference on Systems Engineering Research (CSER 13) (Hg.): Proceedings, 19.–22. März 2013: Elsevier B.V.

Gausemeier, Jürgen; Lanza, Gisela; Lindemann, Udo (2012): Produkte und Produktionssysteme integrativ konzipieren. Modellbildung und Analyse in der frühen Phase der Produktentstehung. München: Carl Hanser Verlag. Online verfügbar unter http://www.hanser-elibrary.com/action/showBook?doi=https://doi.org/10.3139/9783446429857.

Gausemeier, Jürgen; Plass, Christoph (2014): Zukunftsorientierte Unternehmensgestaltung. Strategien, Geschäftsprozesse und IT-Systeme für die Produktion von morgen. 2., überarb. Aufl. München: Hanser.

Gausemeier, Jürgen; Rammig, Franz Josef; Schäfer, Wilhelm (Hg.) (2014): Design methodology for intelligent technical systems. Develop intelligent technical systems of the future. Heidelberg, New York, NY, Dordrecht: Springer (Lecture notes in mechanical engineering).

Graup, Christian (2005): Entwicklung eines innovativen nutzerorientierten Informationsmanagementsystems für KMU: Shaker Aachen.

Grundel, Martin; Abulawi, Jutta; Moeser, Georg; Weilkiens, Tim; Scheithauer, Axel; Kleiner, Sven et al. (2014): FAS 4M—No more: "Please mind the gap". In: Maik S. Maurer, Jutta Abulawi und Sven-Olaf Schulze (Hg.): Tag des Systems Engineering. Bremen, 12.–14. November 2014 ; [TdSE]. München: Hanser, S. 65–74.

Haberfellner, Reinhard; Weck, Olivier L. de; Fricke, Ernst; Vössner, Siegfried (2019): Systems engineering. Fundamentals and applications. 14. überarbeitete Auflage. Cham: Springer International Publishing; Birkhäuser.

Hallerstede, Stefan; Hansen, Finn Overgaard; Holt, Jon; Lauritsen, Rasmus; Lorenzen, Lasse; Peleska, Jan (2012): Technical challenges of SoS requirements engineering. In: 2012 7th International Conference on System of Systems Engineering (SoSE). 2012 7th International Conference on System of Systems Engineering (SoSE). Genova, 16.07.2012–19.07.2012: IEEE, S. 573–578.

Hartmann, Christine; Winzer, Petra (2011): DeCoDe+X in KitVes. Using the Demand Compliant Design in the Development of a Solution for Harvesting High-Altitude Winds for Energy Gereration on Vessels. Pamplona, Spain: Servicios de Publicaciones Universidad de Navarra: Proceedings 14.QMOD Conference on Quality and Service Science (QMOD 2011).

Heinke, Jonas; Mistler, Marian (2019): Agiles und modellbasiertes Projektmanagement in der Produkt- und Dienstleistungsentwicklung. In: Nadine Schlüter und Markus Reiche (Hg.): Herausforderungen im Umgang mit Anforderungen in Zeiten des industriellen Wandels. 1. Auflage. Düren: Shaker (Berichte zum Generic-Management, 2019,1), S. 1–26.

Heinrichsmeyer, Marius (2020): Entwicklung eines zielgerichteten Fehlerursachensuch- und Lösungsalgorithmus [FusLa]. Dissertation. Wuppertal: Bergische Universität Wuppertal.

Heinrichsmeyer, Marius; Ansari, Amirbabak; Schlueter, Nadine; Boehmer, Christian (2020a): Investigation of Problems with High Initial and Update Efforts in the Modeling of Production Systems. A Review on System Modeling Approaches. In: *International Journal on Advances in Systems and Measurements* 13 (3 & 4), S. 264–274. Online verfügbar unter http://www.iariajournals.org/systemsandmeasurements/, zuletzt geprüft am 30.04.2021.

Heinrichsmeyer, Marius; Schlüter, Nadine; Kösling, Fynn; Ansari, Amirbabak (2020b): Validation of a Failure-Cause Searching and Solution-Finding Algorithm in Production based on Complaint Information from the Use Phase. In: ICONS 2020. The Fifteenth International Conference on Systems, Lisbon, Portugal, 23.–27.02.2020. Lisbon: IARIA.

Hering, Ekbert (1989): Petri-Netze. In: Ekbert Hering und Harald Schumy (Hg.): Software-Engineering. Wiesbaden: Vieweg+Teubner Verlag, S. 85–98.

Herrmann, Joachim; Fritz, Holger (2016): Qualitätsmanagement. Lehrbuch für Studium und Praxis. 2., überarbeitete und aktualisierte Auflage. München: Carl Hanser Verlag. Online verfügbar unter http://www.hanser-elibrary.com/isbn/9783446440227.

Hornby, Gregory S. (2007): Modularity, reuse, and hierarchy: Measuring complexity by measuring structure and organization. In: *Complexity* 13 (2), S. 50–61. DOI: https://doi.org/10.1002/cplx.20202.

Huber, Markus; Kölbl, Christian; Lorenz, Robert; Wirsching, Günther (2008): Ein Petrinetz-Modell zur Informationsübertragung per Dialog.

Jockisch, Maike; Holzmüller, Hartmut H. (2009): Ergebnisbericht der Arbeitsgruppe A2: Kundenanforderungen an Industriegüter. Terminologie, Klassifikation und Forschungsfelder. Technical Report 0902. Hg. v. Technische Universität Dortmund. Sonderforschungsbereich 696. Dortmund. Online verfügbar unter http://www.sfb-696.de/uploads/media/Technical_Report_0902_01.pdf, zuletzt geprüft am 23.09.2011.

Join GmbH (2021): IT's all about people. Online verfügbar unter https://www.join.de/, zuletzt aktualisiert am 12.08.2021, zuletzt geprüft am 02.09.2021.

Kaiser, Lydia (2014): Rahmenwerk zur Modellierung einer plausiblen Systemstruktur mechatronischer Systeme. Paderborn, Universität Paderborn, Diss., 2013. Universitätsbibliothek, Paderborn.

Kamenický, Lukás; Markulik, Stefan; Sinay, Juraj (2014): Transformation of Product Characteristics in Terms of an Integrated Management System. In: *Acta Mechanica Slovaca* 18 (2), S. 50–55. DOI: https://doi.org/10.21496/ams.2014.019.

Kanie, K. (2009): Project Management System for Adaptive Product Development. In: Matthias Kreimeyer (Hg.): Proceedings of the 11th International DSM Conference. Greenville, SC, 12 and 13 October 2009. München: Hanser.

Künne, Bernd; Richard, Tim (Hg.) (2009): Sonderforschungsbereich 696. Forderungsgerechte Auslegung von intralogistischen Systemen—Logistics on Demand. Finanzierungsantrag (Fortsetzung) 07/2010 bis 06/2014. Dortmund.

Lamm, J. G.; Weilkiens, T. (2014): Method for Deriving Functional Architectures from Use Cases. In: Maik S. Maurer, Jutta Abulawi und Sven-Olaf Schulze (Hg.): Tag des Systems Engineering. Bremen, 12.–14. November 2014 ; [TdSE]. München: Hanser, S. 225–236.

Lex, A. (2004): Mit Methode zum anforderungsgerechten Roboterdesign. In: Petra Winzer (Hg.): Das Wuppertaler Generic-Managementsystem-Konzept. Aachen: Shaker (Berichte zum Generic-Management, 2/2004).

Lill, Raimar (2021): Modellbasiertes Testen kooperierender autonomer Systeme auf Basis farbiger Petri-Netze. Doctoralthesis. Friedrich-Alexander-Universität Erlangen-Nürnberg (FAU).

Lindemann, U. (2005): Methodische Entwicklung technischer Produkte. Methoden flexibel und situationsgerecht anwenden. Berlin: Springer.

Lindemann, Udo (Hg.) (2016): Handbuch Produktentwicklung. München: Carl Hanser Verlag.

Lindemann, Udo; Maurer, Maik; Braun, Thomas (2009): Structural complexity management. An approach for the field of product design. Berlin: Springer.

Luft, Thomas; Ewringmann, Niklas; Wartzack, Sandro (2014): Application and Validation of the Matrix-based Product Description in a Case Study by Using the Software Loomeo. 24th CIRP Design Conference. In: *Procedia Cirp* 21, S. 479–484. DOI: https://doi.org/10.1016/j.procir.2014.02.056.

Luhmann, N. (1980): Komplexität. Enzyklopädie der Betriebswirtschaftslehre. Stuttgart: Poeschel.

Mamrot, M.; Marchlewitz, S.; Nicklas, J.-P.; Riekhof, F.; Schlueter, N.; Seider, G.; Winzer, P. (2012): Begriffe im Kontext des Generic Systems Engineering—Ansatzes. In: Petra Winzer (Hg.): Generic Systems Engineering als Basis für die Weiterentwicklung des WGMK-Modells. Aachen: Shaker (Berichte zum Generic-Management, 2012,2), S. 21–30.

Mamrot, Michel (2014): Entwicklung eines Ansatzes zur modellbasierten Felddatenrückführung in die Produktentwicklung. 1. Aufl. Herzogenrath: Shaker (Berichte zum Generic-Management, 2014,1).

Mamrot, Michel; Marchlewitz, Stefan; Nicklas, Jan-Peter; Winzer, Petra (2014): Using systems engineering for a requirement-based design support for autonomous robots. In: 2014 IEEE International Conference on Systems, Man, and Cybernetics (SMC). IEEE, S. 3115–3120.

Marchlewitz, Stefan; Nicklas, Jan-Peter; Winzer, Petra (2015): Using systems engineering for improving autonomous robot performance. In: 10th System of Systems Engineering Conference (SoSE). San Antonio, TX, USA: IEEE, S. 65–70.

Marques-Lucena, Catarina; Agostinho, Carlos; Marcelino-Jesus, Elsa; Sarraipa, Joao; Jardim-Goncalves, Ricardo (2015): Collaborative Management of Requirements Using Semantic Wiki Modules. In: Ioan Dumitrache (Hg.): 20th International Conference on Control Systems and Computer Science (CSCS 2015). Bucharest, Romania: IEEE, S. 665–672.

Maurer, Maik; Braun, Thomas (2008): The why-matrix Die Why-Matrix. In: Matthias F. Kreimeyer (Hg.): Proceedings of the 10th International DSM Conference. München: Carl Hanser Verlag, S. 35–44.

Microsoft (2021): Microsoft—Cloud, Computer, Apps und Gaming. Online verfügbar unter https://www.microsoft.com/de-de/, zuletzt aktualisiert am 02.09.2021, zuletzt geprüft am 02.09.2021.

Mirson, Alexander; Skrypnyuk, Oleg; Elezi, Fatos; Lindemann, Udo (2011): MDM-Based Software Modularization by Analysing Inter-Project Dependencies. In: *13th International DSM Conference, Cambridge, MA, USA*, S. 143–157. Online verfügbar unter https://www.designsociety.org/publication/30830/MDM-based+Software+Modularization+by+Analysing+Inter-project+Dependencies, zuletzt geprüft am 13.10.2020.

Mistler, Marian (2020a): Analyse von IT-Werkzeugen zum modellbasierten Generic Systems Engineering für Organisationen auf Basis von e-DeCoDe. In: N. Schlüter, M. Reiche und M. Löwer (Hg.): Potentiale der Informationsvernetzung beim Generic Management. Aachen: Shaker.

Mistler, Marian (2020b): Analyse von IT-Werkzeugen zum modellbasierten Generic Systems Engineering für Organisationen auf Basis von e-DeCoDe. In: Nadine Schlüter, Markus Reiche und Manuel Löwer (Hg.): Potentiale der Informationsvernetzung beim Generic Management. Aachen: Shaker Verlag.

Mistler, Marian (2021): Entwicklung eines Vorgehenskonzeptes zum modellbasierten agilen Anforderungsmanagement (Requirements Engineering und Requirements Management) für Organisationen—REMOt. Dissertation. Wuppertal: Bergische Universität Wuppertal.

Mistler, Marian; Schlueter, N.; Löwer, Manuel (2021a): Agile Design of Organizations using the Information Flow Analysis and the Generic Systems Engineering. In: 2021 IEEE International Conference on Systems, Man, and Cybernetics (SMC): IEEE.

Mistler, Marian; Schlueter, Nadine; Löwer, Manuel (2021b): Analysis of software tools for model-based Generic Systems Engineering for organizations based on e-DeCoDe. In: 2021 IEEE International Systems Conference (SysCon): IEEE.

Mistler, Marian; Schlueter, Nadine; Walter, Bastian; Winzer, Petra (2019): Dealing with Legal Requirements in the Planning Phase of Integrated Management Systems for Agile Organizations. In: Su Mi Dahlgaard-Park und Jens J. Dahlgaard (Hg.): 22nd QMOD-ICQSS Conference. Krakow, Poland: Int. QMOD-ICQSS conference proceedings.

Muschik, Sabine (2011): Development of Systems of Objectives in Early Product Engineering. Entwicklung von Zielsystemen in der frühen Produktentstehung. Karlsruher Institut für Technologie (KIT), Diss., 2011. Hg. v. o. Prof. Dr.-Ing. Dr. h.c. A. Albers. Institut für Produktentwicklung (IPEK). Karlsruhe (Forschungsberichte IPEK, 50). Online verfügbar unter http://nbn-resolving.org/urn:nbn:de:swb:90-237686.

Nicklas, J. P.; Winzer, P.; Dahlgaard-Park, S. M.; Dahlgaard, J. J. (2014): Approach for Using Requirements Engineering in Collaborative Networks. In: Entering the Experience Economy from product quality to experience quality, Proceedings of the 17th QMOD-ICQSS International Conference on Quality and Service Sciences, ICQSS.

Nicklas, Jan-Peter (2016): Ansatz für ein modellbasiertes Anforderungsmanagement für Unternehmensnetzwerke. Dissertation. Aachen: Shaker Verlag (Berichte zum Generic-Management, Band 2016, 2).

Nicklas, Jan-Peter; Mamrot, Michel; Winzer, Petra; Lichte, Daniel; Marchlewitz, Stefan; Wolf, Kai-Dietrich (2016): Use case based approach for an integrated consideration of safety and security aspects for smart home applications. In: 2016 11th System of Systems Engineering Conference (SoSE). 2016 11th System of Systems Engineering Conference (SoSE). Kongsberg, Norway, 12.06.2016–16.06.2016: IEEE, S. 1–6.

Ochel, Lennart (2017): Petri-Netz-basierte Simulation biologischer Prozesse mit OpenModelica.

Ott, Stefan (2009): Konzept zur methodischen System-Modellierung in der anforderungsgerechten Produktentwicklung. Aachen: Shaker.

Partsch, Helmut (1991): Petrinetze. In: Helmut Partsch (Hg.): Requirements Engineering: De Gruyter, S. 100–108.

Partsch, Helmuth (2010): Requirements-Engineering systematisch. Modellbildung für softwaregestützte Systeme. 2. Aufl. Berlin, Heidelberg: Springer.

Pavalkis, Saulius; Nemuraite, Lina; Milevičienė, Edita (2011): Towards Traceability Metamodel for Business Process Modeling Notation. In: Tomas Skersys, Rimantas Butleris, Lina Nemuraite und Reima Suomi (Hg.): Building the e-World Ecosystem. 11th IFIP WG 6.11 Conference on e-Business, e-Services, and e-Society. Berlin, Heidelberg: IFIP International Federation for Information Processing; Springer (IFIP Advances in Information and Communication Technology, 353).

Pohl, Klaus (2016): Requirements Engineering. Fundamentals, principles, and techniques. Berlin, Heidelberg: Springer-Verlag.

Pohl, Klaus; Rupp, Chris (2015): Requirements engineering fundamentals. A study guide for the certified professional for requirements engineering exam, foundation level. IREB compliant. 2nd edition. San Rafael, CA: Rocky Nook.

Pregitzer, Gerhard; Blumör, Alexander; Kleiner, Sven; Krastel, Marcus; Neubert, Michael (2014): Model Based Systems Engineering: Einführung und Anwendung der modellbasierten Arbeitsweise in der Maschinenentwicklung. In: Maik S. Maurer, Jutta Abulawi und Sven-Olaf Schulze (Hg.): Tag des Systems Engineering. Bremen, 12.–14. November 2014 ; [TdSE]. München: Hanser, S. 235–246.

REDPOINT.TESEON (2021): management consultants I digital engineers | smart business apps. Online verfügbar unter https://redpoint.teseon.com/, zuletzt aktualisiert am 02.09.2021, zuletzt geprüft am 02.09.2021.

Reisig, Wolfgang (2010): Petrinetze: modellierungstechnik, analysemethoden, fallstudien: Springer-Verlag.

Riekhof, F.; Winzer, P.; Wörner, L.; Kulig, S. (2012): Funktionsorientierte Auslegung eines Linearantriebs. In: Conf. Entwurf komplexer Automatisierungstechnik EKA. Madgeburg, S. 121–138.

Rudolf, Stefan (2013): Produktionsgerechte Baukastengestaltung. 1. Aufl. Aachen: Apprimus-Verl. (Produktionssystematik, 30/2013).

Rupp, Chris (2021): Requirements-Engineering und -Management. Das Handbuch für Anforderungen in jeder Situation. 7., aktualisierte und erweiterte Auflage. München: Hanser.

Sage, Andrew P.; Rouse, William B. (Hg.) (2009): Handbook of systems engineering and management. 2. ed. Hoboken, NJ: Wiley (Wiley series in systems engineering and management).

Salehi, Vahid; Florian, Gross; Taha, Jihad (2018): Implementation of system modeling language (SysML) in consideration of the CONSENS approach. In: Proceedings of the DESIGN 2018 15th International Design Conference: Faculty of Mechanical Engineering and Naval Architecture, University of Zagreb, Croatia; The Design Society, Glasgow, UK (Design Conference Proceedings), S. 2987–2998.

Schapiro, Seth B.; Henry, Matthew H. (2012): Engineering agile systems through architectural modularity. In: 2012 IEEE International Systems Conference SysCon 2012. 2012 6th Annual IEEE Systems Conference (SysCon). Vancouver, BC, Canada, 19.03.2012–22.03.2012: IEEE, S. 1–6.

Scheer, August-Wilhelm; Cocchi, Andrea (2006): Prozessorientiertes Product Lifecycle Management. Mit … 3 Tab. Berlin, Heidelberg [u. a.]: Springer.

Scheithauer, D. (2014): Qualität im System-Design. In: Maik S. Maurer, Jutta Abulawi und Sven-Olaf Schulze (Hg.): Tag des Systems Engineering. Bremen, 12.–14. November 2014 ; [TdSE]. München: Hanser, S. 225–234.

Schlueter, Nadine (2016): Der DyNamic-Ansatz. Entwicklung eines Ansatzes zur verlässlichen Gestaltung von Unternehmensnetzwerken und ihren Produkt-Service-Systemen. Habilitationsschrift. Wuppertal. Online verfügbar unter http://elpub.bib.uni-wuppertal.de/edocs/dokumente/fbd/maschinenbau/habi2016/schlueter ; http://nbn-resolving.de/urn/resolver.pl?urn=urn%3Anbn%3Ade%3Ahbz%3A468-20171009-111216-2.

Schlueter, Nadine; Heinrichsmeyer, Marius; Bielefeld, Ovidiu; Mistler, Marian; Ansari, Amirbabak (2019): ReMaiN-Concept for Requirements Management and Engineering in R&D business networks in Germany. In: Proceedings of 14th International Conference on System of Systems Engineering. Anchorage, Alaska: IEEE, S. 308–312.

Schlueter, Nadine; Winzer, Petra; Ansari, Amirbabak; Bielefeld, Ovidiu; Dransfeld, Hendrik; Heinrichsmeyer, Marius (2018): KAUSAL. A New Methodological Approach for Model Based Analysis of Complex Failure Chains by Example of an Electromobility Concept. In: 2018 IEEE International Conference on Systems, Man, and Cybernetics (SMC). Miyazaki, Japan: IEEE, S. 947–952.

Schlund, Sebastian (2011): Anforderungsaktualisierung in der Produktentwicklung. Entwicklung einer Methodik zur Aktualisierung von Anforderungen durch die Einbindung anforderungsrelevanter Ereignisse. Aachen: Shaker.

Schmidt III, Robert; Deamer, Jason; Austin, Simon (2011): Understanding adaptability through layer dependencies. In: *DS 68–10: Proceedings of the 18th International Conference on Engineering Design (ICED 11), Impacting Society through Engineering Design, Vol. 10: Design Methods and Tools pt. 2, Lyngby/Copenhagen, Denmark, 15.–19.08.2011*, S. 209–220. Online verfügbar unter https://www.designsociety.org/publication/30752/UNDERSTANDING+ADAPTABILITY+THROUGH+LAYER+DEPENDENCIES, zuletzt geprüft am 13.10.2020.

Schnieder, Eckehard; Schnieder, Lars (2013): Verkehrssicherheit. Maße und Modelle, Methoden und Maßnahmen für den Straßen- und Schienenverkehr. Berlin, Heidelberg: Springer Vieweg (VDI-Buch).

Schuhmann, Holger; Wendel, Heinrich; Braukhane, Andy; Berres, Axel; Gerndt, Andreas; Schreiber, Andreas (2010): Concurrent Systems Engineering in Aerospace: From Excel-based to Model Driven Design. In: *8th Conference on Systems Engineering Research*. Online verfügbar unter https://elib.dlr.de/66373/, zuletzt geprüft am 13.10.2020.

Sendler, Ulrich (2009): Das PLM-Kompendium. Referenzbuch des Produkt-Lebenszyklus-Managements. Dordrecht, New York: Springer (Xpert.press).

SFB 696 (2010): DFG—GEPRIS—SFB 696: Forderungsgerechte Auslegung von intralogistischen Systemen—Logistics on Demand. Online verfügbar unter https://gepris.dfg.de/gepris/projekt/14782203%26context=projekt26task=showDetail%26id=14782203 &, zuletzt aktualisiert am 07.09.2021, zuletzt geprüft am 07.09.2021.

Shimomura, Yoshiki; Kimita, Koji (2013): The Philosopher's Stone for Sustainability. Proceedings of the 4th CIRP International Conference on Industrial Product-Service Systems, Tokyo, Japan,

November 8th–9th, 2012. Berlin, Heidelberg: Springer-Verlag (Lecture Notes in Production Engineering).

Simon, Fritz B. (2009): Einführung in Systemtheorie und Konstruktivismus. 4. Aufl. Heidelberg: Carl-Auer (Carl-Auer Compact).

Sitte, J.; Winzer, P. (2005): Demand Compliant Design of Robotic System. In: Jason Gu und Peter X. Liu (Hg.): 2005 International Conference on Mechatronics and Automation. July 20 to August 1, 2005, Niagara Falls, Ontario, Canada : conference proceedings. Piscataway, NJ: IEEE, S. 1953–1958.

Sitte, J.; Winzer, P. (2011a): Systemmodellierung im Fokus von Generic Systems Engineering. In: Gesellschaft für Systems Engineering e.V. (Hrsg), Tag des Systems Engineering.

Sitte, Joaquin; Winzer, Petra (1999): Bottom-Up Framework for Enterprise Optimisation and Control. In: *Proceedings of the International Enterprise Modelling Conference (IEMC'99)*. Online verfügbar unter https://eprints.qut.edu.au/148008/.

Sitte, Joaquin; Winzer, Petra (2006): Evaluation of a new complex system design method on a mechatronic automotive product. In: 2006 IEEE International Engineering Management Conference. IEEE, S. 278–282.

Sitte, Joaquin; Winzer, Petra (2007): Methodic design of robot vision systems. In: 2007 International Conference on Mechatronics and Automation. IEEE, S. 1758–1763.

Sitte, Joaquin; Winzer, Petra (2011b): Demand-Compliant Design. In: *IEEE Transactions on systems, man, and cybernetics-part A: Systems and Humans* 41 (3), S. 434–448.

Slovak, Roman (2006): Methodische Modellierung und Analyse von Sicherungssystemen des Eisenbahnverkehrs. Hg. v. TU Braunschweig, Fakultät für Maschinenbau und Elektrotechnik. Braunschweig (tech. Diss.; Braunschweig 2006). Online verfügbar unter http://rzbl04.biblio.etc.tu-bs.de:8080/docportal/servlets/MCRFileNodeServlet/DocPortal_derivate_00003799/Dissertation-Slovak-18-12-2006.pdf, zuletzt geprüft am 19.03.2015.

Suh, Nam P. (1998): Axiomatic Design Theory for Systems. In: *Research in Engineering Design* 10 (4), S. 189–209. DOI: https://doi.org/10.1007/s001639870001.

Suh, Nam Pyo (2001): Axiomatic design. Advances and applications. New York, NY [u. a.]: Oxford Univ. Press.

Taha, Jihad; Salehi, Vahid; Abraham, Frank (2018): Development of a Low Powered Wireless IoT Sensor Network based on MBSE. In: 4th IEEE International Symposium on Systems Engineering. October 1–3, 2018, Rome Marriott Park Hotel, Roma, Italy : 2018 symposium proceedings. 2018 IEEE International Systems Engineering Symposium (ISSE). Rome. IEEE International Symposium on Systems Engineering; Institute of Electrical and Electronics Engineers; Systems Council; IEEE International Systems Engineering Symposium; ISSE. Piscataway, NJ: IEEE, S. 1–8.

Two Pillars (2021): Bereit für die Digitalisierung im Anlagenbau? Online verfügbar unter https://www.two-pillars.de/, zuletzt aktualisiert am 02.09.2021, zuletzt geprüft am 02.09.2021.

Weilkiens, T. (2007): Systems engineering with SysML. Modeling, analysis, design. Amsterdam: Morgan Kaufmann OMG Press/Elsevier.

Weilkiens, Tim (2019): Systems Engineering mit SysML/UML. Anforderungen, Analyse, Architektur. 3., überarb. und aktualisierte Aufl. Heidelberg: dpunkt.verlag.

Winzer, P.; Fa. Vossloh Kiepe (2008): Prozesskettenorientiertes Regelkreismodell für ein nachhaltiges robustes Design mechatronischer Systeme. Methodischer Ansatz zur Erhöhung zur Zuverlässigkeit mechatronischer Systeme über den PLC dargestellt am Beispiel des Stromabnehmers. VDMA, 07.11.2008.

Winzer, Petra (2012): PromeSys. Abschlussbericht im Rahmen des Verbundforschungsprojektes "Prozesskettenorientiertes Regelkreismodell für ein nachhaltiges robustes Design mechatronischer Systeme" ; Projektträger für das BMBF—Forschungszentrum Karlsruhe, Produktion und Fertigungstechnologie (PTKA-PFT), Förderkennzeichen 02PG1323. Aachen: Shaker.

Winzer, Petra (2015): Generic System Description and Problem Solving in Systems Engineering. In: *IEEE Systems Journal* 11 (4), S. 2052–2061. DOI: https://doi.org/10.1109/JSYST.2015.2428811.

Winzer, Petra; Braunholz, Helge (2000): Chances and Risks of Process-Oriented Integrated Management Systems. In: *Proceedings of the Human Factors and Ergonomics Society Annual Meeting* 44 (10), S. 277–280. DOI: https://doi.org/10.1177/154193120004401039.

Wynn, David C.; Kreimeyer, Matthias; Eben, Katharina; Maurer, Maik; Lindemann, Udo; Clarkson, John (Hg.) (2010): Managing Complexity by Modelling Dependencies-Proceedings of the 12th International DSM Conference. Cambridge, UK, 22.–23. July 2010. München: Hanser.

Zensen, Andre; Kuster, Jochen (2018): A Comparison of Flexible BPMN and CMMN in Practice: A Case Study on Component Release Processes. In: 2018 IEEE 22nd International Enterprise Distributed Object Computing Conference (EDOC). Stockholm, 16.10.2018–19.10.2018. Stockholm, Sweden: IEEE, S. 105–114

5 The Modules of the GSE Procedural Concept—Mastering Complexity Using Simple Rules

The literature analysis demonstrated that SE currently uses a variety of approach concepts (see Sects. 2.4 and 3.4). For this purpose, SE approaches that serve product development, factory design, safety engineering, and software development were specifically analyzed. However, coping with the complexity of the present and the future requires a general, standardized, modular procedural concept that is usable transdisciplinarily and transparently and traceably allows the actions of the approach to be traced (see Sect. 3.4). Furthermore, it could be derived that an procedural concept should only be developed on the basis of a thinking model and that the thinking model and procedural concept must be synergistically connected. It was also found that SE must be a universal approach to purposefully design systems. However, every general approach of SE should be able to be coupled with specific methods and procedures to enable specialist detailed solutions in addition to the basic, generalist problem-solving.

This is illustrated by the design process of a pantograph in Fig. 5.1. Although a common thinking model in the form of a metamodel was created (see Sect. 3.4), specific design tasks have to be solved with certain methods.

For example, the statics of a pantograph on an O-bus must be tested with different methods and procedures than the functionality test of its control. This insight is not new (Lindemann et al. 2009; Gausemeier et al. 2009; Gausemeier et al. 2014; Lindemann 2016).

Today's products are developed by interdisciplinary teams, which agree on an approach, but must work out their special detailed solutions with specialist methods. They have to be linked with the SE procedural concept. Since there are still various procedural concepts in SE (see Sect. 2.4), the assignment of methods, procedures, and (IT) tools to the respective, specialist SE approach logically follows (see Winzer 2015; Gausemeier et al. 2014; Weilkiens 2019; Lindemann 2016; Lindemann et al. 2009; Sage and Rouse 2009; Haberfellner et al. 2018). (Table 5.1)

N. Schlüter, *Generic Systems Engineering*, https://doi.org/10.1007/978-3-662-67994-4_5

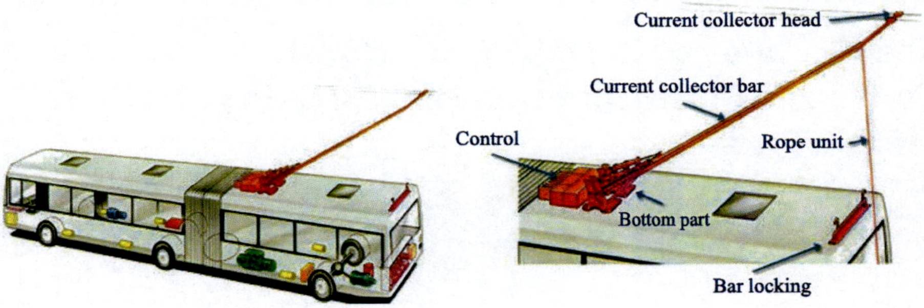

Fig. 5.1 Pantograph (Winzer 2012)

Table 5.1 Overview of methods and procedures according to (Franke 2002, p. 83)

Analysis of existing and new variant-rich product ranges: • Variant modelling, representation • Determination of the appropriate complexity	**Synthesis** of variant-optimized product structures: • Variant-optimizing design of product structures	**Evaluation** of existing and new solution variants	**Transfer** to sales and order processing
• Requirement list • Function analysis • Conjoint analysis • Product structure • Assembly network • Cost structure • Gozinto-/precedence paragraph • Variant tree • Bill of materials comparison/analysis • Assembly-oriented product breakdown • Production-oriented product breakdown • Production overview • Product pyramid • ABC analysis • …	• Series/modules type groups • Standardization/norming • Construction methods • Platforms • Package formation • Parametrization • Modularization • Interface optimization • Classification • Design rules • Same/repeated part matrix • …	• Material cost method • Similarity designations • Relative costs • Search calculation • Feature-based standard costing • Variant cost accounting • Resource-oriented process cost accounting • Variant-oriented process cost accounting • Cost estimation • …	• Enable package offers • Early high-quality sales information • Variant-reducing pricing policy • Early system-supported configuration with customers • Process-oriented processes with structures • Order-neutral pre-assembly groups • Optimization of production depth • Variant-transparent bill of materials systems • Flexibility through late variant emergence

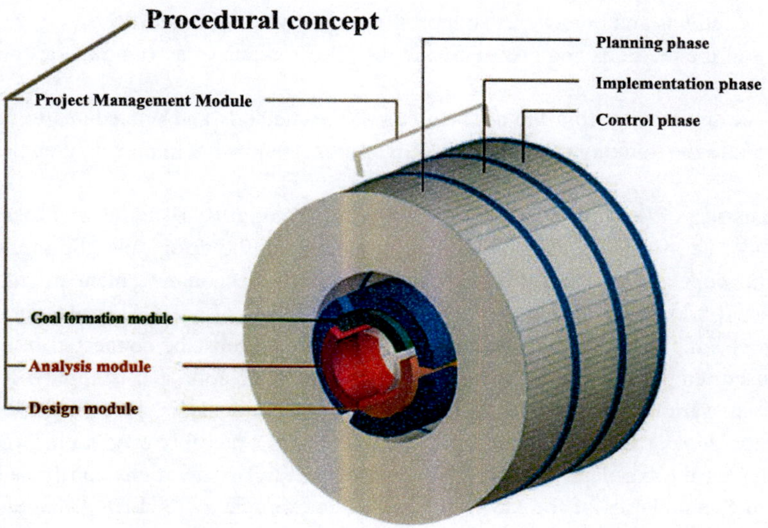

Fig. 5.2 The GSE procedural concept

To counteract this and to offer the transdisciplinary teams a simple, universal problem-solving approach, a ***GSE procedual concept*** was developed for the GSE approach. The GSE approach is a **philosophical SE approach** and consists of an ***analysis, target formation-, as well as design-and project management module.*** The goal of the approach is to standardize all problem-solving processes and also to create comparability across different project management approaches. In this process, ***methods, procedures, and tools are assigned to the GSE modules*** and the resulting ***temporally logical sequence of problem-solving actionsacross*** the phases of ***project management*** for efficient ***planning, implementation, and monitoring*** (see Fig. 5.2). This also meets the often-expressed demand for a connection between SE and project management (Scheithauer 2014; Haberfellner et al. 2018). Furthermore, the modularity of the GSE approach should also contribute to providing the necessary agility for problem-solving in technical and sociotechnical systems (Mistler 2021) (see Sects. 4.4 and 4.5).

The following sections of Chap. 5 are intended to demonstrate that the GSE procedural concept is universal. However, when it is coupled with specialist methods and procedures, special, problem-focused modules result. The following explanations should also prove that the GSE procedural concept is in synergistic connection with the GSE thinking model in the GSE approach. First, the individual universal modules of the GSE procedural concept, i.e., the analysis, the goal formation, and the design module are described. Based on the basic structures of these modules outlined in Sect. 3.3, methods and procedures are introduced and assigned in Sects. 5.1 to 5.3 that provide action-oriented support. The focus is on questions such as:

- Which methods and procedures support the respective GSE modules?
- How can the methods and procedures in the GSE module or across modules be combined?
- How are the results from the applications of the methods and procedures to be integrated into the thinking model of the GSE approach and what impact do they have?

The discussion of possible methods and procedures to be used should first take place per module, i.e., for the analysis, the goal formation and the design module in the GSE procedural concept, and then for the phases of the project management module (see Sect. 5.4). It was found that when applying the modules of the GSE procedural concept, the specific methods and procedures to be selected must be connected in a problem-focused temporal and logical sequence when the GSE approach is applied to solve problems in various areas. Thus, the combination of the analysis, goal formation, and design module does not occur sequentially, but iteratively, possibly even multi-cyclically depending on the problem to be solved (see Sect. 5.5). The iterations can be achieved through the modularity of the GSE approach. For example, MISTLER (Mistler 2021) shows that the multiple integration of the analysis and goal formation module in specific phases is suitable for an iterative approach. Furthermore, it is also assumed from the application of the modules that a multi-cyclic application of the approach concepts created with the GSE approach for new or modified problem situations is also possible. This is indicated by the results of various research works (Mamrot 2014; Nicklas 2016, 2018; Mistler 2021; Mamrot et al. 2014). This chapter will also explain using examples how the GSE approach can interact in its synergistic effect between the GSE thinking model and the GSE procedural concept over a defined period, i.e., the product life cycle.

Suggestions should be given as to which methods and procedures could be used when and how in the GSE procedural concept, without describing them in detail, because this can already be found in EHRLENSPIEL and MEERKAMM, LINDEMANN and PFEIFER and SCHMITT (Lindemann 2016; Ehrlenspiel and Meerkamm 2017; Pfeifer and Schmitt 2021). In addition, a rough sketch should be given of what effects results from methods and procedures have on the GSE thinking model. This aspect is of particular importance because it illustrates ***the interaction between the GSE thinking model and the GSE procedural concept in the GSE approach*** (see Fig. 5.3).

This was partly already investigated in Chap. 4. However, the focus was on the effects of insights from the GSE thinking model on the planning of the procedural concept in the GSE approach. These emerged when creating technical and socio-technical thinking models:

- for the drive, design approaches for the rotor rod,
- for the mechatronic system, a problem-oriented solution approach,
- for the KitVes system, design approaches for minimizing the risk of interaction between the cable and the drum, and
- the agile organizational development of a shoe production.

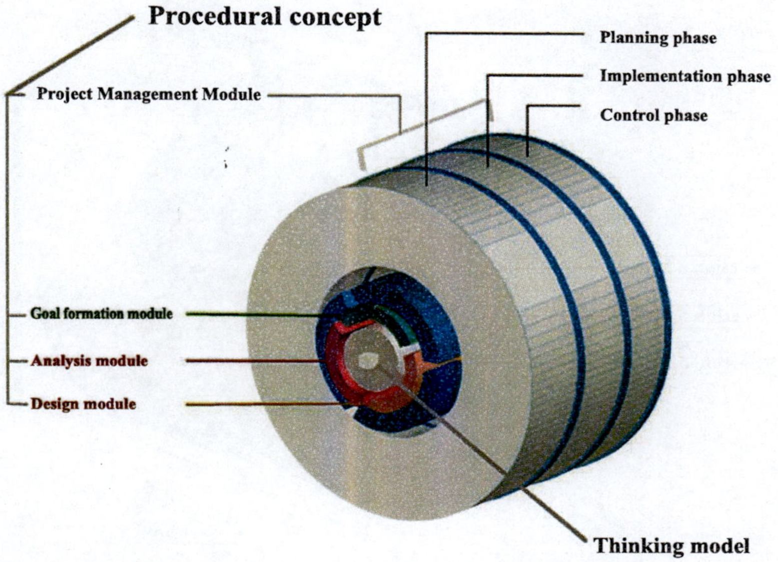

Fig. 5.3 The genesis of the GSE thinking model and the GSE prodecural concept

In Chap. 5, these examples from Sect. 4.4 and 4.5 are revisited to show the effects of refining the GSE procedural concept on the GSE thinking model.

In summary, the aim of Chap. 5 is to present the modules of the GSE procedural concept in more detail. It is intended to illustrate how specific methods and procedures can be selected, linked together, and applied in such a way that a specific, discipline-related problem-solving approach emerges from the universal GSE procedural concept. This is gradually considered in the following sections, as shown in Fig. 5.4, i.e., for the analysis module (red) in Sect. 5.1, for the goal formation module (green) in Sect. 5.2, for the design module (brown) in Sect. 5.3, for the project management module (gray) in Sect. 5.4 and the interaction of the GSE thinking model with the GSE procedural concept in Sect. 5.5.

5.1 The GSE Analysis Module

Basically, the GSE analysis module serves the problem definition, which must in principle be possible through the SE (Arlt 1999). The analysis of the system elements, the interrelationships of the elements with each other, and the interaction with the system environment, as HÄUSLEIN envisages, is not the subject of the GSE analysis module (Häuslein 2004). This system analysis must be part of the workflow for creating the

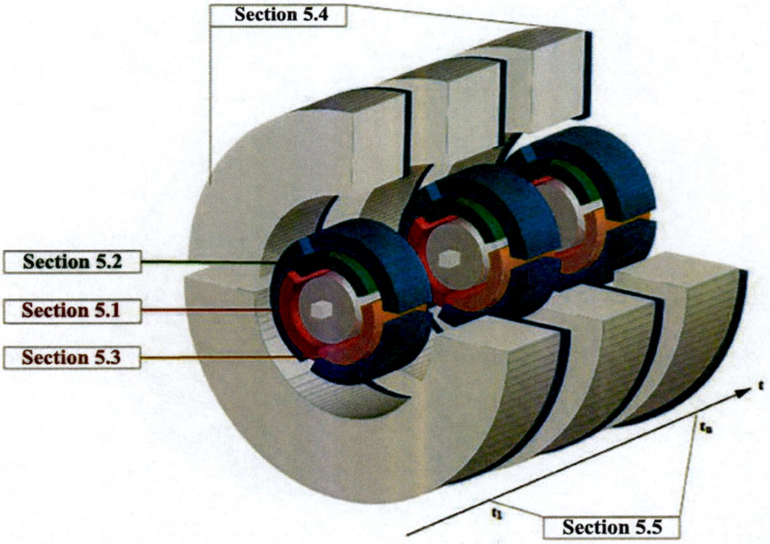

Fig. 5.4 The structure of Chap. 5 for describing the method- and procedure-supported GSE procedural concept in interaction with the GSE thinking model

thinking model. The purpose of the analysis is, according to HABERFELLNER, to make the situation comprehensible, i.e., to understand the problem and its causes, to assign this to a system, and to collect structured information for problem-solving (Haberfellner et al. 2018).

The GSE thinking model is on the one hand the basis for the analysis. On the other hand, the analysis module (red) must allow the analysis results to flow into the GSE thinking model itself as well as into the target formation and the design module, as shown in Fig. 5.5.

The example of the sliding train on the leaves, already presented in Sect. 3.1, illustrated how important it is to clearly define the system boundaries in order to purposefully fix a solution space that allows optimal solutions to be developed. Thus, ***problem finding presupposes an image of the system***—here in the specific case: train, track, and environment—at least as a black-box model***as a prerequisite.*** This rough image of the system can be the basis for the analysis. With the help of the GSE analysis module, it is subsequently possible to clarify which possible causes led to the train sliding in the considered system. For the ***detailed problem recognition*** of the train sliding on leaves, which is the ***task of the GSE analysis module***, various analysis methods and procedures can be used (see Fig. 5.5). Some selected methods and procedures will be explained in the following, and their interaction with the GSE thinking model will be outlined as an example.

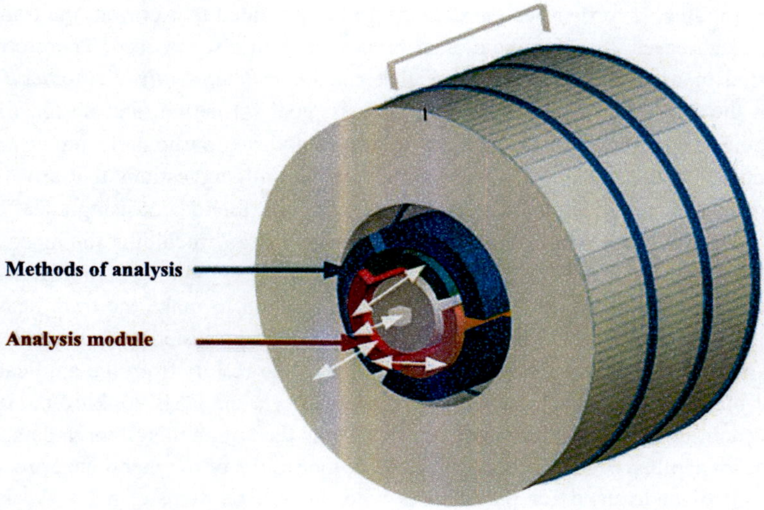

Fig. 5.5 The interactions of the GSE analysis module with the other modules of the GSE approach

Basically, a distinction can be made between:

- universal methods and procedures, and
- specific methods and procedures (Mamrot et al. 2015).

The term ***"universal methods and procedures"***subsumes that these are applicable in the analysis, goal formation, and/or design module and are not related to a specific object, such as errors or scenarios. ***Procedures*** are referred to as systematic approaches. ***Methods*** include a system of rules for goal-oriented realization of theoretical and practical activities (see Winzer 1997; Jochem et al. 2015). The method is a planned approach to achieving a specific goal (Ehrlenspiel and Meerkamm 2017), which can be supported by procedures, i.e., in this example, the cause analysis for the sliding of the train. Universal procedures include brainstorming, brainwriting or mind mapping. To process and document the results, certain ***instruments/tools/tools*** can be used (see also Chap. 4), such as special software for creating mind maps. All three procedures in conjunction with instruments serve the structured collection of opinions of individual group members involved in problem-solving. Referring to the aforementioned example "train", the mentioned procedures can be purposefully connected via a method in order to systematically analyze possible causes for the train sliding. However, if the problem-solving process for the sliding train questions what the most important goals are to prevent the trains from sliding on the leaves, these three mentioned procedures, coupled in a different methodological approach, can also be used in the goal formation process, i.e., in the GSE goal formation module. If further design ideas are to be generated that prevent the train from sliding on the leaves, the aforementioned procedures can also be used. Therefore, they are referred to as ***universal procedures*** that can be ***systematically methodically*** supported in the GSE approach both in the analysis, goal formation, and design module. Establishing the relationship to the GSE thinking model is particularly important. For the example "sliding train on leaves", this means that all suggestions that arose in the brainstorming must be related to the system image. Thus, the idea of using a leaf blower must be assigned to both the "train" and the "railway tracks" including the noise reduction analysis and questioned again as to what effects their implementation has on the respective subsystem. The same applies to the suggestion to make the train heavier or to spread sand on the tracks. In this way, the ideas can be tested and pursued on the thinking model. The transparency and traceability of the results from the application of universal procedures is now systematically generated via the GSE thinking model, i.e., via the system image, which transparently represents the common system understanding of the transdisciplinary team. If this is neglected, the results of the mentioned procedures are difficult to understand for participants who did not participate in the workshops. They are not familiar with the system image that served as the basis. They may have their own models. This can lead to misinterpretations of the results, as the viewer of the results has a different thinking model, which is also not transparent. If, for example, not the train as a subsystem is considered, but only the interaction of the subsystems "wheel" and "track", completely different analysis results are obtained with the aforementioned procedures. Consequently, the results, although the same procedures were used, are only limitedly comparable because they were based on different thinking models.

This example illustrates that the outlined universal procedures on the one hand require a mandatory interaction with the GSE thinking model, and on the other hand can be used in the GSE analysis, GSE goal formation, as well as the GSE design module. These mentioned universal procedures are industry-independent, i.e., they are not only applicable for the design of transport systems, as shown here in the example "sliding train on leaves", but also for the development of new services, for the re-engineering of a mechatronic system or the redesign of a company, to name just a few examples.

In summary, universal methods and procedures are applicable in the GSE analysis, GSE goal formation, and GSE design module, as well as usable regardless of industry or subject matter.

So what are specific methods and procedures?

Specific methods and procedures for the GSE analysis module can be described by three characteristics:

a) They serve only the problem recognition, i.e., they are only applicable in the GSE analysis module. Examples of this are the Q7 tools, such as the histogram, the Pareto diagram, the correlation diagram, or the data collection sheet, to name just a few.
b) They serve only the analysis of a very specific fact or object, e.g.

  - the analysis of failures via failure collection cards, fault trees, or the failure mode and effects analysis (FMEA),
  - the analysis of customer requirements via conjoint analysis or the Kano model,
  - the analysis of cause-effect chains, via the flow diagram or the Ishikawa diagram.
c) They are only applicable in a specific industry. Examples of this are:
  - the analysis of the reliability of seat belts (automotive industry),
  - the analysis of the weather resistance of asphalt (construction industry), and
  - the analysis of the tensile strength of fabrics (textile industry).

In summary, the specific methods and procedures, which are used in the GSE analysis module, either serve problem recognition, or/and the analysis of a very specific fact, and/or they are only applicable for a specific industry.

Also for the specific methods and procedures of the GSE analysis module, it is mandatory that these must be related to the GSE thinking model, as Fig. 5.6 outlines for the ***Q7 tools***.

If the goal is for the analysis results to be reused or to be comprehensible to third parties not involved in the analysis, then the analysis results must be related to the GSE thinking model, as shown in Fig. 5.6. Through the thinking model, the uninvolved person can understand and further use the analysis results. Thus, as shown in Fig. 5.6, the collected data can be clearly assigned to a system element or a system view.

The same applies to the Pareto diagram or the correlation diagram, which can represent interactions between system elements or system views. These too must be clearly defined and possibly refined via the GSE thinking model.

The following will outline the interactions of specific methods and procedures of the GSE analysis module with the GSE thinking model in three factual situations, i.e.:

1. the analysis of failures,
2. the determination of customer requirements, and
3. the analysis of cause and effect chains.

Example 1: The representation of the interactions between the GSE analysis module and the GSE thinking model using the example of special procedures for failure analysis

The analysis of failures in a system can be carried out using a variety of methods and procedures, be it with the ***error collection card***, with the ***fault tree analysis***or with ***FMEA***. Failures and nonconformity are defined as unfulfilled, specified requirements (DIN EN ISO 9000 2015).

Proof of the degree of fulfillment of requirements presupposes a thinking model that correlates the requirements at least with the system, and further with the system elements or their interaction. This is guaranteed by the GSE thinking model, which was created with DeCoDe. If a component or a function of the system under consid-

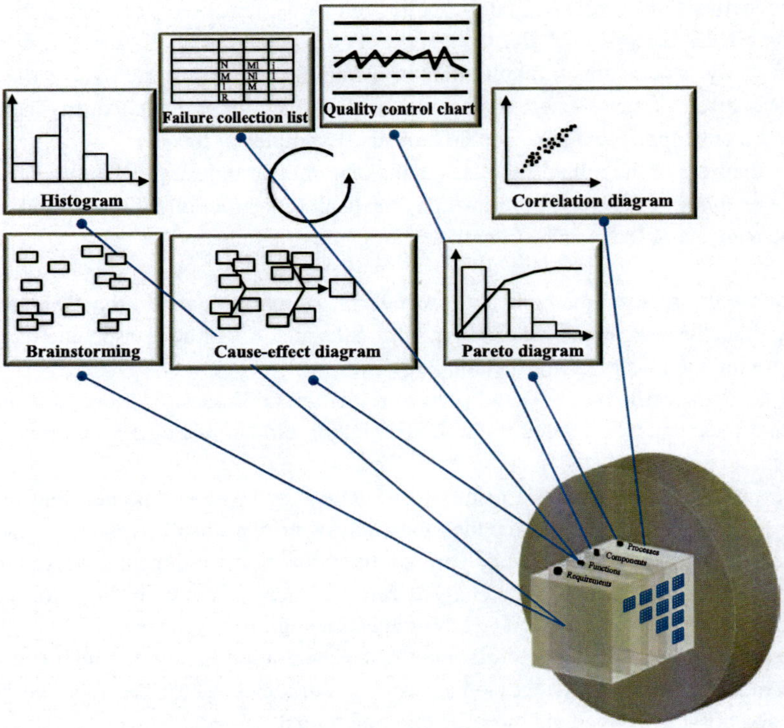

Fig. 5.6 The connection of Q7 tools with the GSE thinking model in reference to GOGOLL (Gogoll 1996, p. 139)

eration does not meet the requirements, the failure can thus be detected and described in more detail. Even when using a different failure concept, the ISO 26262 describes an fault as a failure of an element, i.e., it does not fulfill its required function, the GSE thinking model can be used. The function-component matrix and the function-requirement matrix clearly show which elements of the system have a causal failure. The different failure definition leads to different results of the FMEA and other causalities in the GSE thinking model (Willing and Winzer 2015). For this reason, the failure concept must always be clarified before the method is applied. ◀

What does this look like in the case of FMEA, a method for analyzing, evaluating, and eliminating failures, when the failure concept of DIN EN ISO 9000 (2015) serves as the basis?

The standard for FMEA (DIN EN 60812 2006) requires that the various system components with their characteristics, performances, tasks, and functions must first be documented and the logical link between the system components must be recorded. For this

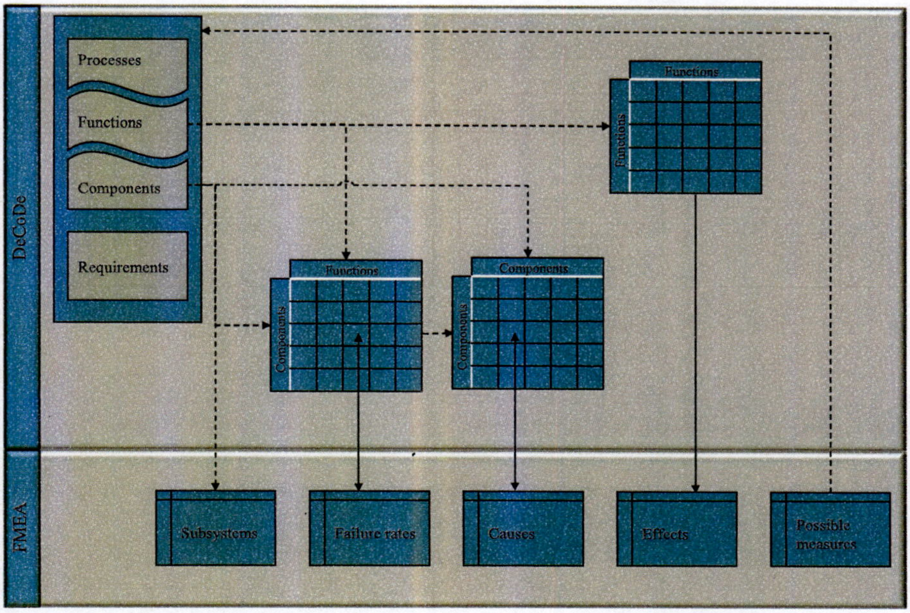

Fig. 5.7 Standardization of the input information into the FMEA through the GSE thinking model based on OTT and WINZER (Ott and Winzer 2007)

reason, a large number of different variants of FMEA have emerged (Pfeifer and Schmitt 2014). So far, the standard does not prescribe a standardized input for this. This can be provided by the GSE thinking model created with DeCoDe (see Fig. 5.7).

If the requirement view in the GES thinking model is faulty or not linked in the system image with the component, function and process view, this is an indication of potential failures. Their causes can be recognized through the relationships between the system views, e.g., function-component view or function-process view, but also within the system view. Consequently, the GSE thinking model forms the input for the mentioned steps, i.e., the failure definition, the determination of potential causes, and the development of ideas for deriving measures, as summarized in Fig. 5.8.

Input information for the FMEA from DeCoDe:

- the description of the characteristics via components or functions (see point 1 from Fig. 5.8),
- the characteristic of potential failures as non-fulfillment of requirements (see point 2 from Fig. 5.8),
- identifying the potential causes of failures via relationships between the system elements (see point 3 from Fig. 5.8).

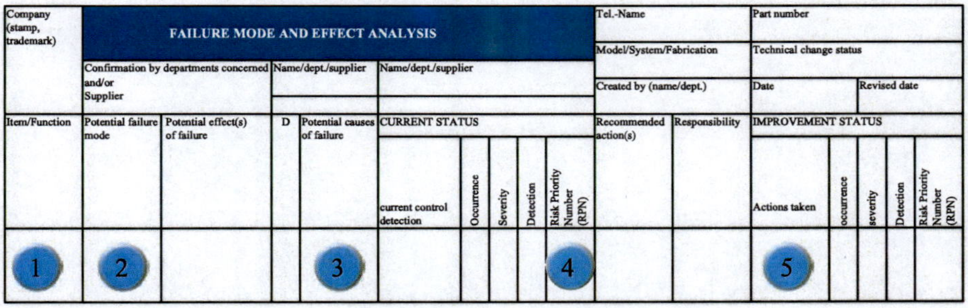

Fig. 5.8 The flow of information between the GSE thinking model and the FMEA (Riekhof and Winzer 2010)

Output information of the FMA for DeCoDe:

- the description of the criticality of functions and components via the risk priority number (RPZ) (see point 4 from Fig. 5.8) and
- the adjustment of the system model according to the actions taken (see point 5 from Fig. 5.8).

Through the FMEA, the criticality of the potential failure with its possible causes is to be evaluated via a risk priority number (see point 4 from Fig. 5.8). This forms the basis for the development of actions for changes to the system. These actions, in turn, must logically lead to an adjustment of the system modeling, i.e., a change in the GSE thinking model. Thus, a more precise GSE thinking model is created, which now avoids the originally identified failures.

This outlined approach of coupling the FMEA with the GSE thinking model makes it clear that this method becomes more efficient through the system-based, standardized input and output information. If a team is to evaluate potential failures using the FMEA, this team does not have to create a system model itself, but can use an already developed GSE thinking model, e.g., of a drive. On this basis, it is now possible to analyze more precisely the plug connection that comes loose during the usage process, which was supposed to secure the power supply for the drive. The results developed by the team can then be fed back into the existing "drive" model. The improvement measures initiated with the help of the FMEA are traceable via the precise system model "drive". Furthermore, if necessary, a consideration of failure sequence chains is possible. However, this requires the combination of further methods and procedures (Willing and Winzer 2015).

Example 2: The analysis of customer requirements in interaction with the GSE thinking model

Customer requirements can be analyzed using the ***Conjoint Analysis*** or the ***Kano Model,*** to name just two methods. These are special analysis methods that can in prin-

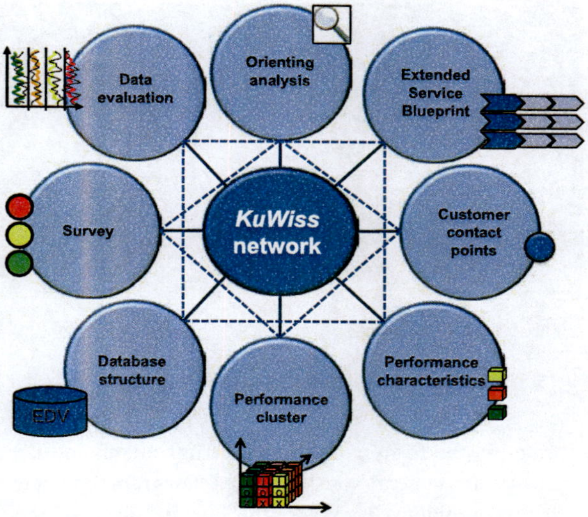

Fig. 5.9 The implementation of the customer voice with KuWISS (Schlüter 2013)

ciple be assigned to the GSE analysis module. The following presents a way in which customer requirements can be analyzed more systematically in conjunction with the GSE thinking model.

Customer requirements are aspects of the requirement view of the generic system model approach. In this specific case, the end customer is understood as the customer. Other stakeholders such as associations, the legislator, etc., can also generate requirements that are also part of the requirement view. These are not considered in the following example.

If it is to be analyzed whether the customer requirement is met or not, it also requires its clear assignment to the respective other views of the system model. Since customer requirements can change over the process of service provision, a time-related assignment to the system approach is necessary, as Fig. 5.9 illustrates.

At the respective customer contact points (see Fig. 5.9), the degree of fulfillment of the customer requirement can be specified at a very specific point in time via the corresponding system feature. This presupposes that the same system model is applied over the entire period of the evaluation of the customer requirement. Only in this way can customer requirements be continuously evaluated (Fiedrich 2010; Schlüter 2011, 2013; Schlüter and Sochacki 2012). ◄

Example 3: The analysis of cause-effect chains in interactions with the thinking model

Cause-effect chains can be represented via ***flow charts,*** but also very simply via the ***Ishikawa Diagram***. With the help of the Ishikawa Diagram, a problem can be characterized and potential causes as well as possible effects on the problem can be illus-

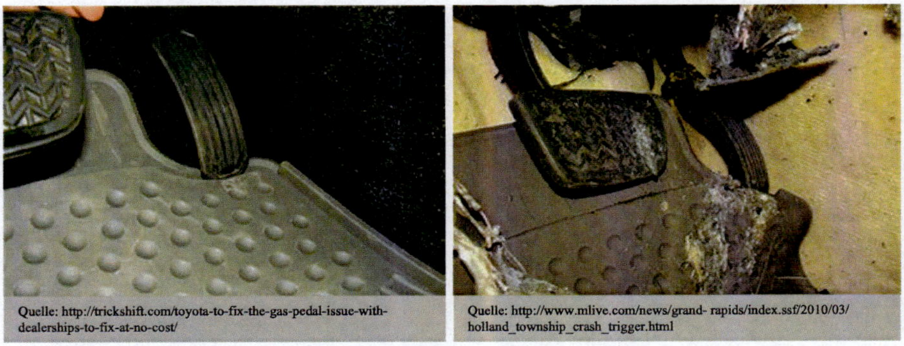

Fig. 5.10 Jamming of the floor mat with the accelerator pedal (Riekhof and Winzer 2010)

trated in the form of fishbones (Brüggemann and Bremer 2012), (Pfeifer and Schmitt 2014). In general, the Ishikawa Diagram characterizes five possible cause fields for the problem. These are the human, the machine, the material, the milieu, or the method. This cause-effect diagram can also be coupled with the thinking model of the GSE approach. Thus, the slipping of the floor mat and its jamming with the accelerator pedal (Focus Online 2011) can be described very simply with the system model (see Fig. 5.10).

Possible causes for the slipping of the floor mat can be starting, sudden braking, slipping when getting in and out, etc. These are all subprocesses of the usage process (process view of the GSE conceptual model). Another cause can be given via the way the floor mat is anchored to the vehicle, or the distance of the accelerator pedal to the footwell. This can be represented via the component-component view in the GSE thinking model. Through the function view, i.e., the networking of brake and strength function, another cause can be derived. The systematic linking of the Ishikawa Diagram with the GSE thinking model is schematically illustrated in Fig. 5.11.

Thus, a more structured analysis is possible. But it also gives ideas for improving the thinking model, as Fig. 5.12 illustrates using the example of Toyota.

In summary, all three examples illustrate how special analysis methods and procedures, which can be assigned to the GSE analysis module, can be carried out more systematically and achieve comprehensible results through their coupling with the GSE thinking model. The GSE thinking model, used as input for the methods and procedures of the GSE analysis module, enables quick interdisciplinary understanding of the initial situation. This is possible because an already existing image of the system, which was worked out with the help of the GSE thinking model, can be used. After the analysis has been carried out, as the examples illustrate, the analysis results can in turn be inserted into the system image. This logically leads to changes in the system image. By comparing the system image at time t_0 before the analysis with the one after the analysis at time t_n, the system change can be made comprehensible.

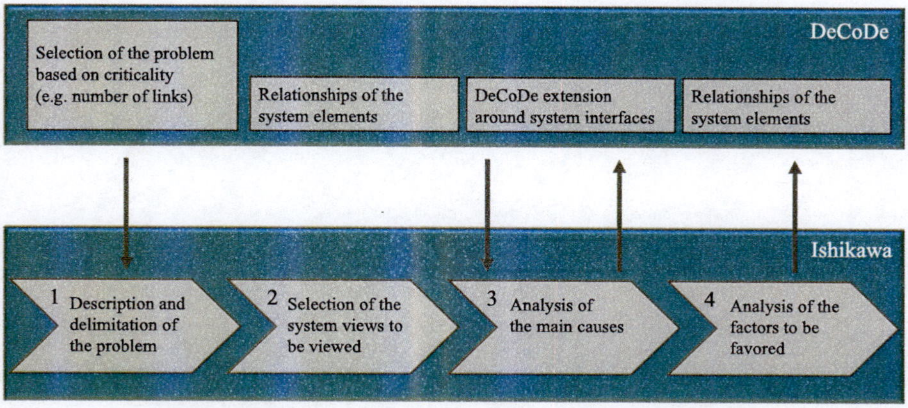

Fig. 5.11 Linking the Ishikawa approach with the GSE thinking model via DeCoDe (Riekhof and Winzer 2010)

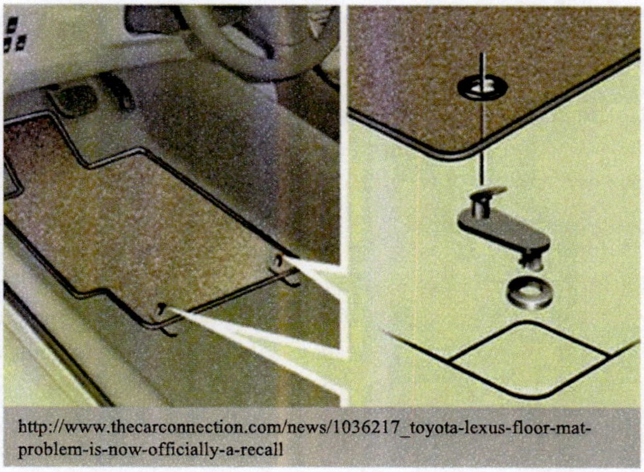

Fig. 5.12 Solution approach for avoiding the slipping of floor mats (Riekhof and Winzer 2010)

5.2 The GSE Goal Formation Module

Objectives arise from prioritized and partly modified requirements for systems. HABERFELLNER characterizes the purpose of objective formation as a systematic summary of intentions that underlie the search for solutions (Haberfellner et al. 2018). ***The taskof the goal formation module module of the GSE approach*** is to derive the most important requirements from the multitude of requirements in order to form goals from them. Goals can, as illustrated in Fig. 5.13, be input information for the analysis module (red).

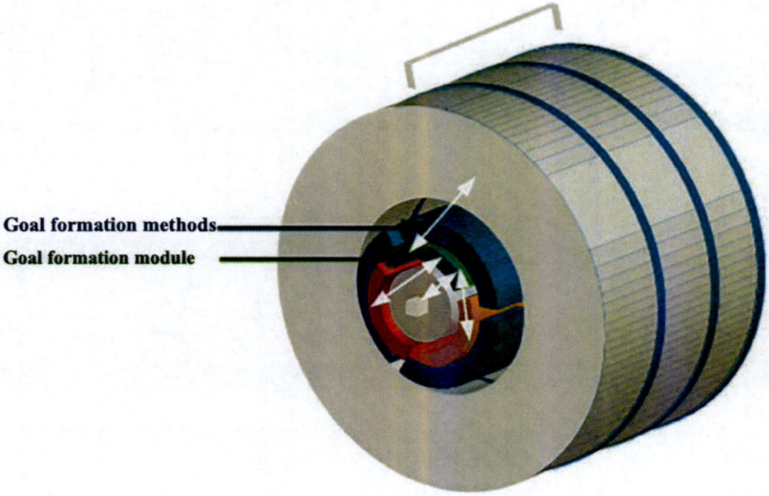

Fig. 5.13 The interactions of the GSE goal formation module with the other GSE modules

But they are also input for the design module (brown) or for the project management module within the scope of project controlling (gray).

Before the selection and use of the methods and procedures schematically represented in Fig. 5.13 within the framework of the GSE goal formation module is described, the essential paths that can lead to the derivation of objectives within the framework of the GSE should first be outlined. Objective formation takes place in interaction with the thought model. Goals, which are the most important result of the Goal formation process, correspond to prioritized and partly specified requirements, which are simultaneously a standardized view in the system model of the GSE approach. Therefore, the dominant interaction of the goal formation module of the GSE procedural concept with the GSE thinking model is justified.

In principle, the ***goal formation in the GSE procedural concept can be achieved via three*** essential, as illustrated in Fig. 5.14, ***paths***:

1. By identifying the stakeholders and their requirements for a system, goals are derived, supported by various methods.
2. By filtering out unfulfilled requirements, goal formation occurs directly. This can:
   a) be directly filtered out from the thinking model or
   b) be analyzed based on methods and procedures over the life cycle of the system.
3. Through requirement-relevant events that can occur during the life cycle of a system, it is possible to derive unrecognized requirements or to specify existing requirements. Special methods and procedures are also required for this.

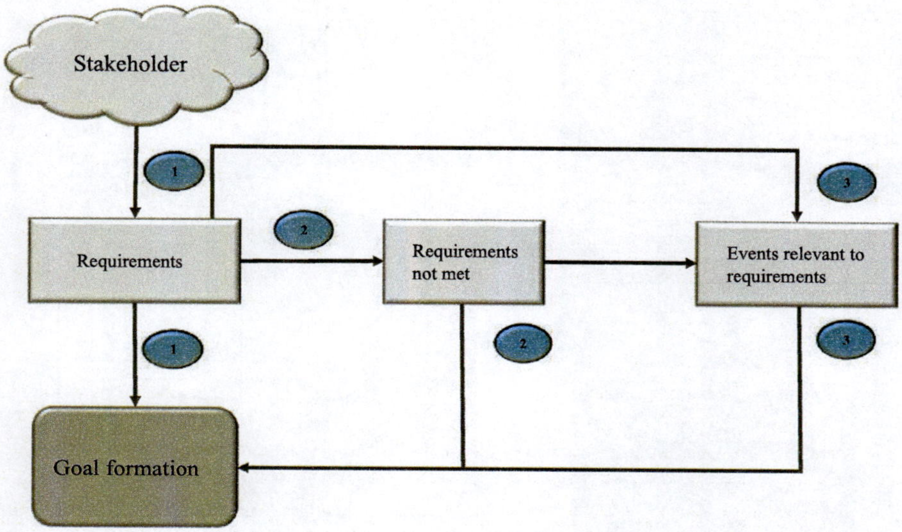

Fig. 5.14 Information input paths into the GSE goal formation module

The following describes these paths individually for generating the necessary input information for the GSE goal formation module with the possible applicable methods and procedures. Finally, a brief characterization of possible methods that can directly generate objectives from the output information of the GSE thinking model in the GSE goal formation module is given.

To 1. Goal formation on the basis of the identified stakeholders and their requirements

This path for deriving input information for the GSE objective formation module includes the following important steps:

1. Identifying and prioritizing the stakeholders.
2. Collecting the requirements of the stakeholders.
3. Comparing the requirements and recognizing contradictions and duplications.
4. Weighing and evaluating the requirements.

The results from these steps are incorporated into the GSE thinking model shown on the right in Fig. 5.15. They serve to create the requirement view.

On the other hand, these results are input information for the GSE goal formation module in order to evaluate them and to fix the degree of fulfillment of requirements, i.e. the objectives.

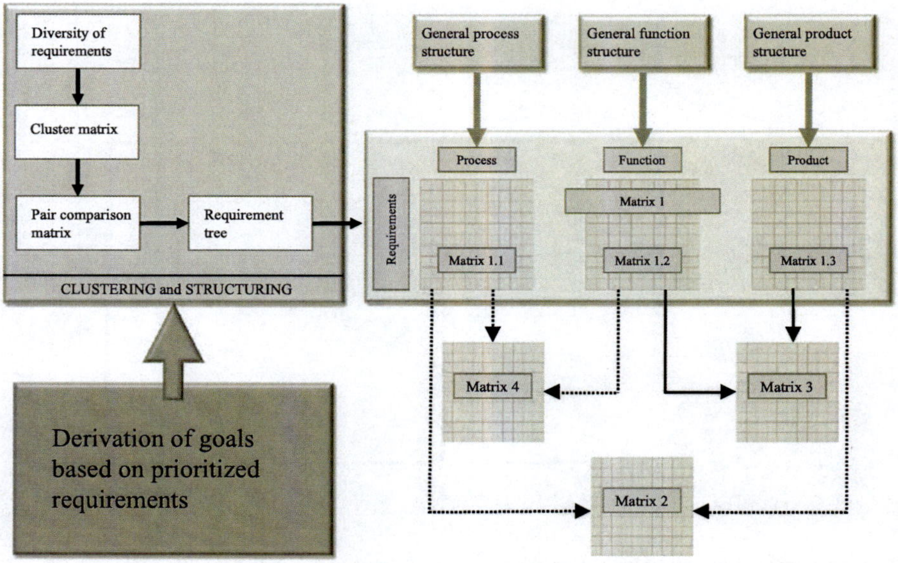

Fig. 5.15 The requirement determination as input information for the GSE thinking model. (Adapted from Lex et al. 2004)

In principle, current and future, even unspoken requirements of stakeholders can be input in this process. The methods and procedures suitable for this often only cover partial aspects. Thus, current and future stakeholders are identified via the ***Stakeholder Radar*** (see Fig. 5.16).

On the other hand, existing and future requirements are collected, for example, ***through market and focus group analyses.*** Other methods and procedures, such as the ***requirement management,*** not only identify requirements, but also compare, evaluate and weigh them.

Other methods and procedures focus only on one group of stakeholders, such as the customer ***in Kansei Engineering*** (Schütte 2002). Here, unspoken (latent) customer requirements are recorded (see Fig. 5.17). If the goal is to delight the customer with the new product to be developed, it is not enough to identify existing requirements. In this case, potential or unspoken requirements must also be collected, as Kansei Engineering allows.

Another group of methods and procedures predicts requirements and compares them with the current ones. The ***scenario technique*** is mentioned here as an example, as shown in Fig. 5.18. Based on current and future requirements, GAUSEMEIER derives objectives for product engineering using scenarios.

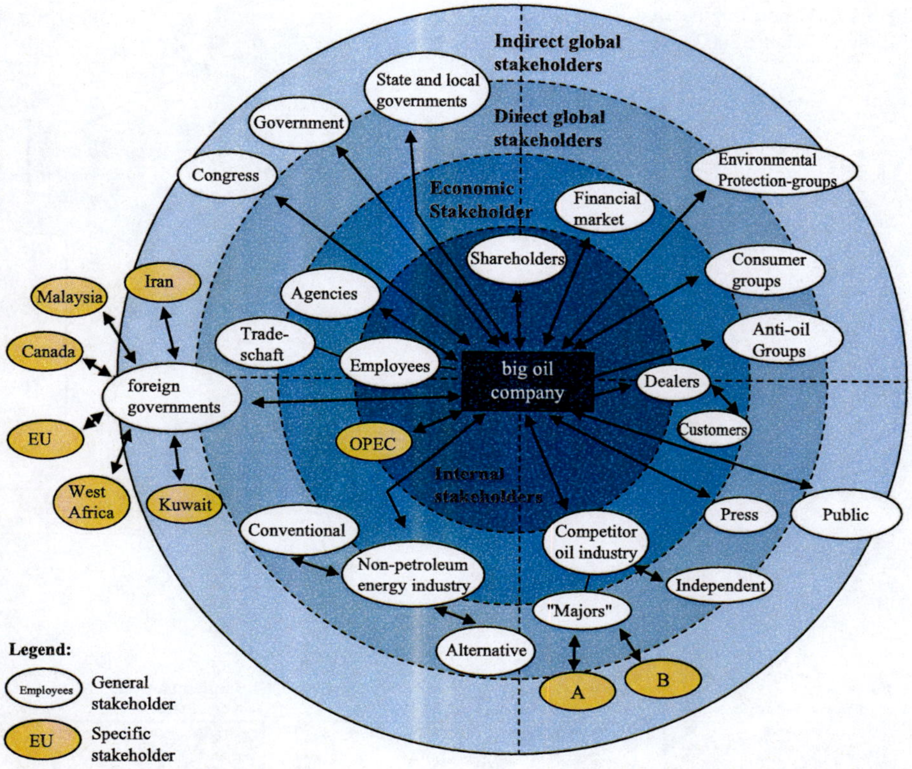

Fig. 5.16 Stakeholder Radar. (According to Gausemeier et al. 2009, p. 172)

None of these mentioned methods and procedures simultaneously captures all stakeholders and their current and future, latent or existing requirements. Thus, all this information cannot be viewed in comparison, weighed and evaluated in order to be able to directly incorporate them as input information into the goal formation process of the GSE procedural concept. For this reason, it is always necessary to connect various methods and procedures, as outlined in this section, in such a way that an efficient goal formation process becomes possible.

To 2. Identifying unfulfilled requirements as input information for the GSE goal formation module

Unfulfilled requirements can be derived from the GSE thinking model and used directly as input information in the GSE objective formation module. Since the GSE thinking model directly links the requirement view with the component, function and process view, it is recognizable which requirements are not fulfilled by components and/or functions as well as processes. By selecting exactly these requirements, input information can

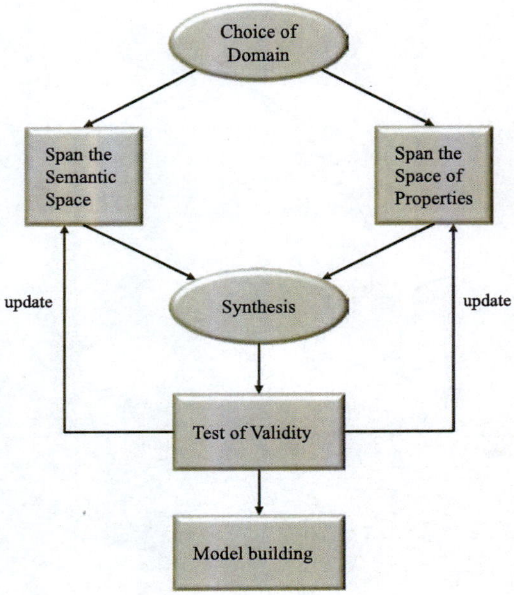

Fig. 5.17 Kansei Engineering procedure (Schütte 2002)

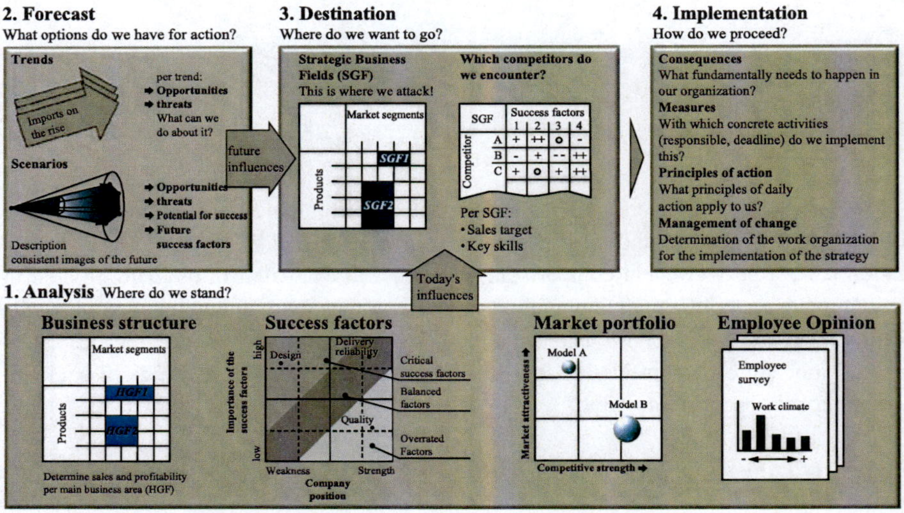

Fig. 5.18 Procedure in strategic product engineering. (After Gausemeier and Plass 2014)

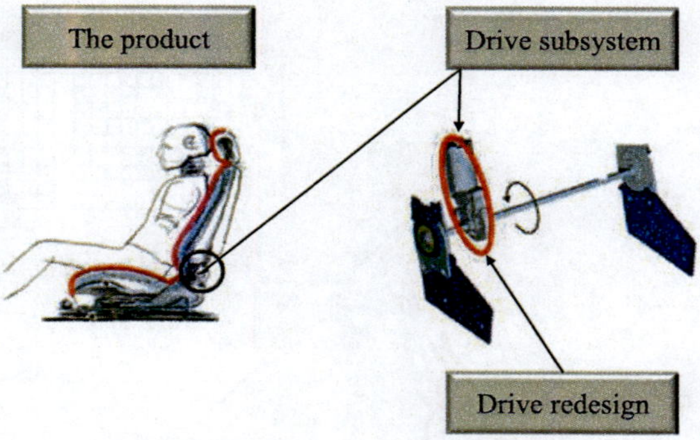

Fig. 5.19 The relationship of the system—seat back—with the subsystem—drive—(Sitte and Winzer 2006)

be obtained directly from the GSE thinking model for the GSE goal formation module. There are two basic variants for this:

A) identifying unfulfilled requirements directly when creating the GSE thinking model and
B) the targeted search for unfulfilled requirements (failures) during the life cycle of a system.

Both variants will be illustrated in the following using selected examples.

Variant A

Unfulfilled requirements can be specifically identified via the GSE thinking model if the thinking model was created using the DeCoDe tools (Sitte and Winzer 2011). They are input information for the GSE goal formation module. The degree of fulfilling these requirements must also be fixed in the goal formation process, as the following example of the re-design of a seat back demonstrates. One of its manufacturers had the task of developing a new subsystem for the backrest angle adjustment (see Fig. 5.19).

The new subsystem to be developed was to convert electrical energy into rotary mechanical energy. When creating the GSE thinking model, it was found that there was no subsystem or components for this conversion function, as Fig. 5.20 illustrates.

For this reason, a search was made across industries for a suitable subsystem for exactly this requirement, as Fig. 5.21 illustrates.

The new subsystem was inserted into the model of the system. As a result, it was possible to demonstrate through the refined thinking model that the requirements are now met.

In summary, this example shows that it is possible, with the help of the GSE thinking model, through the direct networking of the requirement view with the component, process and function view, to demonstrate which requirements are already met, or which are not yet met. Unfulfilled requirements automatically enter the GSE goal formation process, as the example of the drive for the seat back illustrates.

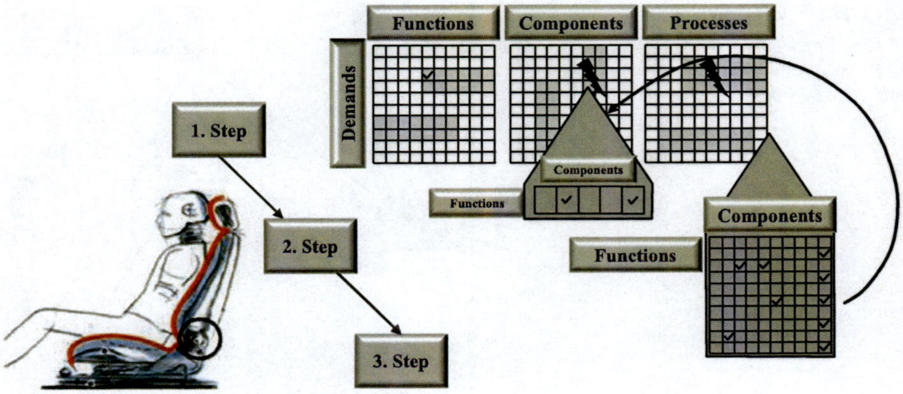

Fig. 5.20 Identifying the unfulfilled requirements using the GSE thinking model using the example of the car seat back (Sitte and Winzer 2006)

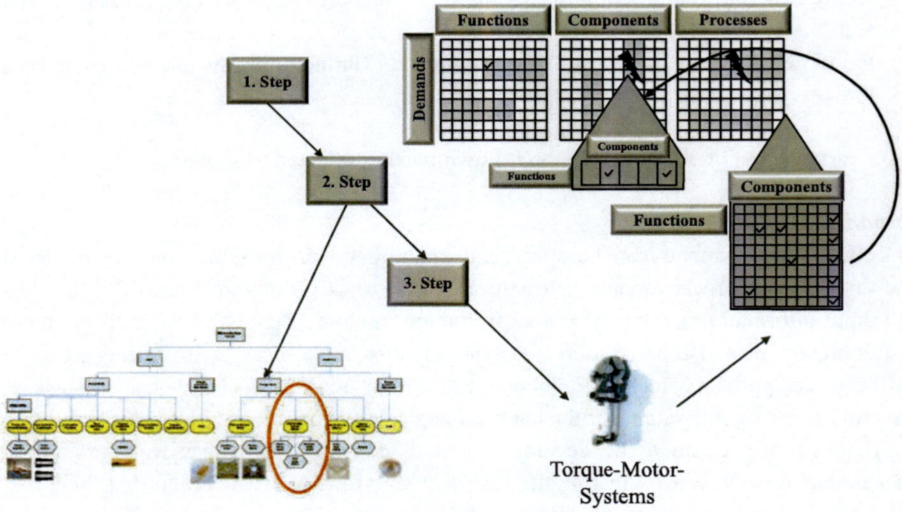

Fig. 5.21 The search for a new subsystem for the car seat back that meets the stated requirements (Sitte and Winzer 2006)

Variant B

Unfulfilled requirements can be detected not only, as just demonstrated, in the product development process, but throughout the entire lifecycle of a product or system. These are then referred to as ***failures*** or complaints. The aim of failure or complaint management is to eliminate the failures or complaints, i.e. the unfulfilled requirements, that are identified in the lifecycle of a system. Its basic principle is shown in Fig. 5.22.

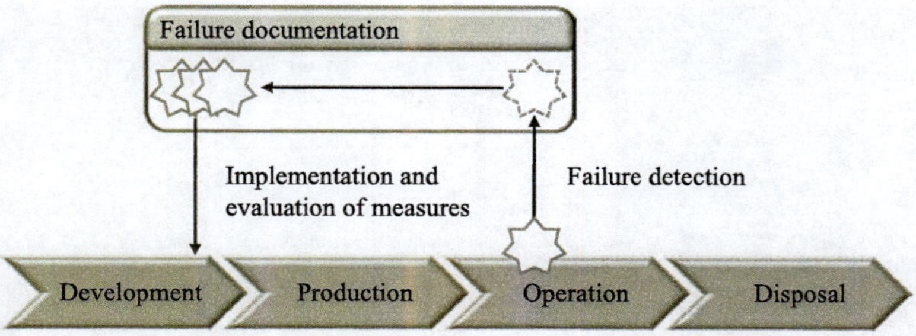

Fig. 5.22 The basic idea of failure management. (According to Schlund 2011, p. 50)

Unfulfilled requirements in the lifecycle of a system can be quickly identified if they can be compared with the GSE thinking model. However, the model of the system must also be refined over the lifecycle.

To 3. The formation of goals based on the identification of new or the refinement of existing requirements caused by the occurrence of requirement-relevant events

So far, only cases have been described in which unfulfilled requirements directly entered the GSE goal formation process. During a system's life, events (requirement-relevant events) can occur from which completely new, previously ***unrecognized requirements*** arise or requirements need to be refined or expanded (Schlund 2011). They must not be directly derived from objectives because they must first be compared with the previous goals. Only based on this comparison can it be determined whether the original goals need to be refined or fundamentally changed. Consequently, they only represent further input information for the GSE goal formation module, which must therefore be reinitiated and run through again. This case corresponds to the third way to obtain input information for the GSE goal formation module, as shown in Fig. 5.23.

The unrecognized requirements or the refinement of existing requirements can occur again and again over the product lifecycle, supported among other things by the ***idea management*** or by ***Poka Yoke,*** as illustrated in Fig. 5.23.

During the life of a product, problems can arise again that result from disturbances, complaints, warranty claims or grievances. If these events can derive new requirements or contribute to the refinement of existing requirements, they are referred to as requirement-relevant events. In contrast to SCHLUND (Schlund 2011), the failure is not referred to as a requirement-relevant event. The failure can only be defined if a requirement was previously fixed, which already entered the goal formation process and it had to be determined during the lifecycle of the system that this requirement is not fulfilled. Consequently, the failure must directly enter the GSE goal formation process and then be eliminated. The situation is completely different with the new or refined requirements identified during the product lifecycle, as illustrated in Fig. 5.23.

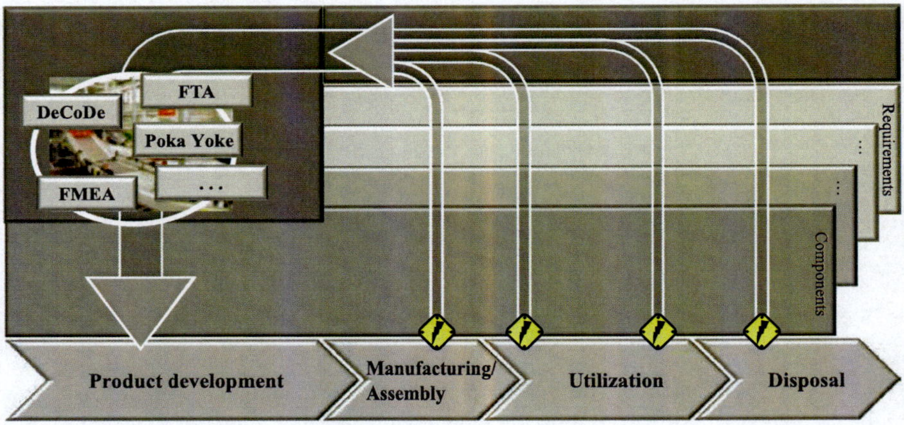

Fig. 5.23 The GSE approach and its application possibilities over the product lifecycle (Müller et al. 2010)

If a drive of a logistics system for the transport of goods up to 10 kg is designed and a customer wants to transport goods up to 25 kg with this system, the manufacturer must now decide whether a redesign of the system is necessary. This can be decided through the GSE goal formation process by comparing it with the other, already existing requirements. However, it always requires a reflection of the requirement-relevant events on the GSE thinking model, which was described for this example in Sect. 4.3. The input, which for the objective formation from the evaluation process of changed systems—output information of the design module—arises, is explained in more detail in Sect. 5.3.

So far, the three possibilities for determining the input information for the GSE goal formation module have been described. These include:

1. the identification of stakeholders and their requirements,
2. the elimination of unfulfilled requirements, and
3. the recognition of new or the refinement of existing requirements based on requirement-relevant events.

The methods and procedures required for this have also been presented in this context. However, further methods and procedures can be directly integrated into the GSE goal formation process in order to prioritize, weight, evaluate the requirements and derive objectives based on them. A corresponding selection for this will be outlined below.

Methods and procedures that can be used within the GSE goal formation module
Classic methods,which are used in the objective formation process, include:

- the Delphi method,
- the conjoint analysis,
- the consistency matrix,
- the mind map,
- the Quality Function Deployment (QFD) or the House of Quality, and also
- methods and procedures of requirements engineering.

While the ***Delphi method*** allows the prioritization of requirements and objectives through expert surveys, target conflicts are filtered out via the ***consistency matrix***. Contradicting requirements or target conflicts can also be identified via the classic ***mind map*** (Lindemann 2005). Prioritizing requirements is also possible via the ***House of Quality*** (see Fig. 5.24).

The initially formulated requirements can be checked via product comparisons, but also via the status of implementation into technical features. This is the basis for a corresponding objective formation. However, QFD focuses exclusively on customer demands and only reflects them in technical features in order to thus represent the degree of objective achievement.

The classic Delphi method and the consistency matrix focus on prioritizing the requirements and thus on the goal formation process.

Also, the ***methods and proceduresof requirements engineering*** can be used within the objective formation module. In general, requirements engineering includes the capturing, systematizing, structuring, evaluating, implementing, and controlling of requirements (see also Sect. 2.4). While the implementation and control of requirements is to be classified in the design phase, the identification, structuring, systematizing, and evaluation of stakeholders and their requirements belong to the goal formation process of the GSE approach. Requirements engineering is comparable to the classic objective formation process and the methods and procedures used there, as described by LINDEMANN (Lindemann 2016). However, it is generally known that requirements change during the course of a product lifecycle. How is this change recorded? What impact does this have on the goal formation process and what on the design process? How are new requirements that arise during the product lifecycle collected, fixed, and integrated into the goal formation? The answer to these questions presupposes that the GSE goal formation module must be in constant interaction with the GSE analysis and design module as well as the GSE thinking model over the product lifecycle. How this can be done will be further explained in Chap. 6.

In this section, methods and procedures have been presented that serve on the one hand to obtain input information for the GSE goal formation module and on the other hand can be used directly in the goal formation process. The following section will go into more detail on how the output information of the GSE goal formation module is now further processed in the GSE design module.

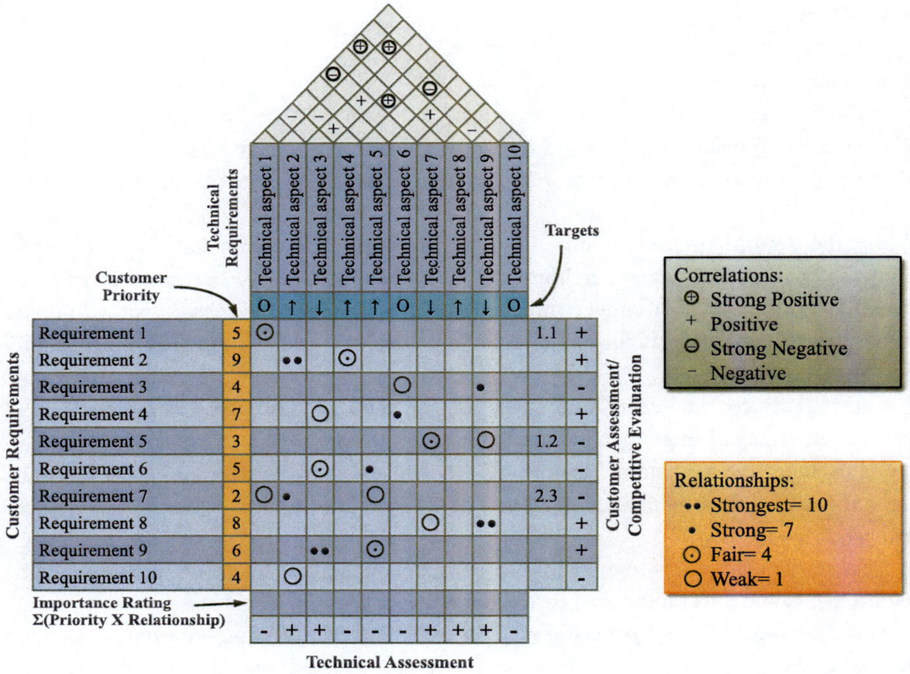

Fig. 5.24 House of Quality information blocks. (According to ffpt.de 2016)

5.3 The GSE Design Module

The task of the ***GSE design module*** is to develop new systems or modify existing ones. HABERFELLNER divides system design into architecture design, i.e. the development of the basic system architecture, and concept design, which includes the concrete design of the selected architecture (Haberfellner et al. 2018). Consequently, design results always lead to a change in the system model and thus presuppose a system model, whether it is a black box or a model created using DeCoDe tools.

The GSE design module is not only connected to the GSE thinking model, but also to the analysis and goal formation module of the GSE approach, as shown in Fig. 5.25.

The input information for the GSE design module can be generated via several paths:

1. They are directly derived from the GSE thinking model.
2. They result from the analysis module.
3. They are generated via the goal formation module.
4. They arise during the project realization as necessary project adjustment.

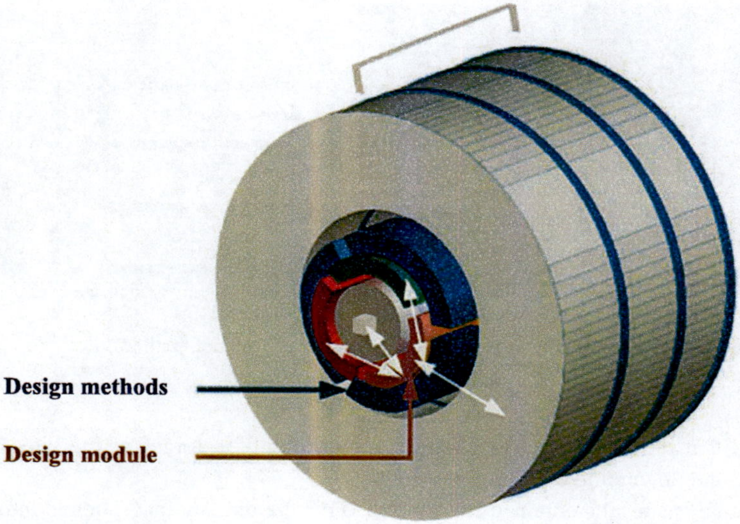

Fig. 5.25 The interactions of the GSE design module with the GSE thinking model and other modules of the GSE approach

The first two possibilities for deriving input information for the GSE design module have already been described in the previous sections. Nevertheless, two examples should briefly recall this.

To 1. The derivation of input information for the GSE design module from the GSE thinking model

If contradictions arise between functions when creating the GSE thinking model using the DeCoDe tools, these serve directly as input information for the GSE design module, as the following example shows.

The data processing function in a PC is realized via a processor. This generates heat. It can disturb the processing function, i.e. the processing of data. Consequently, a new function for heat dissipation is required. This can directly be the input information for a design process, for example for the design of a fan that takes over this heat dissipation function. Thus, when creating the GSE thinking model using the DeCoDe tools for the PC, a function-function matrix generates input information for the GSE design module (Hartmann and Winzer 2011).

To 2. The generation of input information for the GSE design module via the GSE analysis module

Based on the representation of the KitVes system using the GSE model (see also Sect. 4.3), a detailed analysis of the interactions between components using the MTTF

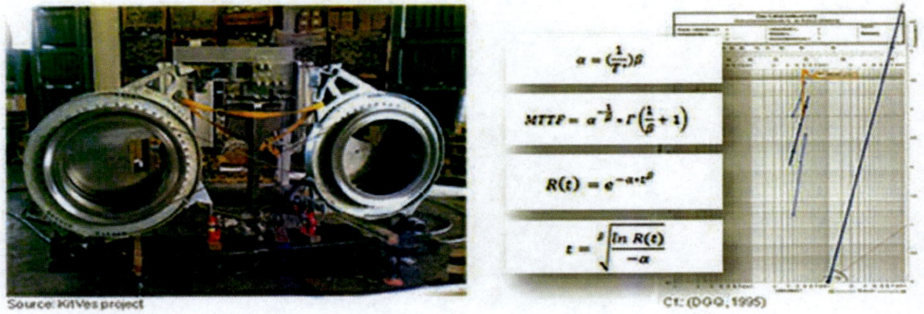

Fig. 5.26 Analysis of the KitVes system using the MTTF (Riekhof et al. 2011)

revealed that the rope is not most stressed by the pulling forces of the kite, but by its winding and unwinding on drums (see Fig. 5.26).

The analysis results obtained from the MTTF were directly incorporated into the GSE design process, where new solution variants for the design of the drums were sought; one of them is shown in Fig. 5.27.

To 3. Deriving input information for the GSE design module, which arises from the GSE goal formation process

This corresponds to the idealized path, i.e., the GSE design module is only initiated when goals have been fixed via the GSE goal formation module. For this variant, selected methods and procedures that can be used here are described below. There are methods and procedures that can be used in the analysis and goal formation module as well as in the design module, such as the ***Delphi method,*** the ***Mind Map,*** the ***Brainstorming*** or the ***Brainwriting,*** to name just a few. Other methods and procedures can be used across the analysis, goal formation, and design module. For example, the ***FMEA*** sequentially enables the analysis of failures, the derivation of goals via the risk priority number, and the derivation of design actions as a result of the design process. The ***QFD*** can also be used for both the analysis and the goal formation and the design. If the roof of the House of Quality shows contradictions between technical features, this is important input information for the design process, which can be initiated with special methods and procedures, such as for example ***TRIZ,*** (Wang et al. 2005).

In summary, it can be assessed that ***the gaining of input information for the GSE design module is possible in a case-specific manner via the following three paths,*** i.e., via:

- the GSE thinking model,
- the GSE analysis module or
- the GSE goal formation module.

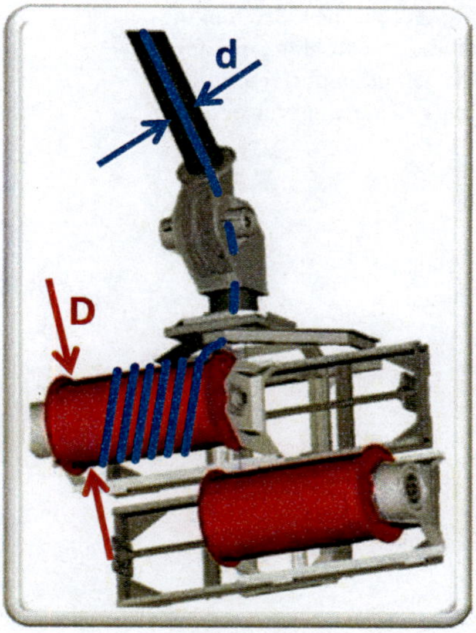

Fig. 5.27 Design solution for the drums of the KitVes system (Riekhof et al. 2011)

If the basic principles of systematic thinking and action, such as ***the basic principle of structuring*** or ***the basic principle*** of ***from the whole to the detail*** are incorporated into the GSE approach, the system should first be analyzed and goals for the design of the system derived before the design process itself begins. However, in practice, this often cannot take place chronologically, especially since—as proven in Sect. 4.1—some analysis methods, such as the FMEA, contain specific analysis, goal formation, and design approaches in one method complex. Thus, the design module of the GSE approach receives input via the GSE thinking model. But it can also receive input information directly via the analysis and also directly via the GSE goal formation module. The input via the GSE goal formation module should be the preferred access to information for the GSE design module. The GSE design module itself also uses special methods and procedures. These can be combined task-specifically using the basic principles of systematic thinking and action. Now the GSE design module, as briefly outlined in Sect. 5.3, can be supported by numerous methods and procedures. Comprehensive explanations on this can also be found in FRANKE, EHRLENSPIEL and MEERKAMM, BENDER and GERICKE, LINDEMANN, GAUSEMEIER and HABERFELLNER (Franke 2002; Ehrlenspiel and Meerkamm 2017; Bender and Gericke 2021; Lindemann 2016; Lindemann et al. 2009; Gausemeier et al. 2014; Haberfellner et al. 2018). For this reason, selected methods and procedures are briefly introduced below, which can be used in the individual sub-steps of the GSE design module.

But first, the steps that are part of this module should be described.

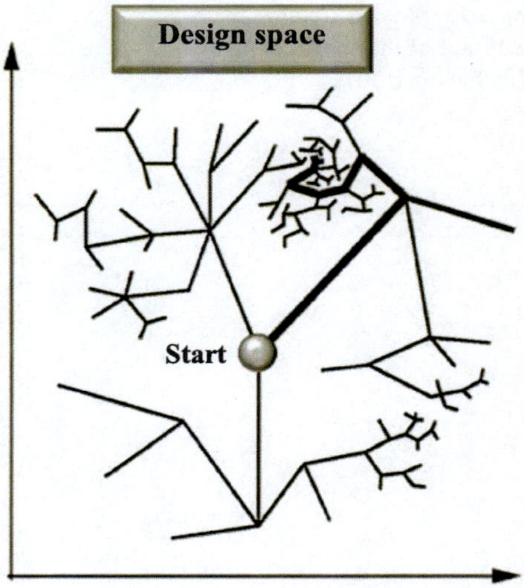

Fig. 5.28 The GSE thinking model as a starting point for the definition of the solution space (Sitte and Winzer 2005)

The GSE design module includes the following sub-steps:

- the definition of the solution space,
- the generation of solution ideas,
- the development, selection, and implementation of solution variants, and
- the evaluation of the implementation process.

The definition of the solution space takes place in strong interaction with the GSE thinking model. This is the starting point for the delimitation of the solution space, as shown in Fig. 5.28.

Building on this, the solution space must be made transparent (see Fig. 5.29), so that the solution ideas to be developed remain manageable.

Only after that is it advisable to select special methods and procedures for developing, selecting, implementing, and evaluating solution ideas and apply them in the GSE design module. This is outlined below for these sub-steps of the GSE design module.

Methods and procedures for generating solution ideas

The classic procedures such as Brainstorming, Brainwriting, and Mind-Mapping are not only suitable for the analysis or goal formation module, they are also suitable for generating new design solutions. If, for example, the goal is to develop the computer for the year 2040, then initial ideas can be generated using the ***progressive abstraction*** or the ***Synectics*** as to what a possible product system could look like. However, this must fundamentally be modeled in its four views, i.e., in the requirement, component, function, and process view.

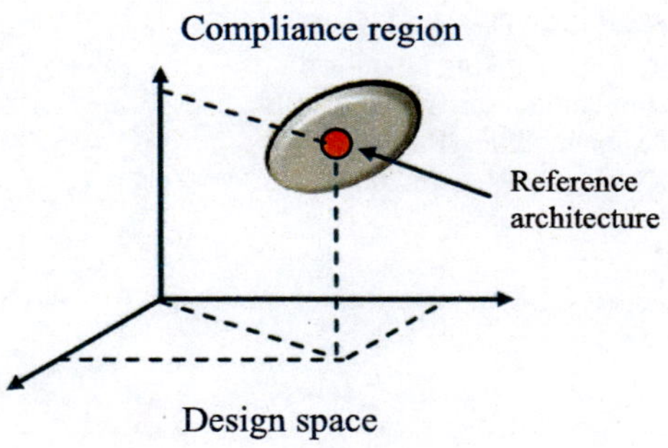

Fig. 5.29 The transparent delimitation of the solution space (Sitte and Winzer 2005)

Methods and procedures for developing and selecting solution variants
Using the the ***635 method*** (Dahms 2010), the individual solution variants can now be roughly evaluated and using the the ***Six Hats*** (Jensen et al. 2000) in detail. On this basis, the best design solution can be selected and the associated GSE thinking model can be quickly specified. If there are contradictions between functions or components, these can be solved using the the ***TRIZ methodology*** according to ALTSCHULER (Yamashina et al. 2002). This results in new or improved solution variants. They are to be evaluated again. The evaluation can also be done using the the ***Delphi method*** (Häder 2009) or by using the the ***Morphological Box*** (Eversheim et al. 2006).

Methods and procedures for implementing and evaluating design solutions
During the implementation and evaluation of the selected design solutions, various methods and procedures can also be used, which have already been used in the GSE analysis and/or GSE goal formation module, such as the the ***Ishikawa diagram,*** which can represent the cause-effect chain of the design solution. The ***FMEA*** is also suitable for systematically eliminating failures found during the evaluation of the design solution. Simulation tools or the ***Rapid prototyping*** as well as modelling and ***DoE*** (Design of Experiments) are methods and procedures that can help evaluate design solutions.

The methods and procedures mentioned or presented in the literature (Ehrlenspiel and Meerkamm 2017; Bender and Gericke 2021; Lindemann et al. 2009; Haberfellner et al. 2018) cannot be clearly assigned to one of the four sub-steps in the GSE design module. By applying the basic principles of systematic thinking and action in the design module, these can be combined depending on the design task. How this can be realized in detail is demonstrated by the following example: As part of the SFB 696 "Logistic on Demand", the goal of subproject B3 was to generate a method workflow to increase the reliability of mechatronic systems in the early phases of product development (Künne and Richard 2009). This is a pure design task. As a result of the research work, the system model for mechatronic systems, which was generated using the GSE thinking model using the DeCoDe tools (see also Sect. 4.3), could be purposefully connected with simulation tools that served the design of the system model. This basic principle is shown in Fig. 5.30.

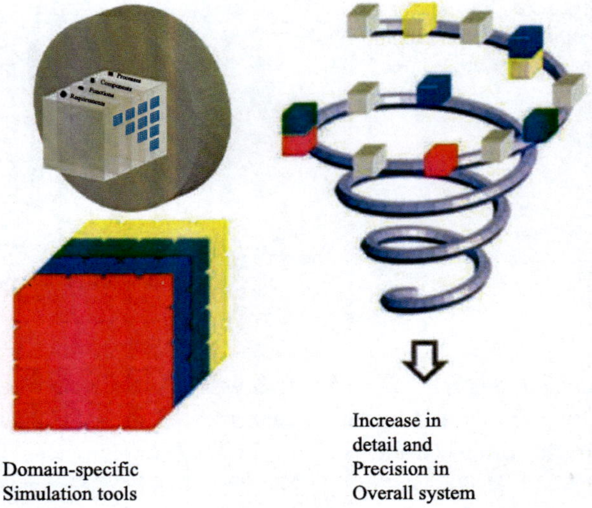

Fig. 5.30 The coupling of the GSE thinking model with simulation tools from the GSE design module. (Based on Künne and Richard 2009)

The basic method workflow, which couples the simulation tools of the GSE design module with the GSE thinking model in such a way that the reliability of the drive can be tested in certain situations, is shown in Fig. 5.31.

This method workflow was used to optimize the drive for a roller conveyor. To do this, the drive was first depicted in interaction with the roller conveyor in the form of a system model using the DeCoDe tools. In the early stages of product development, it had to be proven that the drive could generate a specific torque when carrying a certain load that transports the conveyor belt. The degree of fulfillment of this requirement can only be demonstrated in the early stages of product development using appropriate simulation tools. These results can be found in the following graphic (see Fig. 5.32).

The simulation tool—"ESB"—proves that the realization of the corresponding requirements, i.e., generating torque and starting under full load, becomes possible through the developed drive. At the same time, the simulation was able to prove that the starting torque can be changed by adjusting the rotor bars (see Fig. 5.33).

The torque-slip characteristic (M/s characteristic) of the example machine shows that the starting torque M_A is less than the required nominal torque M_N. This makes it clear that this motor is not capable of delivering the required torque. Instead of replacing the current machine type with a more powerful one, which would lead to even greater oversizing for normal operation, changes to the rotor (deep bar or double cage rotor) can adjust the machine's characteristic curve in such a way that it runs higher in the starting range, thus increasing the starting torque (new starting torque M_A', see Fig. 5.33) and fulfilling the requirement $M_A > M_N$. This change in the rotor bars then had to be depicted again in the GSE thinking model. During the modification of the roller conveyor, a new requirement was added, namely the generation of shock-free operation. This is to be bal-

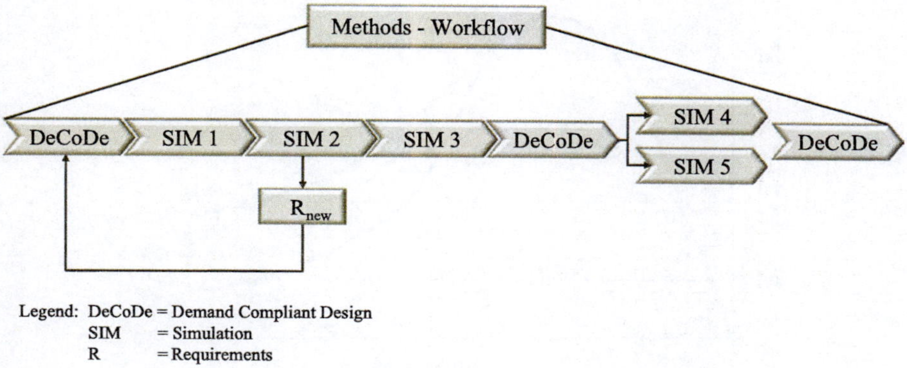

Fig. 5.31 The method workflow for increasing the reliability of mechatronic systems—a planned interaction of the GSE thinking model created and specified using DeCoDe tools and simulation procedures of the GSE design module. (According to Rosendahl et al. 2009)

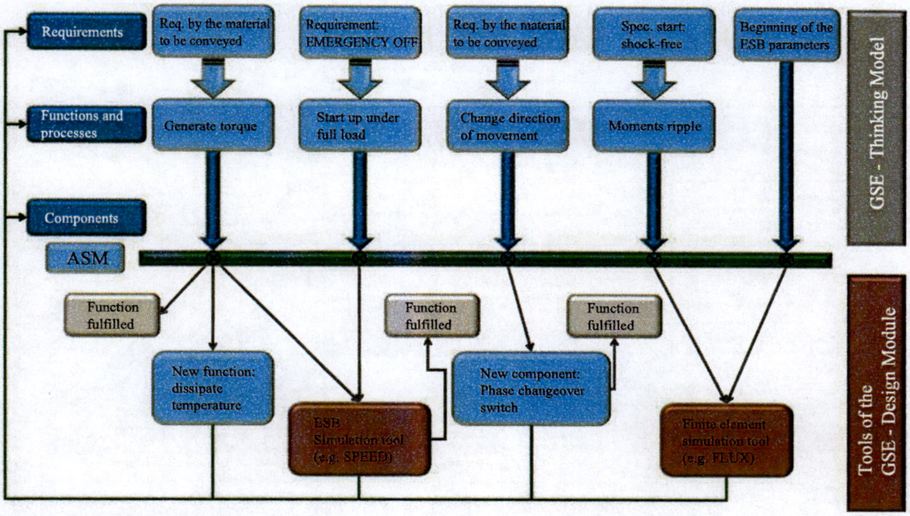

Fig. 5.32 The combination of simulation tools of the GSE design module with the GSE thinking model according to (Rosendahl et al. 2009) (work results interim presentation SFB)

anced with the torque ripple over the drive. For this, an appropriate simulation tool—FLUX—is necessary to check whether exactly this requirement has been realized in the early stages of product modification.

The step sequence outlined using the example can be generalized using the following workflow from Fig. 5.34.

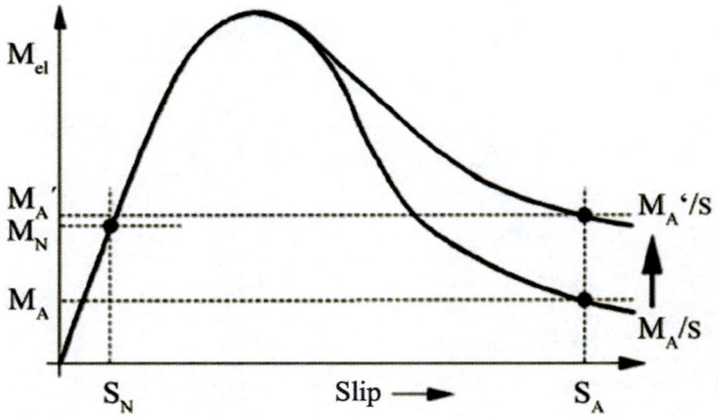

Fig. 5.33 Torque—slip characteristic—result of the simulation of the behavior of an asynchronous machine. (According to Künne and Richard 2009)

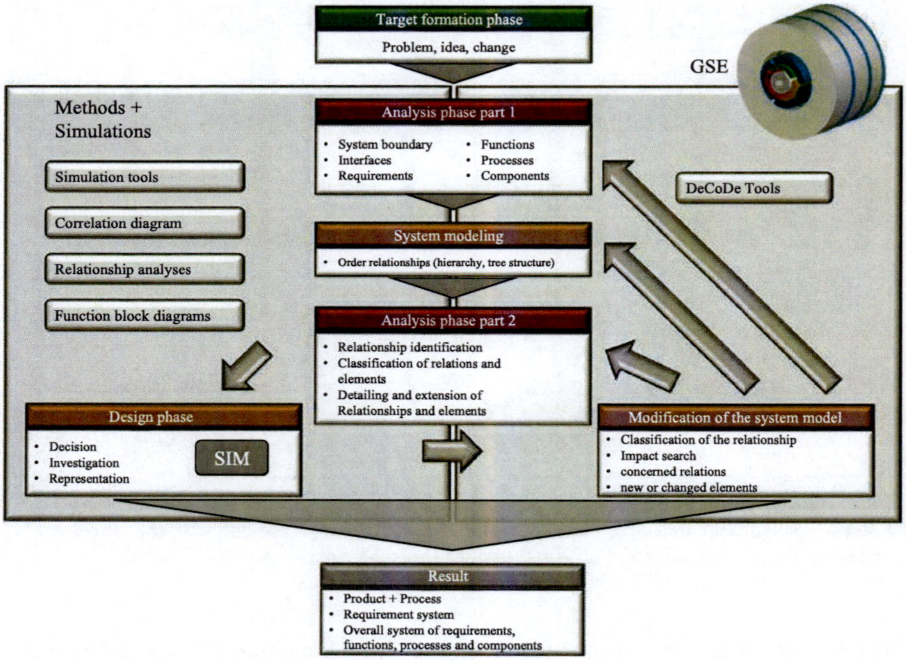

Fig. 5.34 The method workflow using the GSE approach. (Based on Schlund and Winzer 2010)

This workflow illustrates that an image of the drive was initially generated using the GSE thinking model. This provided corresponding inputs for the use of simulation tools. The results of the simulation tools, such as the change in the rotor bars, subsequently led to new requirements or changes in the system model. New requirements in turn required simulations and led to changes in the GSE thinking model. Thus, it was possible to demonstrably prove why and for what reasons which functions, components, and processes change in interaction with the requirements during the design process. The example shows how simulation tools can be meaningfully coupled with the GSE approach as design instruments in the early stages of product development. This is only possible if the information or data structure of the simulation tools is systematically, structured, and standardized transferred into the thinking model.

For this reason, the GSE thinking model with its four views, i.e., the requirement, component, function, and process view, was used in this case. It was shown that this is a viable way to design mechatronic product systems in a traceable, transparent manner in transdisciplinary teams.

In summary, it can also be stated for the GSE design module that there are numerous methods and procedures that generate input information for the GSE design module on the one hand and specifically support the individual steps in the GSE design module on the other hand. These steps are not cyclical, but are determined by the specific design task and the basic principles of systematic thinking and action used. Consequently, the methods and procedures in the GSE design module must be linked iteratively with each other in order to observe the basic principles of systematic thinking and action, in particular the basic principle of critical reflection, but also the basic principle from the rough to the detail and the basic principle of neutrality. The combination of design methods with the GSE thinking model can be expanded as desired and modified depending on the industry or use case. This is proven by various studies by OTT, RIEKHOF, and MÜLLER (see Schlund 2011). These sources prove that the results of these methods systematically complete the system model, provided they are successfully stored in the thinking model of the GSE approach. Within the framework of the GSE design module, it is not necessary to change the entire system model, but only the corresponding parts need to be changed or specified. This possibility is illustrated by the example of the seat backrest presented in Sect. 5.2, where the "seat backrest adjustment" subsystem was designed. The GSE approach necessarily requires that in addition to the changes to a subsystem, necessary adjustments to the overall system must also be examined and implemented. The GSE thinking model allows a quick check and selection of the connection of the subsystem with the overall system. Consequently, the design module is in constant interaction with the thinking model of the GSE approach.

How this design process can be systematically planned, implemented, and executed will be explained in the next section.

5.4 The GSE Project Management Module

In Sect. 3.3, it was shown that within the framework of the GSE approach, the GSE analysis, the GSE target formation and the GSE design module are to be planned, realized, and controlled in their logical sequence through the ***GSE project management module***. In Sects. 5.1 to 5.3, the description of the universal standardized modules of the GSE approach in their basic interaction with selected methods and procedures was carried out. However, how these modules are to be combined in their ***logical sequence*** using the GSE project management module and its selectable methods and procedures will be outlined in this section in principle. It is the task of the GSE project management module, as shown in Fig. 5.35, to plan and successfully implement the interactions between the GSE thinking model and the standardized, universal modules of the GSE approach in a problem-specific manner. Thus, this module includes all fields of project management, as described by HABERFELLNER. He sees project management as the sum of all organizational and dispositive actions used for planning, management, monitoring, and control of problem-solving projects in terms of content, time, and cost (Haberfellner et al. 2018) This must include the handling of resources. This view coincides with the systemic project management of MÜLLER (Müller et al. 2012). He considers the project as "projecting" from a current state to a state to be achieved in the future according to the target setting (Müller et al. 2012, p. 54). The planning and execution of all necessary activities to achieve the goal is given by the project. Exactly at this point is the establishment of a ***connection between the SE and the project management,*** which is demanded in the SE and only partially realized (Welge et al. 2014; Scheithauer 2014), possible. The general goal of the SE is to solve problems (see also Sect. 2.4). Since each problem is unique in its specific kind, the problem-solving process must be systematically adapted using the SE, prepared, and the resulting activities planned, executed, and controlled by means of project management. This also meets the requirement of DIN 69901-5, which defines the project as "an undertaking that is essentially characterized by the uniqueness of the conditions in their entirety, such as target specification, temporal, financial, personnel, and other limitations, a demarcation from other projects and a project-specific organization" (DIN 69901-5 2009). According to MÜLLER, the activities in a project include: "both all technical and all administrative, steering, and auxiliary activities that are necessary. Compliance with temporal, financial, personnel, and other boundaries is explicitly a requirement" (Müller et al. 2012, p. 56). This complements the problem-solving process in the SE. Consequently, the GSE project management module has exactly this function to fulfill. It can be realized with various methods and procedures. Their suitability will be considered in the following section for the phases of planning, implementation, and control of GSE projects.

Thus, project management is a very important component of the GSE approach, which must ensure an efficient problem-solving process. The phases of project management, i.e., the planning, implementation, and control phase, are not additively sequenced

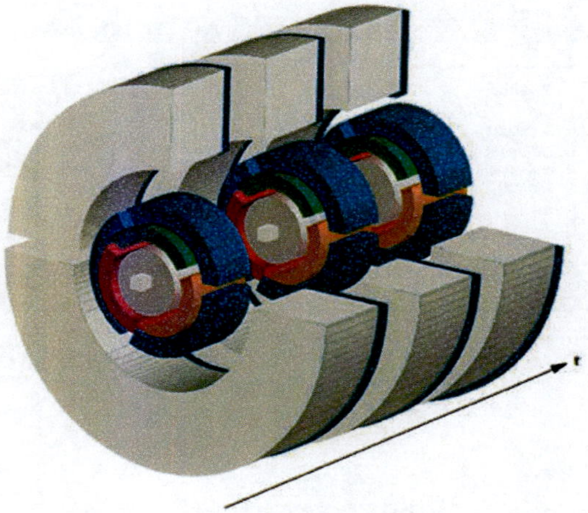

Fig. 5.35 The dynamics of the GSE approach planned, controlled, and implemented via the GSE project management module

one after the other. They overlap, as Fig. 5.36 roughly sketches, over time (Jentsch et al. 2012; Mistler 2018; Eversheim and Schuh 1999b; Lehner 2001; Müller et al. 2012). The project definition phase, which the classic project management envisages, (see also Sect. 3.3) is not necessary, because the project is defined via the GSE thinking model and the project goals are fixed using the GSE goal formation module. The project realization phase and the project completion as well as the follow-up can, as MÜLLER suggests, be part of the project implementation (Müller et al. 2012). With regard to agile project management, MISTLER (Mistler 2021) shows that iterations can be used through the iterative application of the goal formation and analysis module. Thus, in the problem-solving process of project management, the goals can be constantly questioned, analyzed, and adapted (Mistler 2021). From this insight, it follows that the type of project management is determined by the application of the GSE modules.

In each individual phase of project management from Fig. 5.37, information from the GSE thinking model as well as the GSE analysis (red arrow), the GSE goal formation (green arrow), and GSE design module (brown arrow) must therefore be included in a time-differentiated manner. The same applies in the opposite direction, i.e., the information from the phases of project management has a planning, controlling, and controlling effect on the problem-specific, GSE module-combining application of the methods and procedures (see Fig. 5.37).

Therefore, as Fig. 5.37 illustrates, project management is about planning, implementing, and controlling the interaction between the individual modules of the GSE approach, namely the GSE analysis, the GSE goal formation, and the GSE design module, as well as between the thinking model of the GSE approach in their logical sequence. The basic principles of systematic thinking and action (see also Sect. 2.3) must be integrated. The basic principle of going from the general to the detail, which HABERFELLNER favors, for example, allows to maintain an overview of larger projects (Haberfellner et al. 2018).

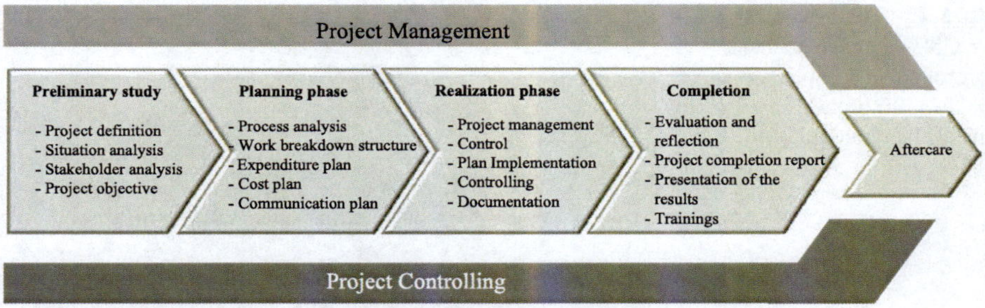

Fig. 5.36 The phases of project management. (Based on Geiger and Pifko 2009)

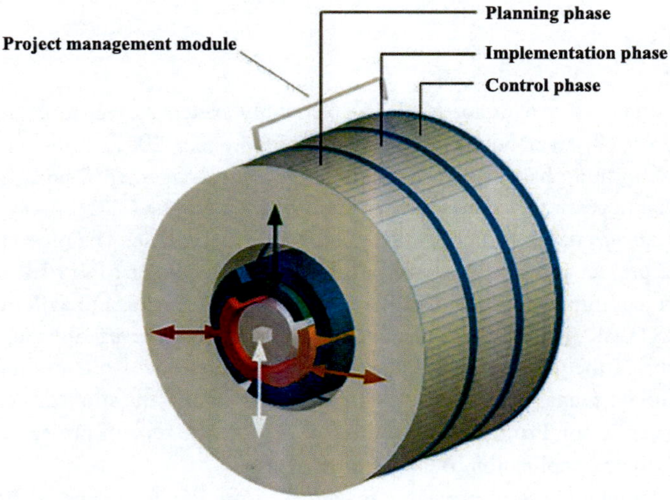

Fig. 5.37 The interaction of the phases of project management with the GSE modules and the GSE thinking model

This also applies to the implementation of the basic principle of minimal models. The latter can, for example, be coupled with the documentation of project results. As already mentioned, the change in the GSE thinking model during the course of the project should be recorded as minimal documentation. In principle, the basic principles of systematic thinking and action presented in Sect. 2.3 depend on the specific problem solution that is to be realized using the GSE approach (see also Chap. 6). These must be selected in a problem-solving-oriented manner during the planning phase and must be strictly observed during project implementation and control. Since these activities in the individual phases of the GSE project management module are to be controlled (managed) in conjunction with various ***basic principles of systematic thinking and action,*** the term "project management" was deliberately chosen for the designation of this GSE module.

The following will first present the ***individual phases of the GSE project management module and briefly introduce possible methods and procedures.*** This will be followed by a description of procedures that are to be applied across all phases of the GSE project management module, as interdisciplinary teams are to be controlled and directed in each of these phases and across phases via information and communication procedures.

The phases of project management in the GSE approach and possible methods and procedures

In the project managementof the ***project planning phase***, it must be decided which modules of the GSE approach will be used when, how, with which methods and procedures and with which capacities. This can determine whether the project should have a classic, agile, or hybrid approach. It is also determined how and when the GSE thinking module will be used or specified and which principles of systematic thinking and action should be applied (see Fig. 5.38).

As a result of the planning phase, a bar chart can be created, as Fig. 5.39 shows in principle.

In this figure, the GSE modules were only coupled with one basic principle of systematic thinking and action—the basic principle of discursive action. Only one planning method, the ***bar chart,*** is used. Other planning methods are the ***network diagram-*** or ***list technique.*** These methods or procedures are assigned to the ***schedule planning*** as a special type of planning, as Fig. 5.40 shows. In addition, Fig. 5.39 exemplifies how iterations could be generated through modular application.

The abstraction of the GSE modules can also be demonstrated using the Scrum approach by SCHWABER and SUTHERLAND (Schwaber and Sutherland 2017) (see Fig. 5.40).

HEINKE and MISTLER (Heinke and Mistler 2019) show that the respective step sequences of the Scrum approach can be transferred to the GSE modules. This means that it is shown when an analysis, target formationor design takes place. Thus, the approach of Scrum and the use of methods and tools can be placed in a larger context. In addition, this abstraction creates comparability between other approaches.

In planning GSE projects, in addition to schedule planning (see Fig. 5.41),the following ***types of planning*** can in principle be used:

- the project structure plan,
- the effort estimation,
- the resource planning,
- the personnel planning and
- the cost planning.

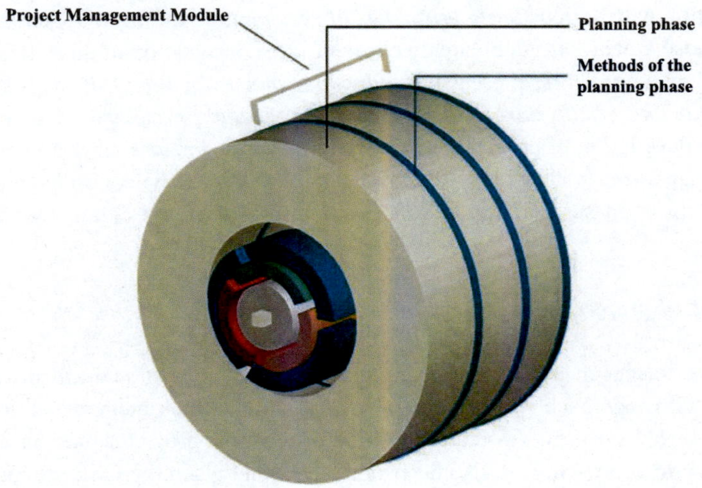

Fig. 5.38 The project planning phase and possible methods and procedures in interaction with the modules of the GSE approach

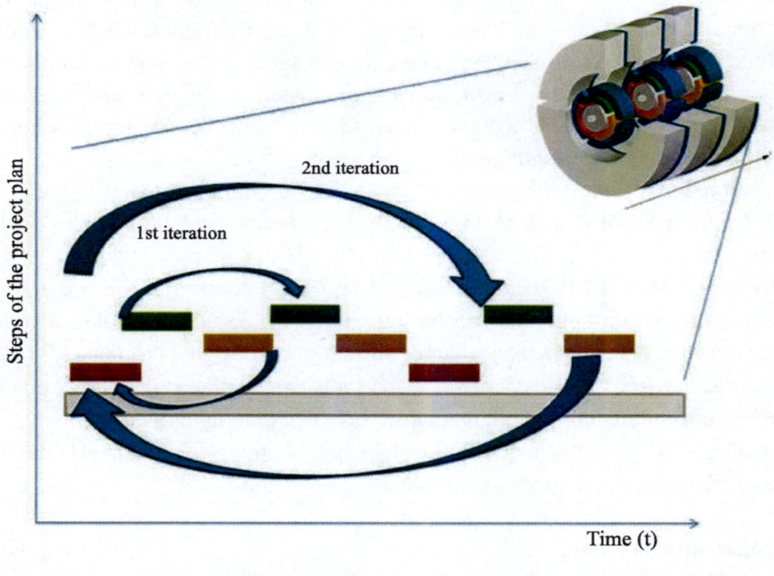

Fig. 5.39 Example of project planning using the bar chart in the GSE approach

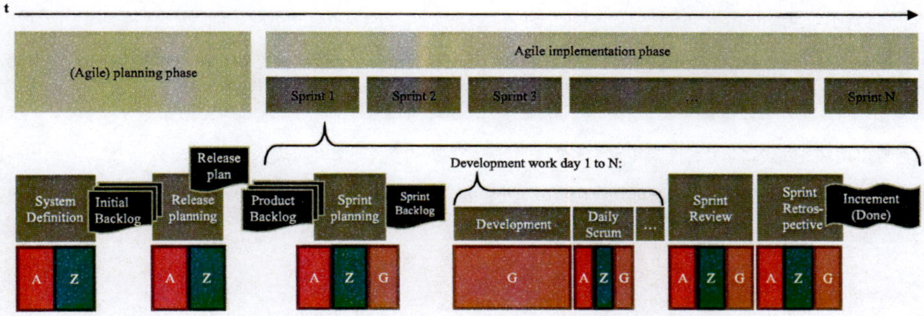

Fig. 5.40 Abstraction of the GSE approach to the Scrum approach. (According to Heinke and Mistler 2019)

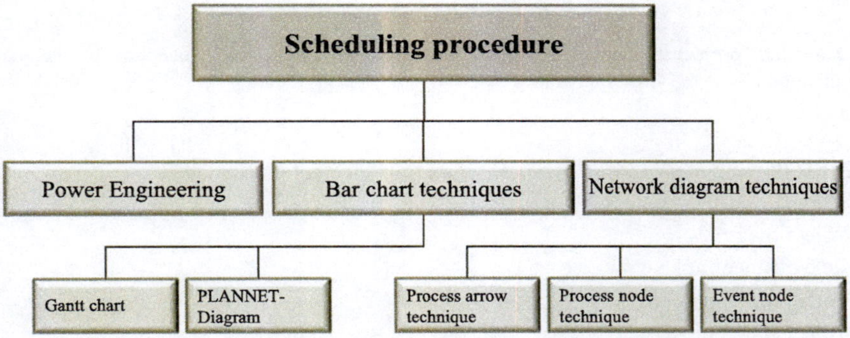

Fig. 5.41 Overview of schedule planning procedures. (According to Zielasek 1995, p. 158)

The use of the types of planning depends on the size of the GSE project. Each type of planning uses specific methods and procedures, as exemplified by schedule planning (Haberfellner et al. 2012; Jentsch et al. 2012).

After the project planning has been completed, the project implementation should begin. However, projects are often planned in a sliding manner, i.e., project planning and project implementation are carried out sequentially.

The goal of ***project implementation*** is the constant search for the best possible way to implement the project plan within the framework of the intended GSE project (see Fig. 5.42).

Project implementation includes the ***project documentation***. In principle, project implementation is permanently linked with the ***project controlling***. During the implementation of the problem-solving project, it must be constantly checked through a goal-actual comparison to what extent the project goals are actually being achieved and to what extent the project plan is being adhered to or needs to be specified.

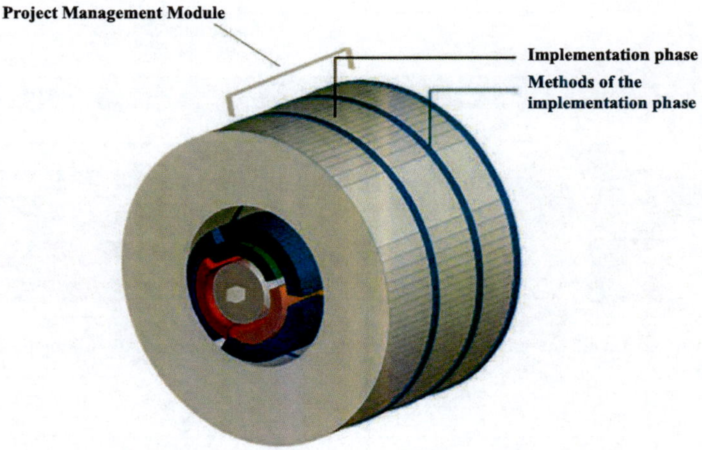

Fig. 5.42 The project implementation phase and possible methods and procedures in interaction with the modules of the GSE approach

The following ***methodsand procedures support project implementation:***

- Project overview,
- Kick-off meeting,
- Project meeting,
- Project status reports,
- Project review,
- Test reports,
- Survey,
- Risk assessment and
- Public relations.

Project meetings, status reports, etc. are to be supported with data and facts. For this, the specific project progress, i.e., the status of the implementation of the planned, temporally and logically coupled GSE modules, must be checked. The following methods and procedures are helpful for this:

- Interview,
- Questionnaire,
- Continuous observation,
- Self-recording,
- External recording and
- Multi-moment recording.

For example, through daily self-recording, it can be traced how a project, which is to be implemented using the GSE approach, was planned, implemented, and realized. Ideas for improving the process, but also problems that arose, for example, when setting up or updating the GSE thinking model, can also be sketched. But as HEINKE and MISTLER (Heinke and Mistler 2019) show, other methods, especially in agile project implementation, can help in the GSE and be used in connection with the other modules: defining persona, user stories, MoSCoW prioritization, cost of delay prioritization, moderated workshops, burn down charts, planning poker, release plan, and task boards (Heinke and Mistler 2019). Fig. 5.44 once again fundamentally outlines the close interlocking of project implementation and project controlling.

Project control is a permanent process that extends over all project phases (see Fig. 5.43). As already highlighted, it serves to monitor the project goals and the project plan. The size of the GSE project determines the methods and procedures of project control to be used.

Project control should be designed to be quantitatively and qualitatively traceable. For this reason, the following things are prerequisites for efficient project control:

- Clearly defined goals that are written down, comprehensible, measurable, non-contradictory, fully realizable, known to all participants, and accepted. This must be guaranteed by the GSE target formation module and the GSE thinking module.
- A well-structured project plan with defined project control points and a complete project documentation.

If these conditions are met, project control can be successfully carried out.

Methods and procedures that support ***project control*** are:

- the project meeting,
- the project inspection,
- the survey,
- the project audit and
- the project review.

Project control within the framework of the GSE approach can be very successful if it is determined for each project phase, with which methodsand procedures the task should be solved in a problem-specific manner. At the fixed key points, the system changes are to be documented via the GSE thinking model. Consequently, at the end of the project duration, it can be precisely traced why and why the GSE thinking model of the specific system was changed and exactly this solution was favored and implemented. This abstract approach will be practically illustrated in Sect. 5.5 using selected examples.

But before this happens, methods and procedures will be discussed that can be used across phases, i.e., across the phases of project planning, implementation, and control.

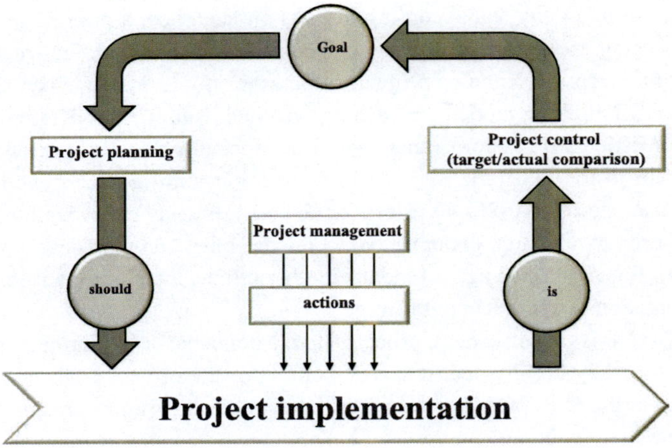

Fig. 5.43 Interaction between project planning, control, implementation, and control, which should be considered when planning GSE projects. (According to Eversheim and Schuh 1999a)

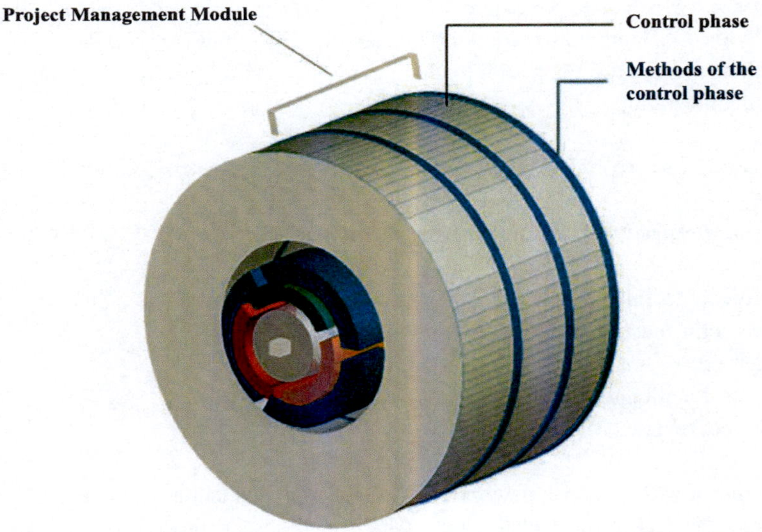

Fig. 5.44 The project control phase and possible methods and procedures in interaction with the modules of the GSE approach

Methods and procedures of project management applicable across phases in the GSE procedural concept

Basically, information needs to be exchanged and communicated in all phases of project management. But decisions also need to be made. These methods and procedures can be used ***across phases in the GSE project management module***, as Fig. 5.45 illustrates.

In all project phases, communication is required both within and outside the project team (see Fig. 5.45 and 5.46).

The methods and procedures summarized in Fig. 5.47 must ensure that the communication results of the various groups working on the GSE project match.

The experience from the KitVes projects (Riekhof 2011) and PromeSys (Winzer 2012) as well as subproject B3 in SFB 696 (Schlund 2011) show that it is particularly helpful when these communication techniques are coupled with the GSE thinking model. This means that when changes occur, it must be determined in the result of the video conference, for example, which relations between the components or functions should be changed in detail. (Fig. 5.48)

In problem-solving projects, always ***decisions*** have to be made. The following methods are recommended for this purpose (Kuster 2011):

- the morphological box,
- bionics,

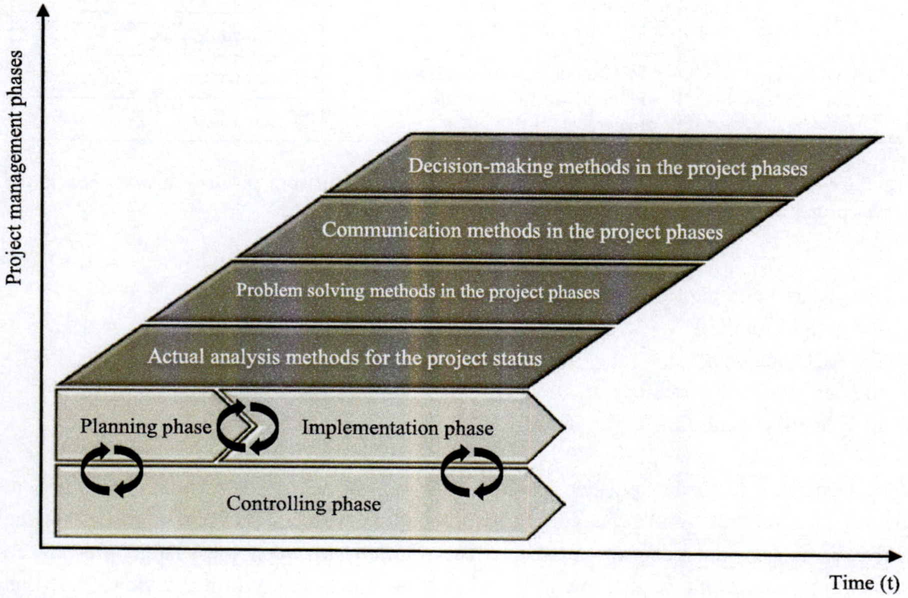

Fig. 5.45 Overview of methods and procedures that can be used across phases in the GSE project management module

Fig. 5.46 Targeted communication in teams (slide from a workshop in the KitVes project)

Character of the situation	Project requirement	Suitable communication channel
High uncertainty	Concrete need for information, concrete need for regulation	Rules
		Forms, reports, plans
↕	↕	E-mail
		Chat Room
		Phone
		Telephone conference
		Video conference
high ambiguity, poor structured situation	Goal definition, clarification of the situation, conflict resolution, assessment	Direct contact
		Group meeting

Fig. 5.47 Methods and procedures to support communication (work result of a workshop in the KitVes project)

- the pro and con game,
- the Delphi method,
- the ABC analysis,
- the decision tree as well as
- the portfolio method.

If it is possible to further perfect the GSE thinking model (see also Sect. 4.4), parts of the decision methods and procedures can be coupled with it. This presupposes that the GSE thinking model not only represents the relations between the system views, i.e., the requirement, component, function, and process view, and within the views, but that these relations are evaluated or attributed via feature values. Since this is currently not

Fig. 5.48 Moderation possibilities depending on the premises (work result of a workshop in the KitVes project)

the case, the projects that are implemented using the GSE approach and coupled with the above-mentioned decision methods and procedures must be documented.

To ensure traceability and transparency, the decision results must be recorded in the GSE thinking model.

The GSE thinking model and GSE procedural concept do not yet automatically match each other. During the course of project implementation and control, the results of the procedural concept must therefore be continuously incorporated into the GSE thinking model and documented for each project phase. The project documentation thus includes the views of the GSE thinking model at different points in time of project progress.

For the cross-phase control of information in GSE projects, the software solutions presented in overview in Table 5.2 can be used.

However, it should be noted with the software tools shown in Table 5.2 that a continuous connection of the IT tools with the GSE thinking model is often difficult to realize. In MISTLER (Mistler 2021), it is shown that the software iQUAVIS is suitable for project management and documentation of the project status via the GSE thinking model (see Sect. 3.3).

In summary, it should be emphasized that the methods and procedures that are universally applicable in project management can also be used within the framework of the GSE approach. It became clear that the project results must be fixed again and again in the system model, i.e., in the GSE thinking model, during the course of the project. Thus, time-determined documents of the GSE thinking model are created during the realization of the GSE procedural concept. How this can be realized in detail will be outlined in the following Sect. 5.5.

Table 5.2 Project management—Guide for planning, monitoring and controlling development projects (Burghardt 2002)

Features / Procedures for project management	Planning data				Actual data				Functions					Techniques				
	Dates	Personnel expenses	Operating resources	Costs	Deadlines	Personnel expenses	Operating resources	Costs	Network planning	Resource planning	Multi-project planning	Cost/turnover	Further allocation	Process network	Batchoritted	Diagonally oriented	Mainframe	Client/Server
EPISTEL		x		x		x		x		(x)			x	x		x	x	
MINIPLAN	x	x	x		x	x			x	x	x					x		x
MS PROJECT	x	x	(x)	(x)	x				x	x	(x)	(x)				x		x
PAC	x	x	x	x	x	x	x	x	x	x	x			(x)		x	x	
PAUS	(x)	x		x		x		x				x	x	x		x		x
SAP R/3 PS	x	(x)	x	x	x	(x)	x	x	x			x	x	x		x		x
SINET-STD	x	x	x		x				x	x				(x)		x	x	
SIPUS	x	x			x	x			x	x	(x)			x		x		x

Legend: x applicable (x) partially applicable

5.5 The Interaction of the Modules of the GSE Procedural Concept and the Consequences for System Modeling

In Sects. 5.1 to 5.4, the individual ***modules of the GSE approach concept*** were presented. In the approach, interactions between the GSE analysis, the GSE goal formation, the GSE design, and the GSE project management module could be outlined. The following two examples are intended to illustrate how all GSE modules of the procedural concept can interact with each other, modified case-specifically.

As already outlined in Sect. 5.3, it was the task of the project "KitVes" to develop a new product system for the use of wind energy and to assess and control the resulting risks.

This was done for:

- the ***drive,***
- the ***mechatronic systems,***
- the ***KitVes system***
- and the ***organizational model of a shoe production.***

The same proof is still missing for the GSE procedural concept. It should be provided in this section for technical systems. These are:

- the low-risk development of the ***KitVes system*** and
- the process of field data feedback into the design and development process using ***mechatronic systems***.

Subsequently, the presentation will be made for a socio-technical system using the example of shoe production.

The interaction of the GSE thinking model and the GSE procedural concept is only roughly described in this section in order to illustrate the effects of the approach on the system model and possible improvements of the GSE approach. The synergies between the two should be the focus of Chapter 6 in detail.

Example 1: Presentation of the interacting modules of the GSE procedural concept for the KitVes system

As already outlined in Sect. 5.3, the task of the "KitVes" project was to develop a new product system for the use of wind energy and to assess and control the resulting risks[1]. A KitVes system to be used on a ship served as a solution variant. For this purpose, high-altitude winds at 500–2000 m were used, with the wind energy being converted into electrical energy by the KiteVes system. The basic principle is shown in Fig. 5.49. ◀

The aim of the project was to design the KitVes system for energy generation on ships in such a low-risk manner that all phases of the product life cycle are taken into account. Thus, at the beginning of the project, the idea arose to use the GSE approach to create a common GSE thinking model for the transdisciplinary development team (see also Sect. 5.3) and on the other hand to capture, systematize and map the requirements for the KitVes system over the phases of the product life cycle the requirements, as illustrated in Fig. 5.50.

At the Quality Gates, the newly emerged, changed or dropped requirements should be integrated into the GSE thinking model of the KitVes system, and the interaction between the components, functions and processes should be specified. However, it soon became apparent that the KitVes system alone consisted of over 200 components and the current computer support for creating the GSE thinking model reached its limits. For this reason, the subject of investigation was narrowed down. The application of basic principles of systematic thinking and action, such as the basic principle from rough to detail and the basic principle of minimal models, helped in this. Thus, the international team

[1] www.kitves.com

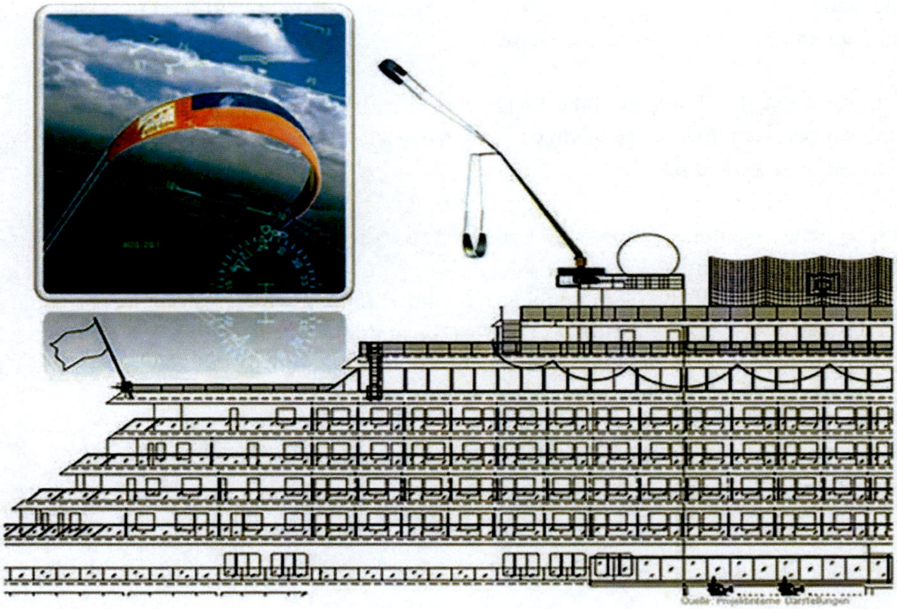

Fig. 5.49 The overall system for energy generation using the KitVes system on the ship. (Based on Riekhof et al. 2011)

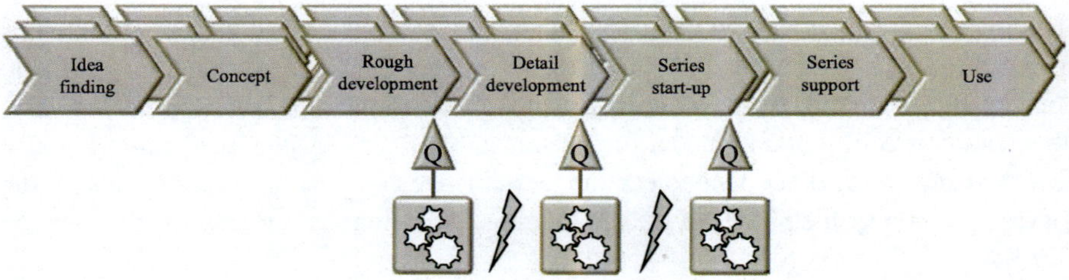

Fig. 5.50 Integration of methods over the product life cycle. (Based on Hartmann et al. 2011)

focused on proving that the kite can indeed convert wind energy into electrical energy and that this happens reliably and with low risk. As a result of the goal specification (first iteration loop in the GSE procedural concept), the subject of investigation was narrowed down and the considered subsystem was outlined (see Fig. 5.51).

After the goal of the reliable design of the KitVes system has been narrowed down to the risk assessment of the kite-steering components, it was now the task of the project team to select methods and procedures that are suitable for risk analysis, assessment and minimization. The methods to be selected should be able to be coupled with the GSE

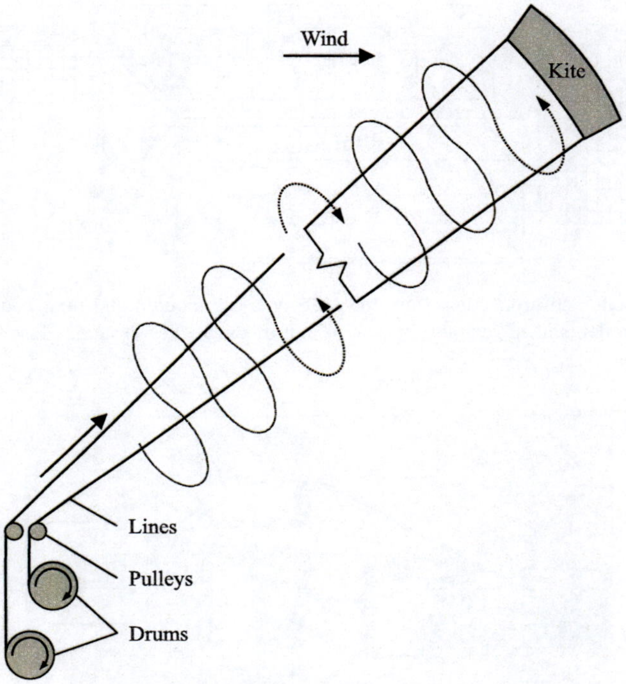

Fig. 5.51 Delimitation of the subject of investigation for risk assessment of the KitVes system on the ship. (Based on Hartmann and Winzer 2011)

thinking model, which was created with the DeCoDe tools. This proof of coupling was achieved by connecting the MTTF with the GSE thinking model for the KitVes system in Sect. 5.3. Another requirement was that the methods and procedures to be selected should be essentially known to the international team, i.e. that all team members should be able to carry out the risk analysis, assessment and design of the Kite-Steering Unit without much training effort. Accordingly, the methods and procedures for risk analysis and assessment shown in Fig. 5.52 were selected. They can be assigned to the GSE analysis module.

Basically, the GSE thinking model provides input for the respective method. Its input, in turn, are the method results (see Fig. 5.52).

In a next step, a logical sequence of combinations of the methodsand procedures for risk analysis and assessment as well as for risk minimization had to be developed as a basic consideration. This is part of the project planning. Fig. 5.53 illustrates the basic principle of the ***logical combination of methods and procedures*** in risk analysis, assessment and design for the Kite-Steering Unit in conjunction with the GSE thinking model.

In a subsequent step, the project team planned the ***temporal*** logical ***sequence*** of the combination of the individual steps. This is also part of the project planning.

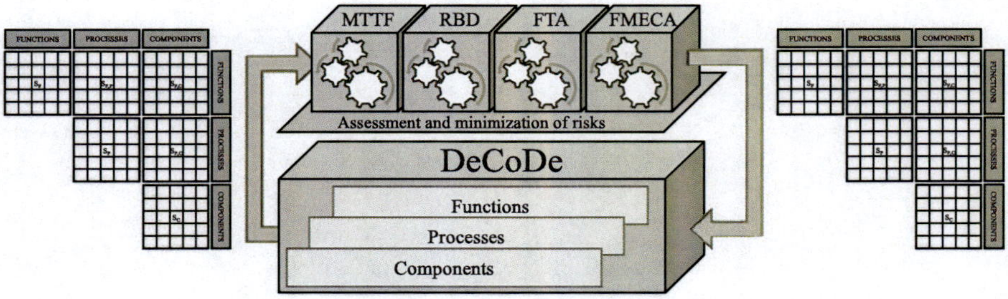

Fig. 5.52 Methods and procedures from the GSE analysis module and their coupling with the GSE thinking model, shown for the low-risk design of the KitVes system. (Based on Hartmann et al. 2011)

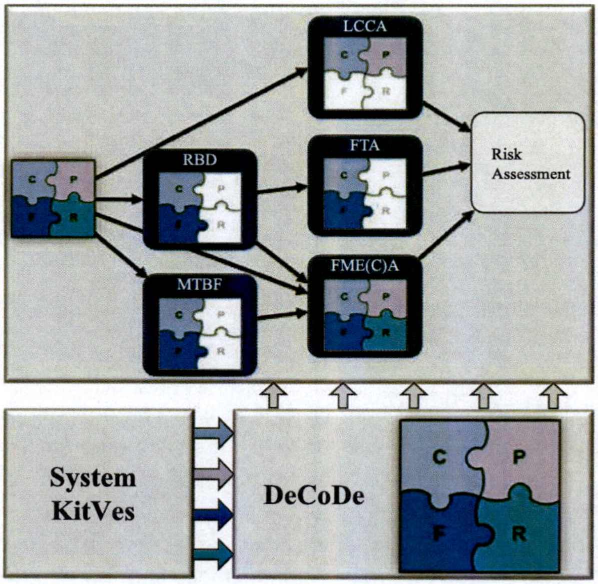

Fig. 5.53 Logical coupling of the methods and procedures for risk analysis and assessment, shown on the KitVes system (Hartmann et al. 2011)

Fig. 5.54 shows how, using simple project planning methods, the temporal logical sequence of the coupling of GES analysis, GSE goal formation and GSE design modules can be carried out. In several workshops, as already described in Sect. 4.3, the individual views of the GSE thinking model for the KitVes system were then generated. For this reason, it is excluded in this section, because this GSE thinking model of the KitVes system (see also Sect. 5.3), as shown in Fig. 5.55, is input for the application of the methods for risk analysis and assessment of the Kite-Steering Unit.

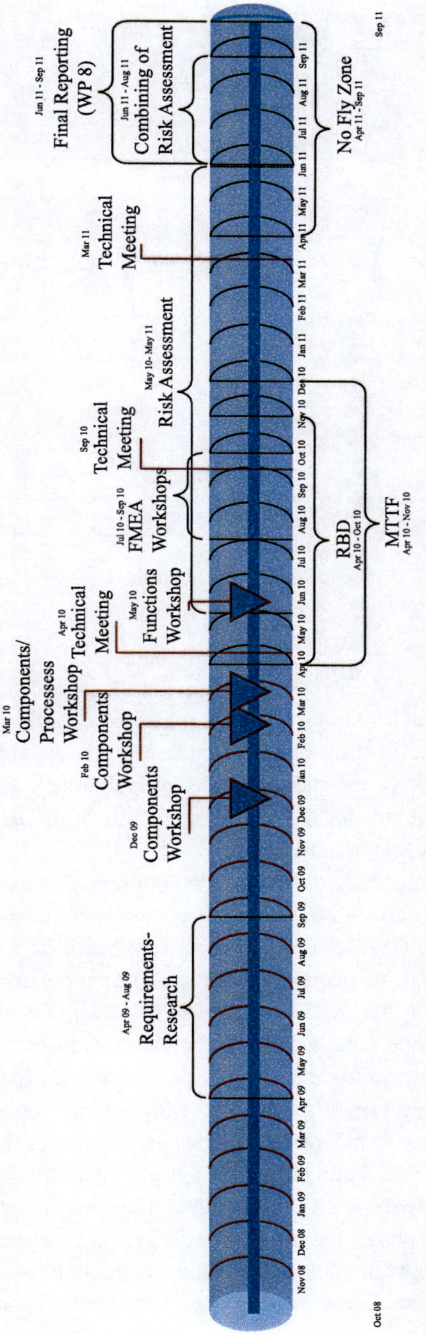

Fig. 5.54 Project planning for the low-risk design of the KitVes system (KitVes project, working document)

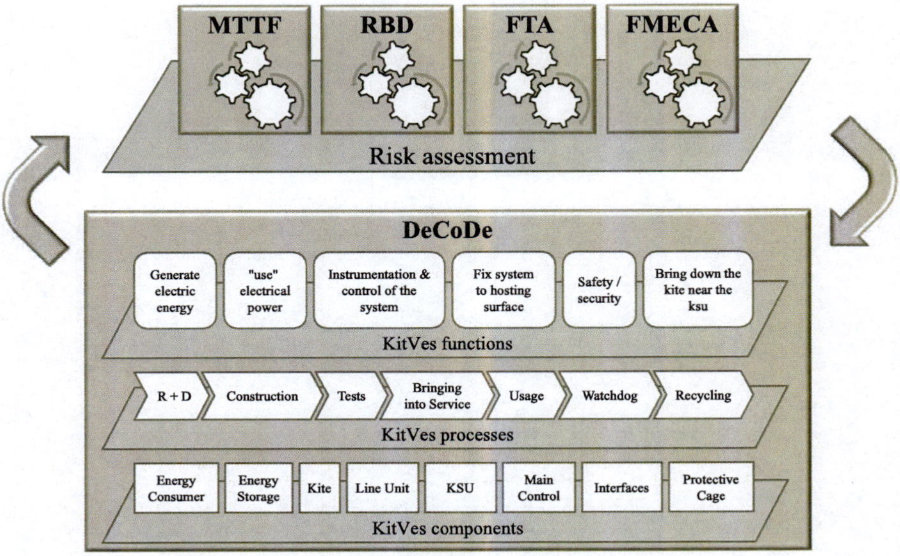

Fig. 5.55 Coupling of the GSE procedural concept with the GSE thinking model, shown on the KitVes system. (Based on Hartmann et al. 2011)

In evaluation of this, the Kite-Steering Unit was designed in such a way that the risks could be minimized. This in turn led to changes in the GSE thinking model. Each design solution changes the image of the original Kite-Steering Unit. Through the application of the life cycle cost analysis, the individual design solutions could now be compared and evaluated, and decision aids for the further perfection of the KitVes system for ships could be provided to the development team.

The project development team, which was composed of representatives from various countries of the European Union, had the opportunity to develop solutions through the joint application of the GSE approach. These were based on a unified and standardized system model. Through this certainly very complex modeling process, it could be ensured that the elements of the KitVes system were clearly designated, described and understood and applied in the same way by team members from the most diverse scientific disciplines. In the course of the application of the methods for risk analysis, assessment and design, it became clear that the effort in data collection can be minimized. Once captured data can be stored in the system image and be immediately available for further use. However, this requires that on the one hand the system images via the DeCoDe tools always refine the GSE thinking model and on the other hand are archived at certain times per project phase. By comparing the system models at each point in time, which is currently only possible manually and not yet computer-supported, the proof of risk minimization can be made comprehensible. The system change is also traceable, i.e. the question of why and why the Kite-Steering Unit had to be changed. This is very quickly possible and well prepared for people not involved through this approach, as the workshops with the EU Commission prove.

In summary, the example of the KitVes system clearly illustrates how the individual modules of the GSE procedural concept interact and can be planned, implemented and realized project-specifically using the GSE project management module. Thus, through the application of the modules of the GSE procedural concept, a specific approach is created, which is precisely adapted to the special problem solution, here the design of a low-risk KitVes system.

Of course, resource and capacity planning was also involved in solving the task. This was omitted in the presentation of the facts in Fig. 5.51 in order to sketch the essentials, i.e. the problem-specific combination of GSE target formation, GSE analysis and GSE design modules, shows Fig. 5.52, that by applying the basic principle from rough to detail or the basic principle of multiple usability, two goal formation phases were switched shortly after each other. Since the risk-minimizing design of the KitVes system over the product life cycle would have exceeded the project framework, a target specification was thus made (2nd phase). It could be estimated in the context of the workshop that the Kite-Steering Unit is the subsystem with the highest risk. For this reason, risk minimization was specifically formulated as a goal for this unit. Consequently, in the further steps during the project duration of two years, the risk analysis, the assessment and the change—i.e. the low-risk design of the Kite-Steering Unit—could take place. The example confirms that the application of the standardized modules in the GSE procedural concept is very helpful in leading project teams, who work on a problem solution in large spatial distances, to the result in a goal-oriented manner. The result also shows that specific methods and procedures from the GSE analysis, GSE goal formation and GSE design modules can be selected and coupled with the GSE thinking model. By connecting with the standardized system model, the information input and output per method is also standardized.

This in turn enables a quick further use of the data generated, but also their traceability is given. Although the interaction of GSE thinking model and GSE procedural concept was only sketched at these examples, it becomes clear that ***it is absolutely necessary to develop a computer-supported GSE thinking model.*** Through a potential computer support of the GSE thinking model, standardized data for processing in the GSE procedural concept can be provided, updated and moreover various points in time of their precision can be designed comparably. Only in this way can continuous traceability and transparency be ensured.

Example 2: Use of the GSE procedural concept for the field data feedback of mechatronic systems into the design and development process

Field data, these are all data and information that arise immediately after the phase of generating the product idea, as Fig. 5.56 shows.

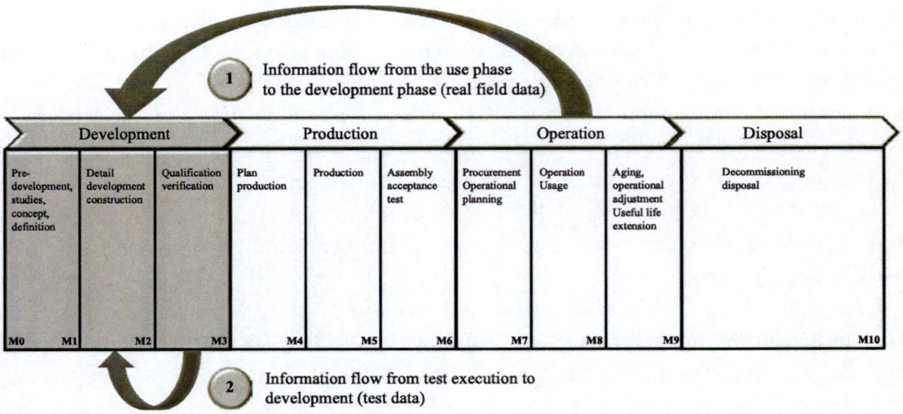

Fig. 5.56 Information flows related to the PLC. (According to [VDI 4003] Riekhof 2010)

From them, statements (characteristics and characteristic values) can be derived that allow conclusions to be drawn as to the extent to which the developed product system meets the requirements. The effective use of field data requires their structured collection and classification. However, the fact is that depending on the purpose of use, the data are already structured differently. This leads to an inaccurate further processing or incomplete collection of field data (Delonga 2007). In addition, there are a large number of different classification characteristics. The structuring of failures, as an example of the subset of field data, shows this in the following Table 5.2.

Table 5.3 makes it clear that the causes of errors are not uniformly standardized. An assignment of failures to the requirements, to the components, to the functions and processes does not take place. However, this is a necessary condition for returning data or information—failures are such—back to the early phases of product development. Edler (Edler 2001, p. 34) holds three types of deficits responsible for the incomplete use of field data, i.e.:

- methodological deficits,
- systematic deficits and
- deficits of integration.

The methodological deficits primarily include the lack of a holistic view of the use of field data. It starts with the collection, structuring and classification of field data and culminates in their targeted evaluation in the corresponding phases of the product life cycle. Often there is ignorance about the purposes for which the collected data can be used. Especially the use of field data by the methods of product development, however, requires a clear information regarding the collected facts (Braunholz 2006). So at pre-

Table 5.3 Failures, their meaning and causes (Winzer and Künne 2009, p. 546)

Classification characteristic	Author/Source
Probability of occurrence	
Probability of occurrence	Bubb und Schmidtke (1993)
Probability of occurrence	Rigby (1970)
Probability of occurrence	Swain, Guttmann (1983)
Probability of occurrence	Ellouze (2007)
Probability of occurrence	Rinne (1995)
Meaning of the Failures	
Meaning of the Failures	Algedri et al. (1998)
Failure severity	Bitkom (2007)
Severity	Telekom Standard TL9000 (1998)
Expected level of damage	Bitkom (2007)
Product status (indirect defects)	IEE 1044-1993 (1993)
Number of users affected	Bitkom (2007)
Risk of damage	DIN 66271 (1995)
Failure sequences	DIN 5350 (Teil 31) (1985)
Causes	
Causes of defects (work defects)	Zimolong (1990)
Causes of defects (work defects)	Hacker (1986)
Origin of defects (sources of product defects)	Algedri et al. (1998)
After action Failure	Algedri et al. (1998)
Sources of Failure	VDI/VDE 2650 (2006)
Reference to requirements	Six Sigma
Experiential knowledge	Ellouze (2007)
Types of Failures in order picking	Lolling (2003)
Source of origin	Ordendi (1993)

sent, especially for mechatronic systems, only incomplete failure evaluation systems are available (Winzer 2012). Due to the time lag between the collection of field data and their utilization within product development, information is often not clearly stored and is therefore difficult to assign to the product model. Approaches such as so-called failure dictionaries (Ebner 1996, p. 92) or mapping tables (Edler 2001, p. 62) for field data collection, address this problem, but only consider partial aspects. Systematic deficits refer to the lack of continuous information and storage systems, which enable the collection and use of field data on a common data basis based on standardized guidelines. The shortcomings of integration refer to the interfaces both between field data and the methods of their use, and between the individual methods themselves. Using the example of "failure", Fig. 5.57 shows various fields of view of selected methods and field data in an excerpt.

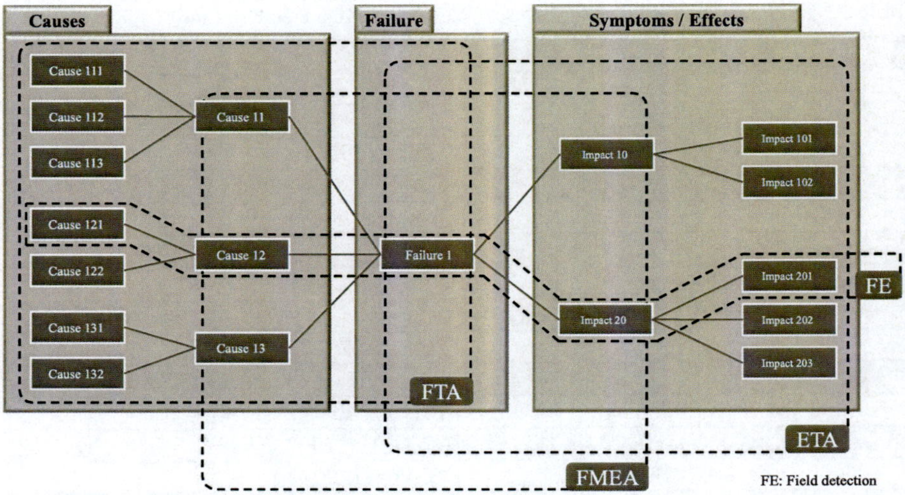

Fig. 5.57 Areas of application of selected QM methods for failure prevention. (After Ebner 1996, p. 74)

None of the presented methods, i.e., FTA, FMEA, and ETA of preventive quality management, allow a comprehensive consideration from the causes to the effects of failures. The Fault Tree Analysis -***FTA*** represents the cause of the error and their necessary combination for failure occurrence including the underlying probabilities of their occurrence . The Event Tree Analysis– Event Tree Analysis –***ETA,*** examines a possible event of a system and its potential effects on the overall system (DIN 25419). The Failure Mode and Effects Analysis—***FMEA***—considers potential failures along with their cause and possible effects, but not the causes and chains of effects. The current practice of collecting field data does not sufficiently take into account the relationship between the various causes and effects. Thus, aspects are overlooked in an isolated application of the mentioned method.

The following example outlines the theoretically described problem from a practical perspective.

The mis-sorting of a conveyed item in a logistics system can have the following causes:

- Failure of the sensor due to loosening of the connections,
- Positioning problems of the conveyed item or
- Blocking of the light beam.

In the next step, it is necessary to examine in detail which requirements:

- for the connections,
- for the positioning and
- for the light beam itself were made.

In this particular case, it was found that the conveyed item was not transported at a certain angle on the sorter, i.e., the conveyed item was outside the intended feature limits. Thus, it became clear in the usage process that the conveyed item was not correctly positioned, resulting in the blocking of the light beam. The requirement "failure-free sorting" in this case is related to the light sensor (component), the function "positioning of the conveyed item" and the process "transporting the conveyed item", which is a sub-process of the usage process. This created the connection between requirement, component, function, and process. This can only be created when the field data are brought into a causal relationship with the GSE thinking model using the DeCoDe tools, as Fig. 5.58 proves.

Based on the hypothesis, i.e., that all field data can be linked to requirements via the GSE thinking model, the following logical sequence was designed for the feedback of field data into the design and development process, which is evident from Fig. 5.59.

The field data, which were recorded, for example, with failure management, QFD or PLM, can thus be assigned to the requirements (red), the components (blue) as well as the functions and processes (each shown in gray in Fig. 5.59) and flow structured into the design and development process. As a result, constructive changes or requirement specifications are possible. The practical proof for this was initially provided for the feedback of test data into the design and development process. The methodological basic approach of the planned procedure (see Fig. 5.60) for a company in the automotive industry was the basis for the concrete project planning.

The measurement results obtained in the course of the experiment were assigned to the product model (GSE thinking model). Since some characteristic values of the experiment did not correspond to the original requirements, cause analyses had to clarify this using suitable methods—here specifically the FMEA. The results of the method in turn flowed into the thinking model and led to its change. Since the thinking model was depicted using the PromeSys portal, this computer support quickly updated the thinking model. Fig. 5.61 illustrates the methodological approach in conjunction with the computer-aided precision of the thinking model via the PromeSys portal in principle.

The modified GSE procedural concept for test data feedback into the early phases of design and development made it clear in summary that the measured values are as structurable as the system model, i.e., the GSE thinking model. In the present case, the GSE thinking model was created using the DeCoDe tools. Thus, the modeling of the system took place in the requirement, component, function, and process view. Measured values always consist of characteristic values in various forms. These are also to be assigned to the system elements or their relation in the GSE thinking model. Currently, the GSE

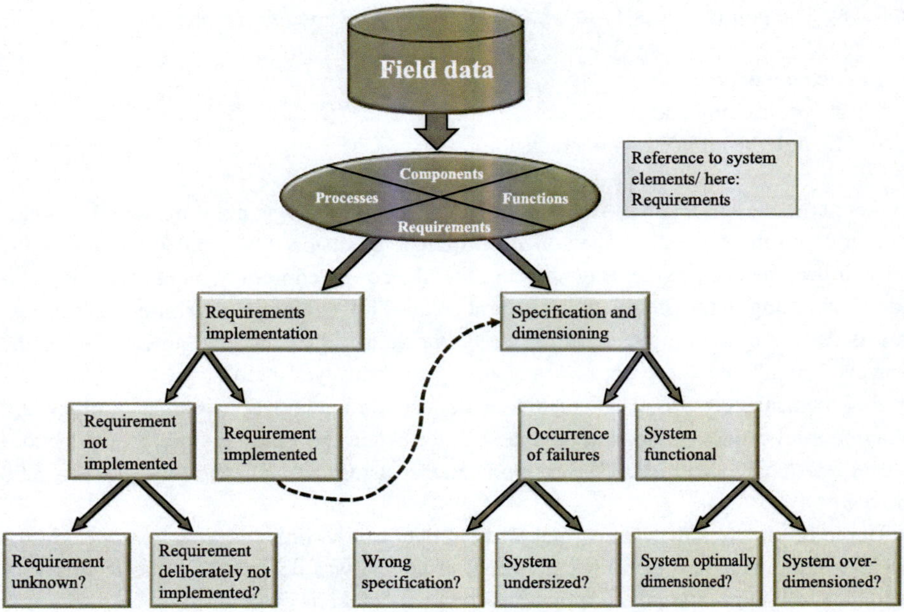

Fig. 5.58 Coupling of the field data with the GSE thinking model, which was created using the DeCoDe tools (Riekhof 2010, p. 12)

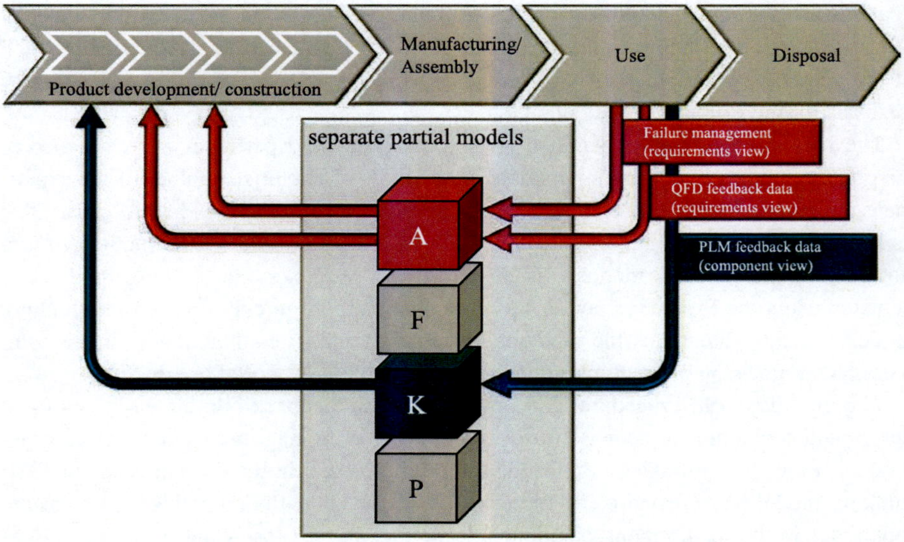

Fig. 5.59 Modified GSE procedural concept for the use of field data in various phases of the product life cycle

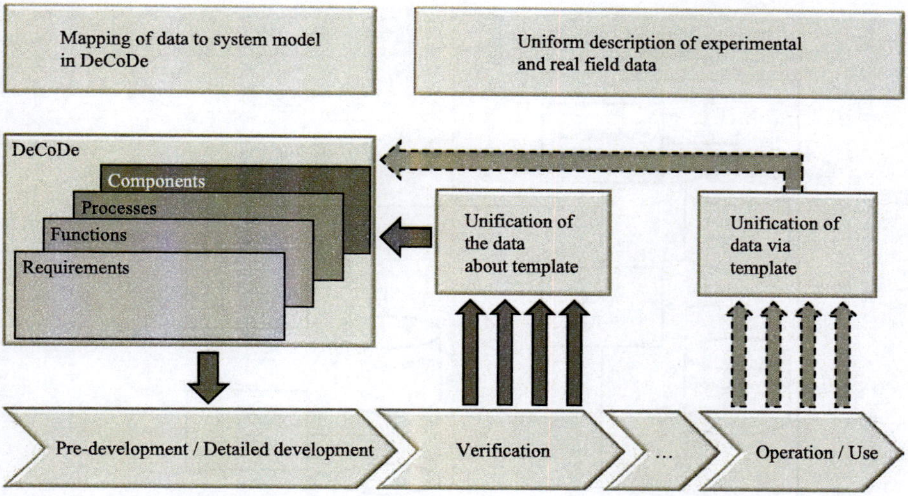

Fig. 5.60 Modified GSE procedural concept for the use of field data in various phases of the product life cycle (Riekhof 2010)

thinking model does not yet have standardized attributions of the system elements and their relations. Therefore, in this particular case, it was necessary to generate ideas for the attribution of the system model via the procedural concept in order to bring about a solution. In the future, this should be predetermined and standardized by the GSE thinking model. The example of test data feedback also shows that the modules of the GSE procedural concept can be combined with each other in different temporal and logical sequences. Table 5.4 abstracts the outlined approach via a bar chart (a method used in project planning).

The comparison of this approach with that of the risk minimization of the KitVes system makes it clear that both GSE procedural concepts differ in the combination of the modules, but the modules are analogous. ***Thus, a procedure for a problem solution with universal, standardized modules can be planned in such a way that an individual problem solution becomes targeted.*** These two examples also show that the standardized modules of the GSE procedural concept can be supplemented problem-specifically with specific methods and procedures. They also illustrate that the basic principles of systematic thinking and acting can be seamlessly and problem-related integrated within the framework of the GSE procedural concept. They help in systematic problem solving and in creating a well-structured project plan. The planning of the type and manner of coupling the individual modules of the GSE procedural concept, their implementation and the control of their effectiveness can be efficiently implemented with the methods of project management. The two examples for the KitVes system and the test data feedback also illustrate this.

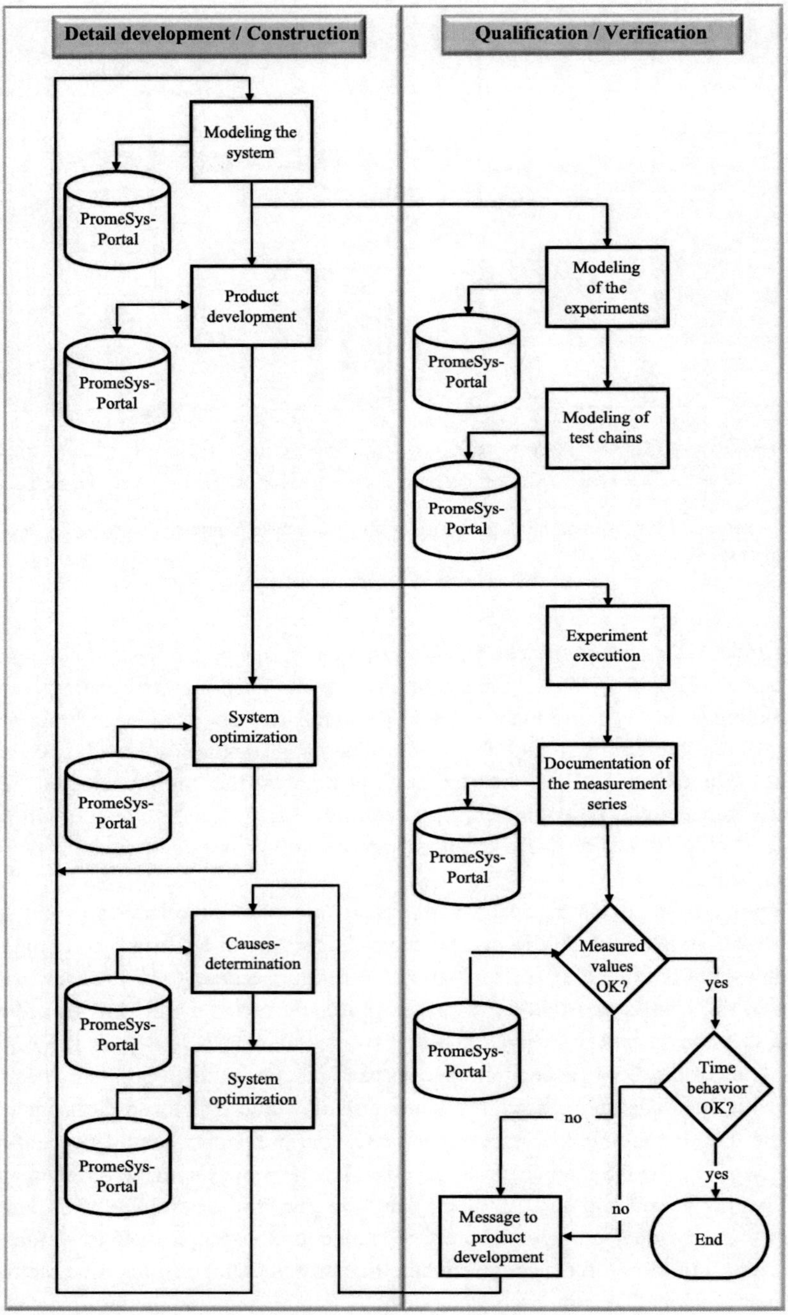

Fig. 5.61 The PromeSys portal as the backbone of IT-supported test data feedback (work results from the PromeSys research project)

Table 5.4 Project planning for the development of a methodological approach to field data feedback

Steps \ Time	1. Year				2. Year			
I. Field data classification and system modeling	■	■	■					
II. Field data feedback			■	■	■			
III. Utilization						■	■	■
IV. Validation			■	■	■	■	■	■

In the later Chap. 6 the interaction between the GSE thinking model and the GSE procedural concept will be examined in more detail using examples of technical systems. ***Thepermanent comparison between GSE thinking model and GSE procedural concept***—so the thesis—***supports the more efficient application of the GSE approach.*** First, the interaction of GSE thinking model and GSE procedural concept in socio-technical systems will be outlined below.

Example 3: Model-based organizational development using the example of a shoe production

To illustrate how the GSE modules can be used to build an procedural concept for organizational development, the following presents the procedural concept for model-based agile requirements management (**R**equirements **E**ngineering and Requirements **M**anagement) for **O**rganisa**t**ions **(REMOt)** developed by MISTLER (Mistler 2021) (see Fig. 5.62).

The REMOt approach concept shown in Fig. 5.62 is based on the GSE approach. It consists of four steps (system definition, survey, weighting, and validation) that can be derived from the PDCA cycle and are underpinned with the GSE modules. In all steps, the analysis and goal formation module is built in, so that agility in the problem-solving process can be ensured. Furthermore, at the center of the REMOt procedural concept is the REMOt organizational model, which is based on the GSE thinking model e-DeCoDe. In addition, the individual steps are underpinned with phases in which methods and (IT) tools can be applied problem-specifically using the developed REMOt kit. These are adapted to the problem-solving process on the one hand and to the thinking model on the other (Mistler 2021).

REMOt Step A: In REMOt Step A, the goal is to achieve a system definition. This results in a rough system delimitation and problem definition. With regard to the ReMaiN approach, REMOt Step A corresponds to the sensitization phase and system definition for the collection of information. The system delimitation is assigned to the GSE analysis module. The project idea and the project benefit should be clarified here. After the system delimitation has been carried out through the analysis, the problem

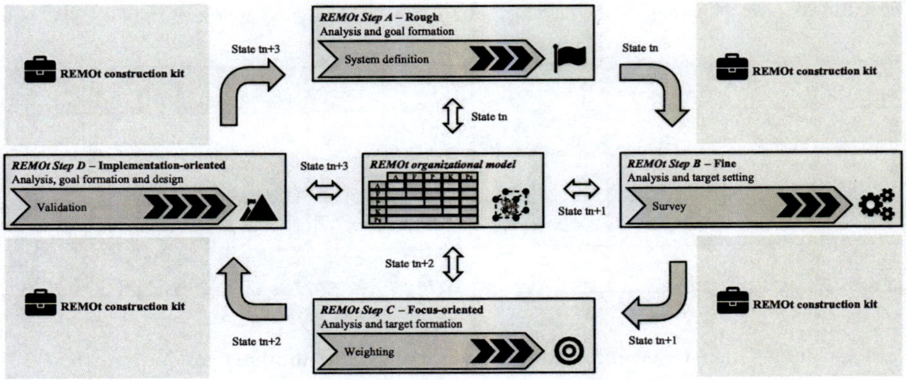

Fig. 5.62 REMOt procedural concept according to (Mistler 2021)

definition specifies the problem by identifying the stakeholders and their requirements. Consequently, the problem definition belongs to the GSE goal formation module. The results from REMOt Step A are documented using the REMOt organizational model. Thus, the state t_n represents the initial situation (Mistler 2021) (see Fig. 5.62).

REMOt Step B: In REMOt Step B, a fine analysis and target formation is aimed for. This is done by means of a structured collection of information on the organizational system. The basis for the collection is the rough system delimitation and problem definition from REMOt Step A. In order for a structured information collection to be enabled, the strategic orientation of the collection is of enormous importance. Therefore, in REMOt Step B, the Information Flow Analysis (IFLA) by WINZER and SITTE (Sitte and Winzer 1999) is used. It is an integral part of the GSE analysis module and has already been used and further developed in various works (Sitte and Winzer 1996, 1999; Winzer and Braunholz 2000, 2003; Mistler 2021; Mistler et al. 2021; Braunholz 2006). First, a structuring of the collection information is carried out in order to achieve a fine target formation for the IFLA. Attributes are determined with regard to the problem that should be queried in the IFLA. The structuring of the collection is therefore to be classified in the GSE goal formation module. After this has been carried out, the implementation and follow-up of the IFLA, which are assigned to the GSE analysis module, take place. The result of the REMOt Step B forms the e-DeCoDe modeling with the state t_{1+n} (Mistler 2021) (see Fig. 5.62).

REMOt Step C: REMOt Step C pursues a focus-oriented analysis and goal setting. For this purpose, a structured weighting is carried out for the problems of the organizational system. The starting point for the weighting are the results of REMOt Step B. Based on this information, it should be examined which problems are most important in terms of organizational system development and whether these are sufficient to evaluate the problems. This structuring for weighting is therefore assigned to the GSE analysis module. The subsequent evaluation of the problems of the organizational system is attributed to the GSE goal formation module. The result of REMOt Step C is to be documented in the REMOt organizational model and marked with the state t_{2+n} (Mistler 2021) (see Fig. 5.62).

REMOt Step D: The purpose of REMOt Step C is an implementation-oriented analysis, goal foramtion and design. For this purpose, a validation is carried out, which triggers a continuous improvement process (CIP). The entire validation of REMOt Step C is assigned to the GSE design module, as a solution space is fixed here, solution variants are designed and a check of the feasibility of the solution variants takes place. First, a solution space is fixed, which forms the basis for the design of the organizational system. Based on this, solution variants are developed. This phase is assigned to the GSE analysis module. Among other things, there is the possibility to check whether enough information has been collected to develop solution variants for the problems of the organizational system. From the solution variants, measures are derived, which are checked for their feasibility. This means whether these measures can meet, partially meet or not meet the requirements of the organizational system. Thus, there is the possibility in this phase to introduce new requirements or modify requirements. The action plan resulting from the interplay of analysis and goal setting finally triggers a CIP. The CIP forms the result of REMOt Step D and is in iterative connection with the REMOt organizational model and is to be marked as state t_{n+3}. The implementation results are finally to be made transparent in the overall organization. Thus, among other things, new projects can emerge, which can build on the previous findings (Mistler 2021) (see Fig. 5.62).

REMOt Organizational Model: As shown in Fig. 5.62 and described in REMOt Steps A, B, C, and D, the REMOt organizational model is iteratively connected with the steps. The results of a step are always documented in the REMOt organizational model. Thus, the storage of the results is assigned to the GSE project management module. To gather the information for the states, it is necessary that the GSE project management module is also connected with the GSE analysis, goal formation, and design module.

The REMOt organizational model is based on the e-DeCoDe approach, which is described in detail in Chap. 4. Since the e-DeCoDe approach is a meta-model for describing socio-technical systems, MISTLER (Mistler 2021) recognized that the approach must be transferred to the description of organizational systems. In addition, the approach must also enable agility in system modeling. This means ensuring a flexible adaptability of the system model in case of changes through modularity (Mistler 2021). How MISTLER transferred the e-DeCoDe approach to the modeling of organizational systems and how agility in system modeling is generated in the process is described below.

For this purpose, the REMOt organizational model with the implemented e-DeCoDe approach is first shown in Fig. 5.63.

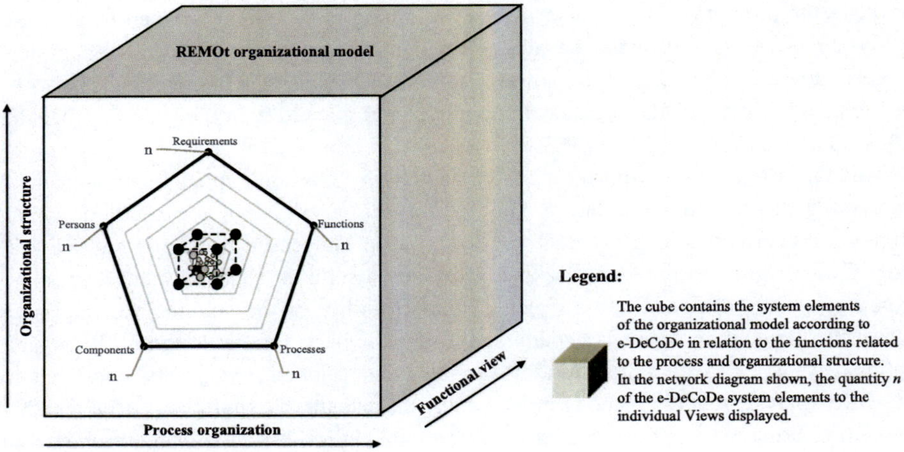

Fig. 5.63 REMOt Organizational Model (Mistler 2021, p. 46)

The REMOt organizational model is represented in the form of a cube. This has three different axes: **Process Organization**[2] (x-axis), **Organizational Structure**[3] (y-axis) and **Functional View** (z-axis). Within the three dimensions, a network diagram is shown, which represents the e-DeCoDe system elements. The letter "n" stands for the number of determined e-DeCoDe system elements in the REMOt organizational model.

The process and organizational structure are closely related. Nevertheless, it can be noted that both the process organization and the organizational structure have a purpose in the organization. If it is recognized which elements of the process organization and the organizational structure work towards the same function, these can be connected via the functional view (Mistler 2021). This connection has been transferred to the e-DeCoDe modeling language and visualized, as can be seen in Fig. 5.64.

In Fig. 5.64, it can be seen that the e-DeCoDe system elements have been transferred to the organizational system view. Thus, according to MISTLER, the following definitions in connection with the e-DeCoDe modeling result for the organizational system modeling with the REMOt organizational model:

[2] "The **process organization** describes the temporal logical coupling of business processes, which can be subdivided into sub-processes taking into account the respective input and output. At the last level, business processes are described by activities and thus determine their purpose (function) in the process organization. In the course of this, the execution of activities is implemented by persons with the necessary tools, resources and information" Mistler 2021, p. 45.

[3] "The **organizational structure** describes the rights, powers (competencies) and responsibility (duties) of persons through their role (function) in an organization. Furthermore, it defines the relationship contexts between organizational units, separates the sub-organizational units from each other and arranges them in a certain hierarchy" Mistler 2021, p. 43.

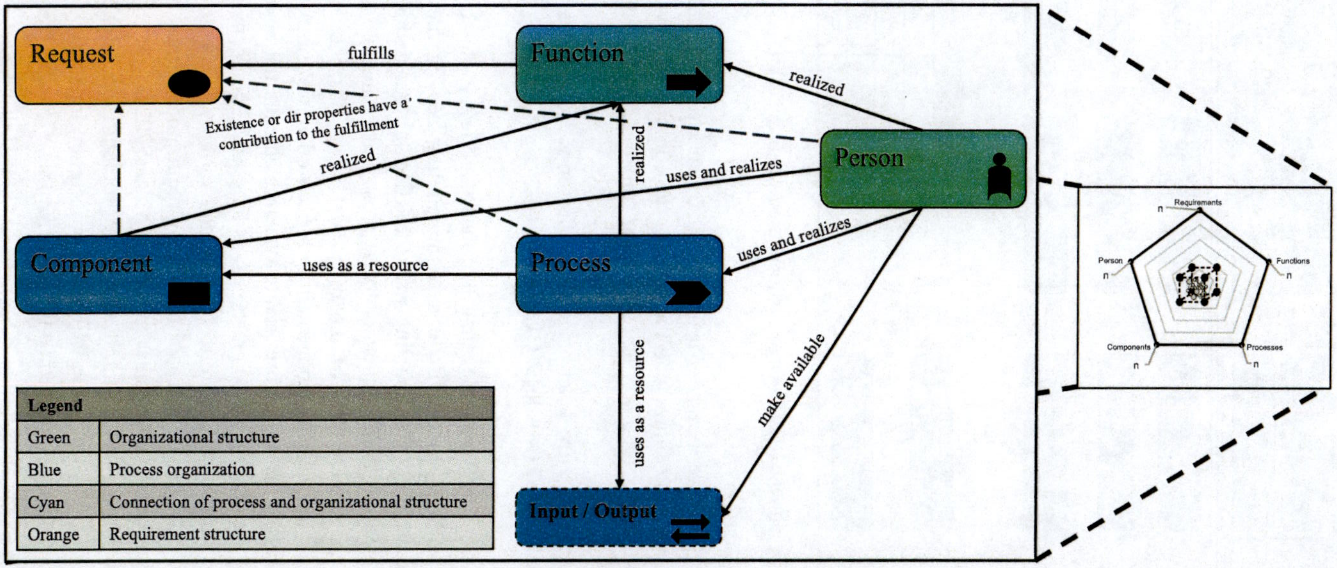

Fig. 5.64 Connection of the REMOt Organizational Model Views (Mistler 2021, p. 47)

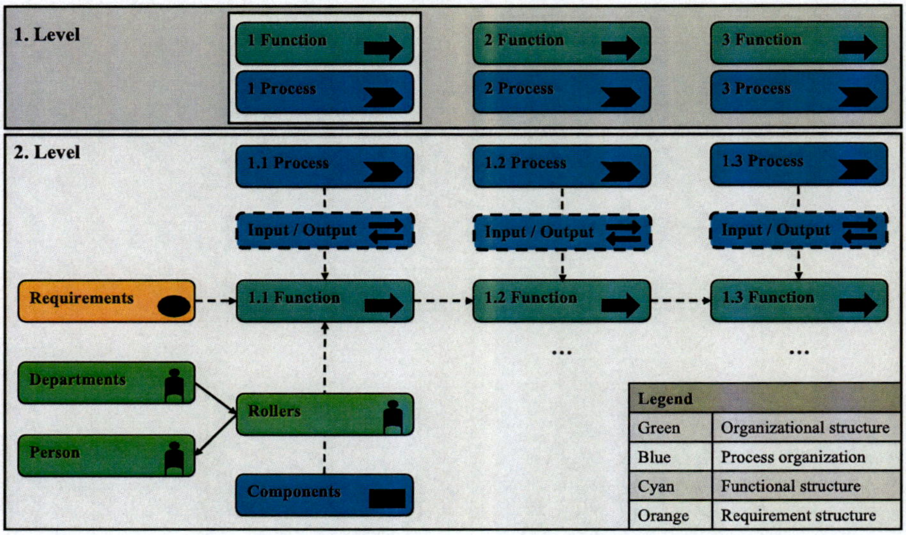

Fig. 5.65 Agile Design with the REMOt Organizational Model (Mistler 2021, p. 49)

- "The **functional view** describes ***what*** is done in the organizational system.
- The **process organization** shows with the process and component view as well as the processing of input into output, ***how*** and ***with what*** something is done in the organizational system.
- The **organizational structure** describes with the person view, ***who*** does something in the organizational system" (Mistler 2021, p. 47).

From this insight, MISTLER concludes that the functional view is suitable for forming modules with regard to the desired agility in system modeling (see Fig. 5.64).

Fig. 5.65 shows the basic agile modeling with the REMOt organizational model. For this, a functional view must first be created that is most understandable for the organization. This forms the lowest level of the functional view. Based on this, all other e-DeCoDe system elements can be connected with each other via the functional view. With regard to agility, the functions form stable system elements that also represent the company's goals and therefore must be preserved, as long as the company's goals do not change. The other system elements are to be designed variably in relation to the function. This means, for example, in the case of changing requirements, that the process and organizational structure can be adjusted accordingly (Mistler 2021).

In summary, this chapter has exemplified how a problem-specific approach can be built using the GSE modules, using the REMOt approach developed by MISTLER. The ReMaiN approach (see Chap. 4) served as the basis for this. How the REMOt approach can be applied in practice is shown in Chap. 6. The symbiosis of the GSE approach with the ReMaiN approach illustrates that other approaches for other problem areas can also be adapted with the GSE approach. Thus, it is possible to retain the core contents of other approaches and to transform them in a comprehensible way for a specific purpose. Such an adaptation could also be achieved in the work of HEINKE and MISTLER (Heinke and Mistler 2019) with the Scrum approach of SCHWABER and SUTHERLAND (Schwaber and Sutherland 2017) (see Sect. 5.4).

5.6 Summary of the Modules of the GSE Procedural Concept

After the GSE modules have been explained in detail in Sects. 5.1 to 5.5, it makes sense to summarize the essential contents in this Sect. 5.6. This is important so that the basic significance of the modules can also be easily conveyed. Therefore, brief statements about the core contents are repeatedly found in various works dealing with the GSE (see Mamrot 2014; Nicklas 2016; Mistler 2021; Mistler et al. 2021). A current overview of the core statements of the analysis, goal formation, design, and project management modules is given by (Mistler 2021), which is why these summaries are reproduced below in a slightly modified form.

Analysis Module: The analysis module aims to define a problem, focus on a system model so that a solution space can be fixed for the development of an optimal solution for a described problem. This means that a problem understanding for the causes of the problem is generated and this can be assigned to the focused system. This is the premise for collecting and structuring information to solve a problem. It should be emphasized that a system representation requires at least one black-box model. The black-box model serves as a rough basis for analyzing causes and problems (Mistler 2021) (see Sect. 5.1).

Goal Formation Module: In the goal formation module, it is about prioritizing and in some cases also modifying requirements for systems. Thus, the purpose of the module is to manage the diversity of requirements for system development and to derive the essential ones from this diversity. Only in this way can a goal be set for system development. This is achieved by prioritizing the stakeholders. Their requirements are collected and compared with each other. This way, contradictory or duplicate requirements can be identified, weighted, and evaluated (Mistler 2021) (see Sect. 5.2).

Design Module: The design module aims to develop new systems or modify existing ones. Therefore, this module serves to define the solution space, generate solution ideas, develop, select and implement solution variants. In addition, the validation of the implementation process of the requirements for the system is an essential part of the module. The solution space definition takes place in close interaction with the GSE thinking model (Mistler 2021) (see Sect. 5.3).

Project Management Module: The project management module is closely related to the GSE thinking model, but also to the other described, standardized, and universal GSE modules. It serves to control the system development in a timely, logical, efficient, and problem-solving oriented manner. The basic principles of systemic thinking and action must be included. Furthermore, the basic principle from the rough to the detail must be particularly observed in order to be able to overlook the goal formation for an entire project, but also to be able to divide it into smaller tasks that serve the entire project. The basic principle of minimal models must also be taken into account so that, for example, only necessary documentation and modeling are carried out. Minimal documentation can be achieved by fixing the change of the GSE thinking model during the course of the project. The project management module thus includes all technical, administrative, steering, and supporting tasks to plan, carry out, and control the problem-solving process. It also includes the overview and compliance with personnel, financial, and time constraints. It should be emphasized that the planning, implementation, and control phase of the GSE project management module overlap and are not sequentially connected. This is necessary to ensure agility in the problem-solving process. The definition of a project is not provided for in the GSE procedural concept. Rather, the project is defined and fixed with the help of goal formation (Mistler 2021) (see Sect. 5.4).

Interaction of the Modules: The interaction of the modules is necessary in order to be able to react agilely to problem situations. For example, if solution variants are developed in the design module, it is still useful to continuously reflect on the objectives of system development, i.e., the requirements for system development. Furthermore, this can leave open the possibility of integrating or modifying requirements in system design. However, to question the objectives more precisely, the analysis module is also needed. This serves the purpose of examining the determined requirements more closely in relation to the existing information about the system. Among other things, it could be determined whether there is not enough information available to evaluate the implementation of the requirements or whether the feasibility of the requirements appears unrealistic (Mistler 2021) (see Sect. 5.5).

To illustrate how the GSE modules can be used to build an procedural concept, the following chapter serves.

References

Arlt, Gregor (1999): Systemansatz eines produkt- und ablauforientierten Qualitätsmanagements durch Integration der Systemtechnik. Univ., Diss. Duisburg, 1999. Als Ms. gedr. Düsseldorf: VDI-Verl. (Fortschritt-Berichte VDIReihe 16, Technik und Wirtschaft 109).

Bender, Beate; Gericke, Kilian (2021): Pahl/Beitz Konstruktionslehre. Methoden und Anwendung erfolgreicher Produktentwicklung. 9th ed. 2021. Berlin, Heidelberg: Springer Berlin Heidelberg; Imprint: Springer Vieweg.

Braunholz, Helge (2006): Werkzeugentwicklung für informationsflussorientierte Prozessmodelle. Aachen: Shaker Verlag GmbH.

Brüggemann, Holger; Bremer, Peik (2012): Grundlagen Qualitätsmanagement. Von den Werkzeugen über Methoden zum TQM. Wiesbaden: Vieweg+Teubner Verlag.

Dahms, Matthias (2010): Motivieren, Delegieren, Kritisieren. Die Erfolgsfaktoren der Führungskraft. 2. Aufl. Wiesbaden: Gabler.

Delonga, Melani (2007): Zuverlässigkeitsmanagementsystem auf Basis von Felddaten. Dissertation. Stuttgart: Universität Stuttgart.

DIN 69901–5 (2009): Projektmanagement_- Projektmanagementsysteme_- Teil_5: Begriffe. Berlin: Beuth Verlag GmbH.

DIN EN 60812 (2006): Analysetechniken für die Funktionsfähigkeit von Systemen_- Verfahren für die Fehlzustandsart- und -auswirkungsanalyse (FMEA) (IEC_60812:2006); Deutsche Fassung EN_60812:2006. Berlin: Beuth Verlag GmbH.

DIN EN ISO 9000 (2015): Qualitätsmanagementsysteme—Grundlagen und Begriffe. Berlin: Beuth Verlag.

Ebner, Claus (1996): Ganzheitliches Verfügbarkeits-und Qualitätsmanagement unter Verwendung von Felddaten. Berlin, New York: Springer.

Edler, Andreas (2001): Nutzung von Felddaten in der qualitätsgetriebenen Produktentwicklung und im Service. Berlin: IPK.

Ehrlenspiel, Klaus; Meerkamm, Harald (2017): Integrierte Produktentwicklung. Denkabläufe, Methodeneinsatz, Zusammenarbeit. 6., vollständig überarbeitete und erweiterte Auflage. München, Wien: Hanser. Online verfügbar unter http://www.hanser-fachbuch.de/buch/Integrierte+Produktentwicklung/9783446440890.

Eversheim, Walter; Liestmann, Volker; Winkelmann, Katrin (2006): Anwendungspotenziale ingenieurwissenschaftlicher methoden für das service engineering. In: Service Engineering: Springer, S. 423–442.

Eversheim, Walter; Schuh, Günther (Hg.) (1999a): Produktion und Management. Betrieb von Produktionssystemen. Berlin: Springer (Produktion und Management, 4).

Eversheim, Walter; Schuh, Günther (1999b): Produktmanagement. Berlin: Springer (VDI-Buch-Wirtschaftsingenieurwesen,/Hrsg. von Walter Eversheim; Günther Schuh; 2).

ffpt.de (2016): Free House of Quality Template for PowerPoint. Online verfügbar unter http://www.free-power-point-templates.com/articles/free-house-of-quality-template-forpowerpoint-qfd-template/, zuletzt geprüft am 15.02.2016.

Fiedrich, Sabine (2010): KuWiss—Einsatz einer unternehmensspezifischen Methodik zur kontinuierlichen Messung der Kundenstimme im Baukastensystem. In: Petra Winzer (Hg.): Entwicklungen im Wuppertaler Generic-Management-Konzept. Aachen: Shaker, S. 19–32.

Focus Online (2011): Toyota Rückruf für 400.000 Hybrid-Autos. Online verfügbar unter http://www.focus.de/auto/news/toyota-rueckruf-fuer-400-000-hybrid-autos_aid:478330.html, zuletzt geprüft am 26.07.2011.

Franke, Hans-Joachim (2002): Variantenmanagement in der Einzel- und Kleinserienfertigung. Mit 33 Tabellen. München [u.a.]: Hanser.

Gausemeier, Jürgen; Plass, Christoph (2014): Zukunftsorientierte Unternehmensgestaltung. Strategien, Geschäftsprozesse und IT-Systeme für die Produktion von morgen. 2., überarb. Aufl. München: Hanser.

Gausemeier, Jürgen; Plass, Christoph; Wenzelmann, Christoph (Hg.) (2009): Zukunftsorientierte Unternehmensgestaltung. Strategien, Geschäftsprozesse und IT-Systeme für die Produktion von morgen. München, Wien: Hanser.

Gausemeier, Jürgen; Rammig, Franz Josef; Schäfer, Wilhelm (Hg.) (2014): Design methodology for intelligent technical systems. Develop intelligent technical systems of the future. Heidelberg, New York, NY, Dordrecht: Springer (Lecture notes in mechanical engineering).

Geiger, Ingrid Katharina; Pifko, Clarisse (2009): Projektmanagement—Zertifizierung nach IPMA(3.0)-Ebenen D und C. Grundlagen und Kompetenzelemente, Methoden und Techniken mit zahlreichen Beispielen. 2., überarb. Aufl., Ausg.: U0039. Zürich: Compendio Bildungsmedien (Betriebswirtschaftslehre).

Gogoll, Alexander (1996): Untersuchung der Einsatzmöglichkeiten industrieller Qualitätstechniken im Dienstleistungsbereich: IPK Berlin.

Haberfellner, Reinhard; Weck, Olivier de; Fricke, Ernst; Vössner, Siegfried (2012): Systems engineering-grundlagen und anwendung. 12. In: *Auflage. Zürich: Orell Füssli,* 978-3280040683.

Haberfellner, Reinhard; Weck, Olivier de; Fricke, Ernst; Vössner, Siegfried (2018): Systems Engineering. Grundlagen und Anwendung. 14. überarbeitete Auflage. Zürich: Orell Füssli Verlag.

Häder, Michael (2009): Delphi-Befragungen. Ein Arbeitsbuch. 2. Aufl. Wiesbaden: VS, Verl. für Sozialwiss. (Lehrbuch).

Hartmann, C.; Winzer, P.; Riekhof, F. (2011): DeCoDe+X Methodenverknüpfung und Systemmodellierung zur Unterstützung der Risikobeurteilung in den frühen Phasen der Produktentwicklung. In: *25. TTZ. Leonberg*.

Hartmann, Christine; Winzer, Petra (2011): DeCoDe+X in KitVes. Using the Demand Compliant Design in the Development of a Solution for Harvesting High-Altitude Winds for Energy Gereration on Vessels. Pamplona, Spain: Servicios de Publicaciones Universidad de Navarra: Proceedings 14.QMOD Conference on Quality and Service Science (QMOD 2011).

Häuslein, Andreas (2004): Systemanalyse. Grundlagen, Techniken, Notierungen. Berlin: VDE-Verl. Online verfügbar unter http://www.gbv.de/dms/hebis-darmstadt/toc/113453388.pdf.

Heinke, Jonas; Mistler, Marian (2019): Agiles und modellbasiertes Projektmanagement in der Produkt- und Dienstleistungsentwicklung. In: Nadine Schlüter und Markus Reiche (Hg.): Herausforderungen im Umgang mit Anforderungen in Zeiten des industriellen Wandels. 1. Auflage. Düren: Shaker (Berichte zum Generic-Management, 2019, 1), S. 1–26.

Jensen, Daniel; Feland, John; Bowe, Marty; Self, Brian (2000): A 6 Hats Based Team Formation Strategy: Development And Comparison With An Mbti Based Approach. In: 2000 Annual Conference, S. 5–9.

Jentsch, David; Trommler, Ullrich; Horbach, Sebastian; Ackermann, Jörg; Müller, Egon (2012): Entwicklung eines interaktiven Planungshandbuches für kompetenzzellenbasierte Produktionsnetze. In: E. Müller und A.C Bullinger (Hg.): Fachtagung "Vernetzt Planen und Produzieren—VPP2012" sowie Symposium "Wissenschaft und Praxis—W&P". Intelligent vernetzte Arbeits- und Fabriksysteme VPP 2012 –. Fachtagung Vernetzt Planen und Produzieren und Symposium Wissenschaft und Praxis. Chemnitz. Chemnitz: Eigenverlag TU Chemnitz, ISSN: 0947–2495 (Wissenschaftliche Schriftenreihe des IBF, Sonderheft 18), S. 53–62.

Jochem, R.; Rößle, D.; Ariza Alvarez, J. E. (2015): Taxonomie von Qualitätsmanagement-Methoden. In: Stefan Bracke, Michel Mamrot und Petra Winzer (Hg.): Qualitätsmethoden im Diskurs zwischen Wissenschaft und Praxis. Bericht zur GQW-Jahrestagung 2015 in Wuppertal. 1. Aufl. Herzogenrath: Shaker (Berichte zum Qualitätsmanagement, 2015, 17).

Künne, Bernd; Richard, Tim (Hg.) (2009): Sonderforschungsbereich 696. Forderungsgerechte Auslegung von intralogistischen Systemen—Logistics on Demand. Finanzierungsantrag (Fortsetzung) 07/2010 bis 06/2014. Dortmund.

Kuster, Jürg (2011): Handbuch Projektmanagement. Dordrecht: Springer.

Lehner, Johannes M. (2001): Praxisorientiertes Projektmanagement. Grundlagenwissen an Fallbeispielen illustriert. 1. Aufl. Wiesbaden: Gabler.

Lex, A.; Winzer, P.; Sitte, J. (2004): Generic Management Design-a Method of Collecting Knowledge Systematically during the Developing Process. In: *VDI BERICHTE,* S. 279–284.

Lindemann, U. (2005): Methodische Entwicklung technischer Produkte. Methoden flexibel und situationsgerecht anwenden. Berlin: Springer.

Lindemann, Udo (Hg.) (2016): Handbuch Produktentwicklung. München: Carl Hanser Verlag.

Lindemann, Udo; Maurer, Maik; Braun, Thomas (2009): Structural complexity management. An approach for the field of product design. Berlin: Springer.

Mamrot, M.; Schlueter, Nadine.; Winzer, P. (2014): Generic Systems Engineering (GSE) in der praktischen Anwendung. In: Petra Winzer (Hg.): Trends zur Handhabung von Komplexität im Qualitätsingenieurwesen. Aachen: Shaker (Berichte zum Generic-Management, 2014, 2), S. 1–18.

Mamrot, M.; Schlüter, N.; Winzer, P. (2015): Wie können wir Qualität auch in Zukunft sichern? In: Stefan Bracke, Michel Mamrot und Petra Winzer (Hg.): Qualitätsmethoden im Diskurs zwischen Wissenschaft und Praxis. Bericht zur GQW-Jahrestagung 2015 in Wuppertal. 1. Aufl. Herzogenrath: Shaker (Berichte zum Qualitätsmanagement, 2015, 17).

Mamrot, Michel (2014): Entwicklung eines Ansatzes zur modellbasierten Felddatenrückführung in die Produktentwicklung. 1. Aufl. Herzogenrath: Shaker (Berichte zum Generic-Management, 2014, 1).

Mistler, Marian (2018): Einheitliches, modellbasiertes und agiles Anforderungsmanagement zur Entwicklung eines softwaregestützten Informationssystems für Organisationen. In: Nadine Schlüter und Markus Reiche (Hg.): Umgang mit Anforderungen in agilen Organisationen. 1. Auflage. Herzogenrath: Shaker (Berichte zum Generic-Management, 2018,2), S. 13–41.

Mistler, Marian (2021): Entwicklung eines Vorgehenskonzeptes zum modellbasierten agilen Anforderungsmanagement (Requirements Engineering und Requirements Management) für Organisationen—REMOt. Dissertation. Wuppertal: Bergische Universität Wuppertal.

Mistler, Marian; Schlueter, N.; Löwer, Manuel (2021): Agile Design of Organizations using the Information Flow Analysis and the Generic Systems Engineering. In: 2021 IEEE International Conference on Systems, Man, and Cybernetics (SMC): IEEE.

Müller, Egon; Jentsch, David; Trommler, Ullrich; Horbach, Sebastian; Ackermann, Jörg (2012): Entwicklung eines interaktiven Planungshandbuches für kompetenzzellenbasierte Produktionsnetze. In: E. Müller und A.C Bullinger (Hg.): Fachtagung "Vernetzt Planen und Produzieren—VPP2012" sowie Symposium "Wissenschaft und Praxis—W&P". Intelligent vernetzte Arbeits- und Fabriksysteme VPP 2012. Fachtagung Vernetzt Planen und Produzieren und Symposium Wissenschaft und Praxis. Chemnitz. Chemnitz: Eigenverlag TU Chemnitz, ISSN: 0947–2495 (Wissenschaftliche Schriftenreihe des IBF, Sonderheft 18), S. 53–62.

Müller, Nico; Schlund, Sebastian; Winzer, Petra (2010): Modellierung komplexer mechatronischer Systeme anhand des Demand Compliant Design. In: E. Schnieder, U. Jumar und C. Diedrich (Hg.): Entwurf komplexer Automatisierungssysteme. Beschreibungsmittel, Methoden, Werkzeuge und Anwendungen; 11. Fachtagung mit Tutorium, 25. bis 27. Mai 2010 in Magdeburg, Denkfabrik im Wissenschaftshafen. Magdeburg: Ifak, S. 79–88.

Nicklas, Jan-Peter (2016): Ansatz für ein modellbasiertes Anforderungsmanagement für Unternehmensnetzwerke. Dissertation. Aachen: Shaker Verlag (Berichte zum Generic-Management, Band 2016, 2).

Nicklas, Jan-Peter (2018): Möglichkeiten der Zusammenführung eines modellbasierten Anforderungsmanagements mit Tools zur Risikoabschätzung. In: Nadine Schlüter und Markus Reiche (Hg.): Umgang mit Anforderungen in agilen Organisationen. 1. Auflage. Herzogenrath: Shaker (Berichte zum Generic-Management, 2018, 2), S. 67–85.

Ott, Stefan; Winzer, Petra (2007): Cultivating Knowledge methodically: Improving analysis resolution with DeCoDe and FMEA. In: Proceedings of QMOD 2007. Helsingborg/Sweden. 18.–20. Juni 2007.

Pfeifer, Tilo; Schmitt, Robert (Hg.) (2014): Masing Handbuch Qualitätsmanagement. 6., überarbeitete Auflage. München: Hanser, Carl.

Pfeifer, Tilo; Schmitt, Robert (2021): Masing Handbuch Qualitätsmanagement. München: Carl Hanser Verlag GmbH & Co. KG.

Riekhof, F.; Winzer, P. (2010): Produktqualität und Q-Gates. AutoUni. AutoUni—Volkswagen Aktiengesellschaft. MobileLifeCampus, Wolfsburg, 27.05.2010. Online verfügbar unter http://

www.sjf.tuke.sk/kbakp/Documents/DeCoDe-PLC-Methodenkopplung.pdf, zuletzt geprüft am 03.05.2013.

Riekhof, F.; Winzer, P.; Willing, M. (2011): Reliability in early product development phases. Using the DeCoDe+ X approach for a data-based discussion of design decisions. In: *Proceedings of QMOD*, S. 29–31.

Riekhof, Florian (2010): Methodische Rückführung von Versuchsdaten in die Produktentwicklung und Validierung der Methodik am Beispiel eines Sitzlehneneinstellers. Masterthesis. [unveröffentlicht]. Fachgebiet für Produktsicherheit und Qualitätswesen. Bergische Universität Wuppertal.

Riekhof, Florian (2011): Ansatz zur systematischen Versuchsdatenrückführung in die Produktentwicklung. In: Petra Winzer (Hg.): Anforderungsgerechte Produkt- und Dienstleistungsentwicklung im Rahmen des Wuppertaler Generic-Management-Konzeptes. 1. Aufl. Aachen: Shaker (Berichte zum Generic-Management, 2011, 2), S. 77–108.

Rosendahl, J.; Kulig, S.; Schlund, Sebastian; Winzer, Petra (2009): Methodenworkflow zur Entwicklung mechatronischer Systeme. In: Bernd Künne, W. Tillmann und H.-A Crostack (Hg.): Forderungsgerechte Auslegung von intralogistischen Systemen. Logistics on Demand. Dortmund: Verl. Praxiswissen, S. 63–79.

Sage, Andrew P.; Rouse, William B. (Hg.) (2009): Handbook of systems engineering and management. 2. ed. Hoboken, NJ: Wiley (Wiley series in systems engineering and management).

Scheithauer, D. (2014): Qualität im System-Design. In: Maik S. Maurer, Jutta Abulawi und Sven-Olaf Schulze (Hg.): Tag des Systems Engineering. Bremen, 12.–14. November 2014; [TdSE]. München: Hanser, S. 225–234.

Schlund, S.; Winzer, P. (2010): DeCoDe-Modell zur anforderungsgerechten Produktentwicklung. In: Gerhard Bandow (Hg.): "Das ist gar kein Modell!". Unterschiedliche Modelle und Modellierungen in Betriebswirtschaftslehre und Ingenieurwissenschaften. 1. Aufl. Wiesbaden: Gabler.

Schlund, Sebastian (2011): Anforderungsaktualisierung in der Produktentwicklung. Entwicklung einer Methodik zur Aktualisierung von Anforderungen durch die Einbindung anforderungsrelevanter Ereignisse. Aachen: Shaker.

Schlüter, N. (2011): Kundenzufriedenheitsmessungen in Netzwerken. In: Petra Winzer (Hg.): Anforderungsgerechte Produkt- und Dienstleistungsentwicklung im Rahmen des Wuppertaler Generic-Management-Konzeptes. 1. Aufl. Aachen: Shaker (Berichte zum Generic-Management, 2011, 2).

Schlüter, N.; Sochacki, S. (2012): Qualitative Netzwerkanalyse hinsichtlich der Anwendbarkeit von KuWiss-Netz. In: *Generic Systems Engineering als Basis für die Weiterentwicklung des WGMK-Modells. Shaker, Aachen.*

Schlüter, Nadine (2013): Entwicklung einer Vorgehensweise zur Implementierung einer forderungsgerechten Kundenzufriedenheitsmessung in Unternehmensnetzwerken: Shaker Verlag.

Schütte, Simon (2002): Designing feelings into products: Integrating kansei engineering methodology in product development. Universität Linköping, Linköping.

Schwaber, Ken; Sutherland, Jeff (2017): The Scrum Guide. The Definitive Guide to Scrum: The Rules of the Game. Online verfügbar unter https://www.scrumguides.org/docs/scrumguide/v2017/2017-Scrum-Guide-US.pdf, zuletzt geprüft am 17.10.2020.

Sitte, J.; Winzer, P. (1996): Measurement of information flow for enterprise model building. In: Proceedings 1996 IEEE Conference on Emerging Technologies and Factory Automation. ETFA'96. 1996 IEEE Conference on Emerging Technologies and Factory Automation. ETFA'96. Kauai, HI, USA, 18–21 Nov. 1996: IEEE, S. 378–384.

Sitte, J.; Winzer, P. (2005): Demand Compliant Design of Robotic System. In: Jason Gu und Peter X. Liu (Hg.): 2005 International Conference on Mechatronics and Automation. July 20 to August 1, 2005, Niagara Falls, Ontario, Canada : conference proceedings. Piscataway, NJ: IEEE, S. 1953–1958.

Sitte, J.; Winzer, P. (2011): Systemmodellierung im Fokus von Generic Systems Engineering. In: Gesellschaft für Systems Engineering e. V. (Hrsg), Tag des Systems Engineering.

Sitte, Joaquin; Winzer, Petra (1999): Bottom-Up Framework for Enterprise Optimisation and Control. In: *Proceedings of the International Enterprise Modelling Conference (IEMC'99)*. Online verfügbar unter https://eprints.qut.edu.au/148008/.

Sitte, Joaquin; Winzer, Petra (2006): Evaluation of a new complex system design method on a mechatronic automotive product. In: 2006 IEEE International Engineering Management Conference. IEEE, S. 278–282.

Wang, Hao; Chen, Guanlong; Lin, Zhongqin; Wang, Haihua (2005): Algorithm of integrating QFD and TRIZ for the innovative design process. In: *International Journal of Computer Applications in Technology* 23 (1), S. 41.

Weilkiens, Tim (2019): Systems Engineering mit SysML/UML. Anforderungen, Analyse, Architektur. 3., überarb. und aktualisierte Aufl. Heidelberg: dpunkt.verlag.

Welge, Matthias; Martinez, Nicolas; Steblou, Katarina; Friedrich, Christian (2014): Einsatz agiler Projektmanagement Methoden zur Erfüllung von Automotive SPICE Anforderungen. Erreichung von MAN. 3 Projektmanagement Level 2 unter Anwendung des Scrum Frameworks. In: *Tag des Systems Engineering*, S. 155.

Willing, M.; Winzer, P. (2015): Fehler vermeiden heißt Fehler verstehen—Anforderungen an eine neue Methodik. In: Stefan Bracke, Michel Mamrot und Petra Winzer (Hg.): Qualitätsmethoden im Diskurs zwischen Wissenschaft und Praxis. Bericht zur GQW-Jahrestagung 2015 in Wuppertal. 1. Aufl. Herzogenrath: Shaker (Berichte zum Qualitätsmanagement, 2015, 17), S. 303–320.

Winzer, Petra (1997): Chancen zur umfassenden Unternehmensgestaltung. Methodischer Ansatz zur qualitäts-, human- und ökologieorientierten Gestaltung von Arbeits- und Fabriksystemen. Techn. Univ., Habil.-Schr. Berlin, 1996. Frankfurt am Main: Lang (Europäische HochschulschriftenReihe 5, Volks- und Betriebswirtschaft, Bd. 2189).

Winzer, Petra (2012): PromeSys. Abschlussbericht im Rahmen des Verbundforschungsprojektes "Prozesskettenorientiertes Regelkreismodell für ein nachhaltiges robustes Design mechatronischer Systeme"; Projektträger für das BMBF—Forschungszentrum Karlsruhe, Produktion und Fertigungstechnologie (PTKA-PFT), Förderkennzeichen 02PG1323. Aachen: Shaker.

Winzer, Petra (2015): Generic System Description and Problem Solving in Systems Engineering. In: *IEEE Systems Journal* 11 (4), S. 2052–2061. DOI: https://doi.org/10.1109/JSYST.2015.2428811.

Winzer, Petra; Braunholz, Helge (2000): Chances and Risks of Process-Oriented Integrated Management Systems. In: *Proceedings of the Human Factors and Ergonomics Society Annual Meeting* 44 (10), S. 277–280. DOI: https://doi.org/10.1177/154193120004401039.

Winzer, Petra; Braunholz, Helge (2003): Die Informationsflussanalyse—eine Grundlage für die Prozessbewertung mittels Transaktionskostentheorie. In: Tilo Pfeifer (Hg.): Prozessorientiertes Qualitätsmanagement—Gestalten, Umsetzen, Bewerten. Aachen: Shaker (Berichte zum Qualitätsmanagement, 5), S. 125–140.

Yamashina, Hajime; Ito, Takaaki; Kawada, Hiroshi (2002): Innovative product development process by integrating QFD and TRIZ. In: *International Journal of Production Research* 40 (5), S. 1031–1050.

Zielasek, Gotthold (1995): Projektmanagement. Erfolgreich durch Aktivierung aller Unternehmensebenen. Berlin, Heidelberg: Springer Berlin Heidelberg.

6 Case Studies—Managing New Dimensions of Complexity With GSE

The *first example,* presented in Sect. 6.1, shows, ***how the increasing diversity of requirements in the product development phase can be managed using the GSE approach.*** The increase in the diversity of requirements is an expression of the new dimensions of complexity management fixed in Chap. 2, which is due to the increasing individualization of products.

The ***second example*** of Sect. 6.2 demonstrates, ***how merging system boundaries,*** which are a consequence of the trend towards miniaturization, ***can be designed by transdisciplinary teams using the GSE approach in product development***.

The ***third example*** from Sect. 6.3 illustrates how, despite increasing division of labor and specialization as a result of globalization, ***the reliability of product systems, especially mechatronic systems, can be ensured transparently and traceably over the lifecycle by using the GSE approach.***

In the ***fourth example*** from Sect. 6.4, ***potential critical failures in the usage process, especially in the product-environment interaction in the early phases of product development, are detected using failure networks.*** The GSE approach allows for a systematic and structured procedure that can be used across disciplines.

The ***fifth example*** from Sect. 6.5 shows what ***support the GSE approach*** provides ***in the feedback of field data into the design and development process,*** in order to learn from past mistakes in a targeted manner (Mamrot 2014).

To what extent not only DeCoDe can be used for error claim tracing, but also e-DeDoDe and the implied assignment of the claim to the corresponding responsible production sites is outlined in the ***sixth example*** (Sect. 6.6, ***failure cause search and solution algorithm).***

The ***seventh example,*** presented in Sect. 6.7 summarizes the ***application of the GSE approach in the early phases of customer-integrated product development in industrial plants*** (Schlüter and Winzer 2014, pp. 267–278). The coordination between the customer

N. Schlüter, *Generic Systems Engineering*, https://doi.org/10.1007/978-3-662-67994-4_6

and the developers has always been a problem. How the communication process between the two can be improved using the GSE approach is illustrated by this example.

The ***eighth example*** (Sect. 6.8) then deals with the use of ***GSE for organizational development.*** Using the example of the further development of a manufacturing company, it is explained how new requirements can be initially integrated into the production processes using GSE, how the processes and information flows are then redesigned, and finally rolled out in the QM system.

While examples one to five relate to technical systems, examples six to eight focus on socio-technical systems. They all illustrate ***how problem-specific modified solutions can be developed using the standardized GSE approach*** and how DeCoDe as well as e-DeCoDe as a standardized model with minimal views can handle complexity. This is only possible by ***implementing thesynergy between GSE conceptual modeland GSE procedural concept and continuously refining the GSE thinking model.*** Certainly, product lifecycle considerations are state of the art in science and technology (cf. Lindemann 2005; Lindemann et al. 2009; Gausemeier 2009; Pfeifer and Schmitt 2014; Abramovici et al. 2014, pp. 140–145), but it remains to be noted that the product lifecycle phases are only partially standardized. In the defined Quality-Gates of Fig. 5.1, the examination of the degree of requirement fulfillment often takes place via checklists or protocolled workshops (Prefi 2014). This creates a multitude of information that cannot be connected and that cannot—or only with difficulty—be assigned to the conceptual model of the respective system under consideration (cf. Fig. 6.1).

The case studies also aim to show that the basic idea of product lifecycle management can be coupled with the GSE approach. Based on the defined product lifecycle and using the Quality-Gates, in conjunction with the GSE approach, i.e. especially with the standardized GSE thinking model, the foundation for a structured information flow can

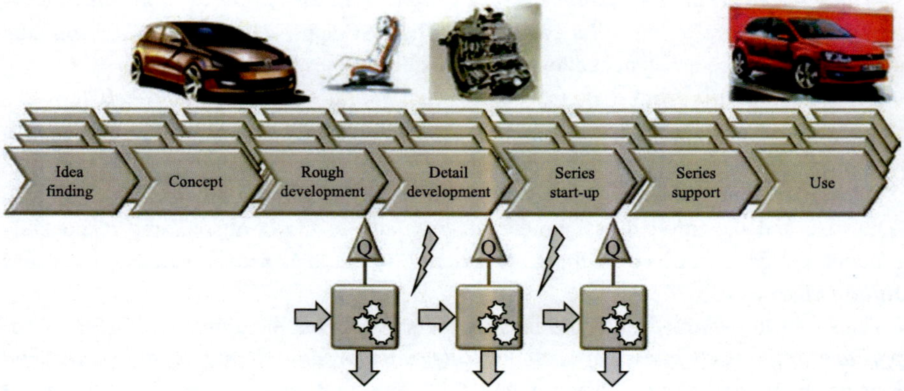

Fig. 6.1 Quality-Gates and the management of the flood of information (Riekhof and Winzer 2010)

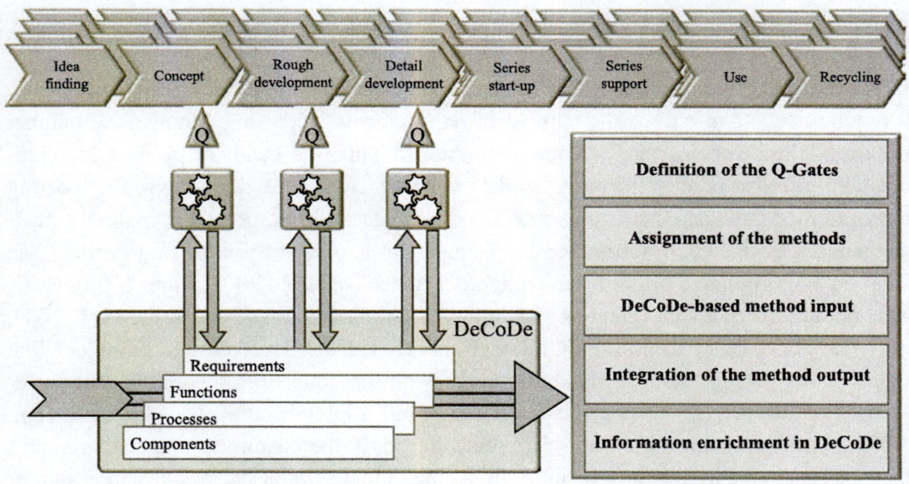

Fig. 6.2 The systematic connection of the GSE thinking model with the GSE procedural concept based on the lifecycle of a system (Riekhof and Winzer 2010)

be laid. This is a mandatory prerequisite for Industry 4.0. At the same time, it offers a solution idea for the topic of Big Data (Bauernhansl et al. 2014).

Figure 6.2 illustrates the basic idea of considering the product lifecycle using the GSE approach. Here, the product lifecycle is a guideline for the GSE procedural concept. At the Quality-Gates, it is decided which methods and procedures are necessary for analysis, goal setting or design and how these flow into the GSE thinking model. Thus, changes in the GSE thinking model always occur at the Quality-Gates, which need to be documented. By comparing the GSE thinking models at each Quality-Gate, traceability of the system's change can be guaranteed.

This basic principle of interlocking the GSE thinking model and GSE procedural concept is also demonstrated in the examples in the following chapters, i.e. in:

- the requirement-based product development (first example, Sect. 6.1),
- the development of mechatronic systems (second example, Sect. 6.2),
- the reliable design of products over the product lifecycle (third example, Sect. 6.3),
- the identification of failures in potentially critical usage processes (fourth example, Sect. 6.4),
- the feedback of field data into the design and development process (fifth example, Sect. 6.5),
- the failure cause search and solution algorithm (sixth example, Sect. 6.6),
- the customer-integrated product development (seventh example, Sect. 6.7) and
- the requirement-based organizational development using the REMOt approach (eighth example, Sect. 6.8).

6.1 Requirement Update in Product Development

A crucial aspect of in managing current and future complexity is mastering the diversity of requirements. Not only is the number of requirements increasing, but also the number of stakeholders who set requirements for systems. This is a result of product individualization, but also of globalization, as elaborated in Chap. 1. Requirements Engineering or requirements management is intended to contribute to this aspect of complexity management. Requirements Engineering & Management is part of Systems Engineering (see Chap. 3) and originated primarily in computer science (see Pohl 2008; Rupp 2009). However, requirements management is also important for marketing (see Davis et al. 2007, pp. 1–31; Jockisch et al. 2009). JOKISCH and HOLZMÜLLER clearly point out that requirements management in marketing focuses on the collection, recording, and evaluation of requirements, but neglects further use and implementation (see Jockisch et al. 2009). In addition, not all stakeholders, but often only the customers and their requirements are included in the consideration. In engineering disciplines, especially in design or construction, the collection, evaluation, and verification of the degree of requirement fulfillment is an integral part of product development (see Ehrlenspiel 2003; Pahl et al. 2005; Lindemann et al. 2009; Prefi 2014, pp. 402–427). The requirement update, according to SCHLUND (see Schlund 2011), i.e., the constant updating of requirements throughout the product development process, requires systematic support, as outlined in Fig. 6.3.

The collection of requirements is done, for example, through research, observations, and conversations. They are documented using requirement lists and specifications. Their specification can be done through the pattern, the prototype, the pre-series, etc. However, new requirements can also arise. For their appropriate structuring and documentation, software support is suggested, such as DOORS[1] .

The evaluation of the requirement is also often only from the customer's perspective and is difficult to understand (see Schuh 2009; Schlund et al. 2009, pp. 54–59). This can also lead to conflicts of objectives (see Lindemann 2005; Lindemann et al. 2009). The requirement update and its modeling are insufficiently implemented. The focus is on a continuous integration of new or changed requirements and thus represents a cross-process with interfaces to the other areas of requirements management. A permanent coupling of requirements management with the phases of product development is demanded, but not consistently implemented (see Pahl et al. 2007; Lindemann 2007; Ehrlenspiel 2009; Schlund 2011). For this reason, SCHLUND developed (see Schlund 2011) a solution proposal for the outlined problem of requirement updating in product development using the GSE approach (see Fig. 6.4).

SCHLUND used the GSE thinking model with the DeCoDe tools to create the image of the logistical system. Consequently, the logistical system is modeled in four prede-

[1] DOORS (English Dynamic Object Oriented Requirements System) is a software for requirements management.

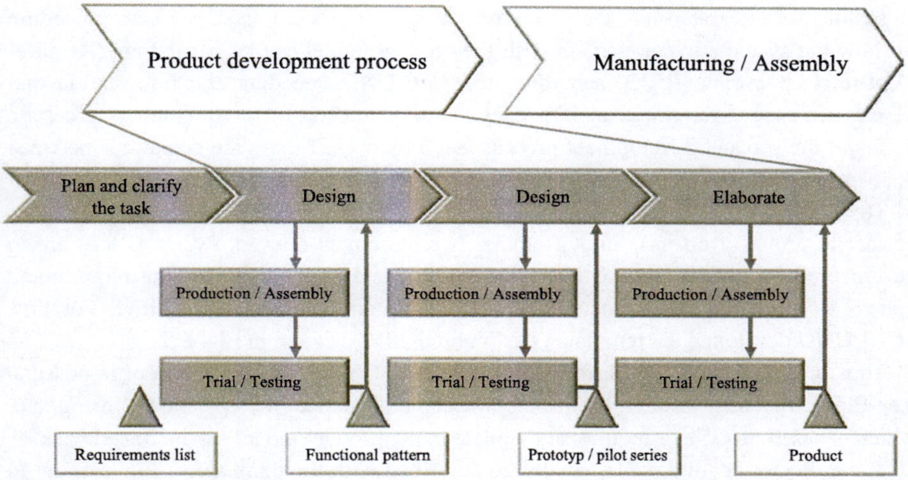

Fig. 6.3 The process of product development and stopping points for requirement updating (see Schlund 2011, p. 10)

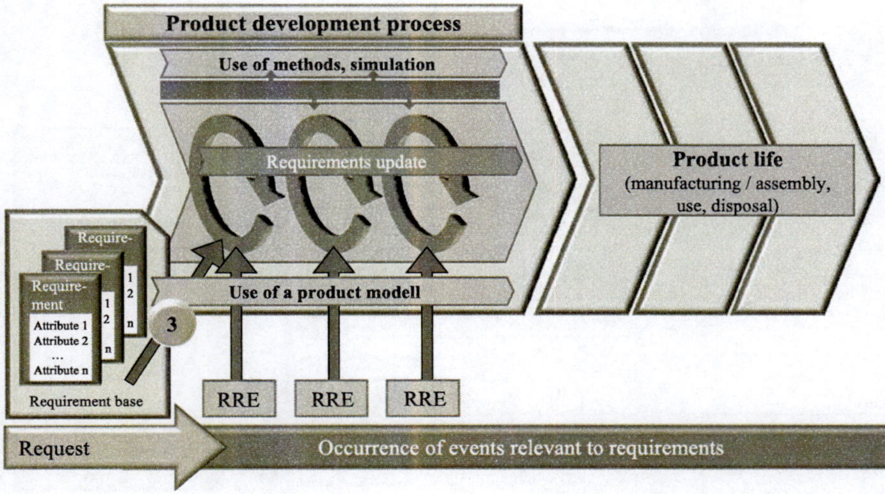

Fig. 6.4 The integration of requirements into the product model (see Schlund 2011, p. 81)

fined views, i.e., the requirement, component, function, and process view (see also Sect. 4.3). To systematically implement the initially defined requirements within the product development process, on the one hand, the product model itself is changed (see point 3 in Fig. 6.4). On the other hand, there can also be a requirement update. This must be seamlessly integrated into the product development process and describe a procedure for how the systematic comparison with the product model can be carried out. Requirement-relevant events (RRE), according to SCHLUND, are data and information that have a reference to requirements (R). They are to be linked to the requirements at certain points of the product development process (see Fig. 6.4). If there are deviations between them, appropriate corrective measures are necessary (see Schlund 2011).

These can include errors, disturbances, complaints, non-compliance with tolerances (a special group of errors), among others. If requirement-relevant events (RRE) are to be mirrored on the requirements (R)—this is a basic prerequisite for the requirement update—a corresponding standardization of the product model is required. For this, SCHLUND developed the template (see Schlund 2011), shown in Fig. 6.5.

Thus, a linking of the GSE thinking model and GSE procedural concept is possible (see Fig. 6.6). Thus, through the problem-specific sequence of steps shown in Fig. 6.6, which is based on GSE, a requirement update in the product model can be traced back.

The collected requirements, which are described with the template of Fig. 6.5, are to be set in relations with the system model, i.e., with the component, function, and process view. If new, changed requirements arise during the product development process

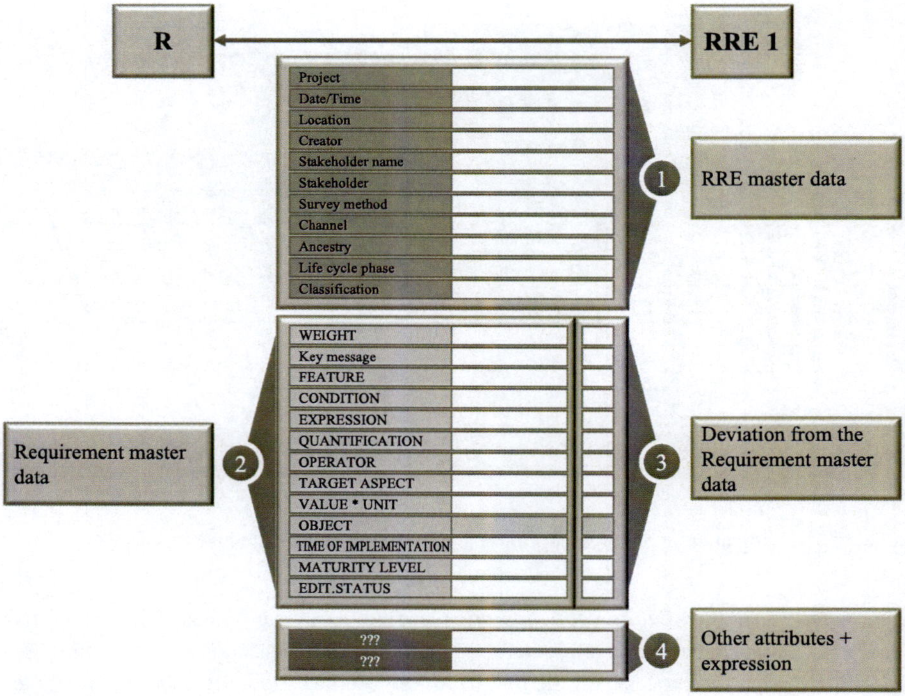

Fig. 6.5 RRE recording using the RRE template (see Schlund 2011, p. 99)

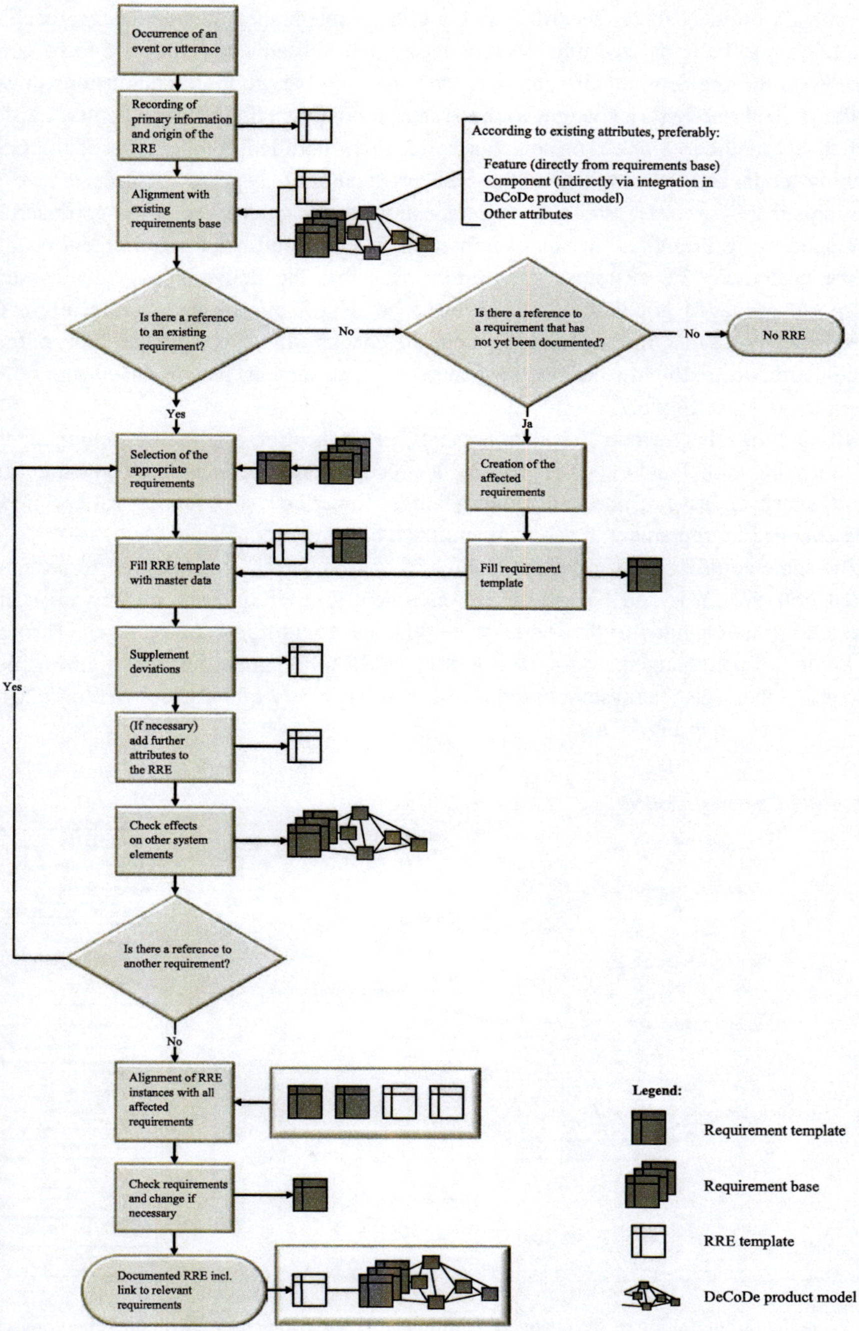

Fig. 6.6 Comparison of potential RRE with an existing requirement base using the requirement template and DeCoDe product model. (After Schlund 2011, p. 101)

or some are omitted, these, according to the GSE problem-specific solution approach of Fig. 4.6, are to be reinserted into the product model. Subsequently, they are to be compared with the previous requirements, re-evaluated, and appropriate conclusions drawn for the possibly necessary changes to the system model. SCHLUND (see Schlund 2011) applied his method, which represents an application-modified combination of the GSE thinking model and the modules of the GSE procedural concept, to the roller conveyor of a logistical system. He was able to demonstrate the practicability of the approach he developed using examples. In one example, SCHLUND outlines a requirement change by the customer. The customer requirement was that the delivered logistical system transports packaged goods at a maximum speed of 1.5 m/s instead of the originally agreed speed of 2 m/s. In the following steps, he checks what consequences this contract specification desired by the customer (change of a requirement) has on the change of the system model (see Fig. 6.7).

By applying the method, which represents an application-modified coupling of the GSE thinking model and the GSE procedural concept, it was demonstrated how the drive train (asynchronous machine, tangential gearbox, lower belt, and support rollers) needs to be changed to implement the above-mentioned customer requirement.

The same could be proven for the change of the requirement, “goods are to be transported both forwards and backwards”. Consequently, the requirement “forward transport” had to be changed to the requirement “transport with direction change”. Through the linking of the changed requirement with the GSE thinking model, shown in Fig. 6.8, changes in the logistical system could also be initiated very quickly (see Schlund 2011, p. 133), as shown in Fig. 6.9.

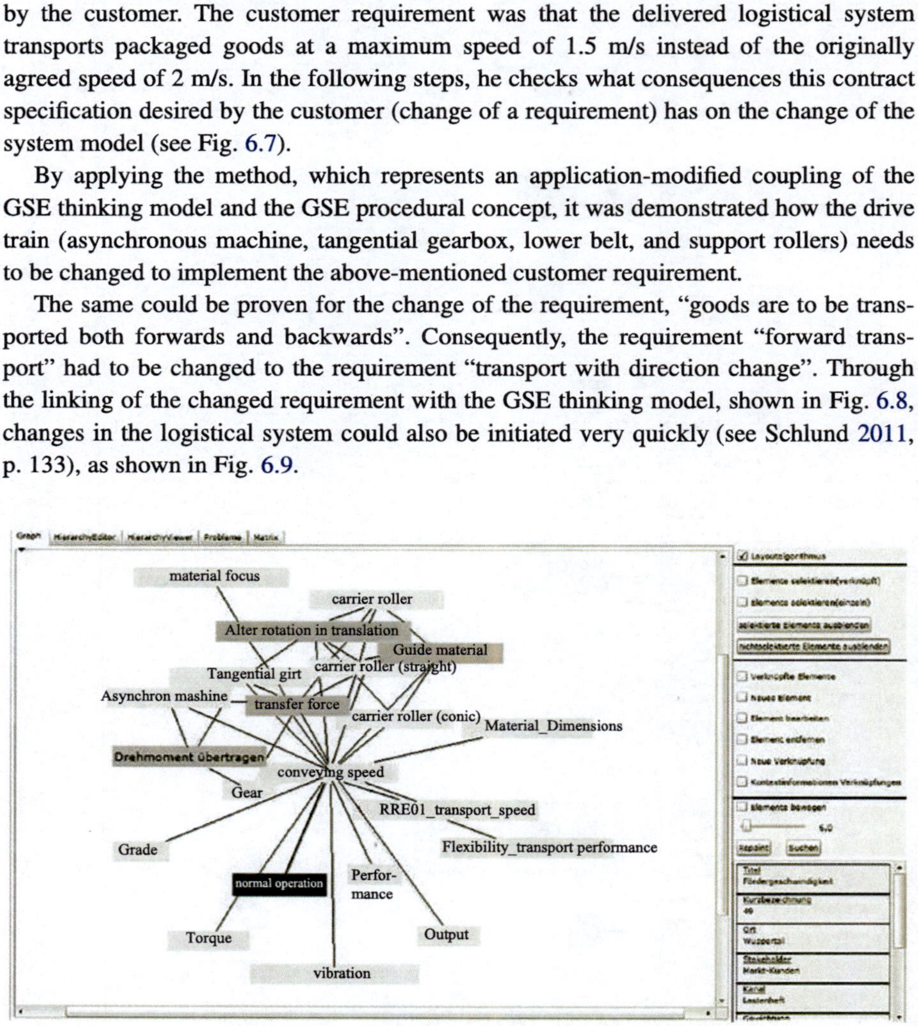

Fig. 6.7 The direct linking of the changed requirement “conveying speed” with the other views of the GSE thinking model using the DeCoDe tools. (After Schlund 2011, p. 132)

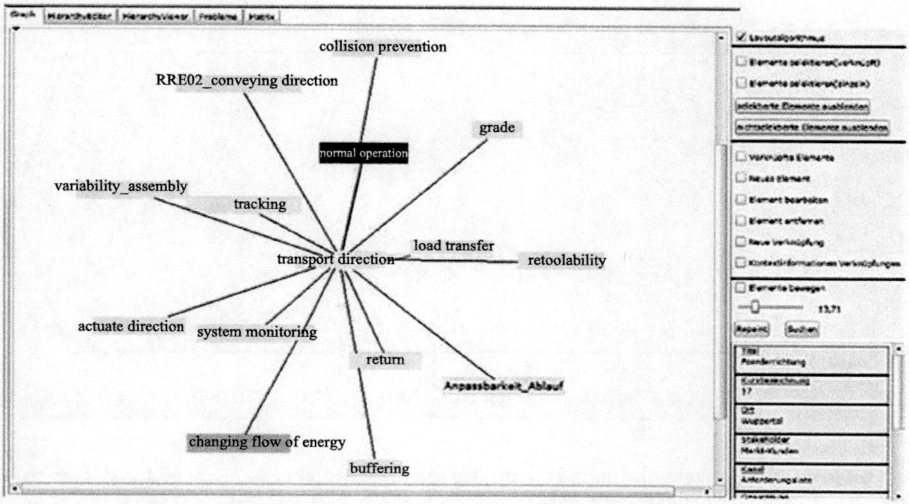

Fig. 6.8 The direct linking of the requirement "conveying direction" with the other views of the GSE thinking model using the DeCoDe tools (see Schlund 2011, p. 133)

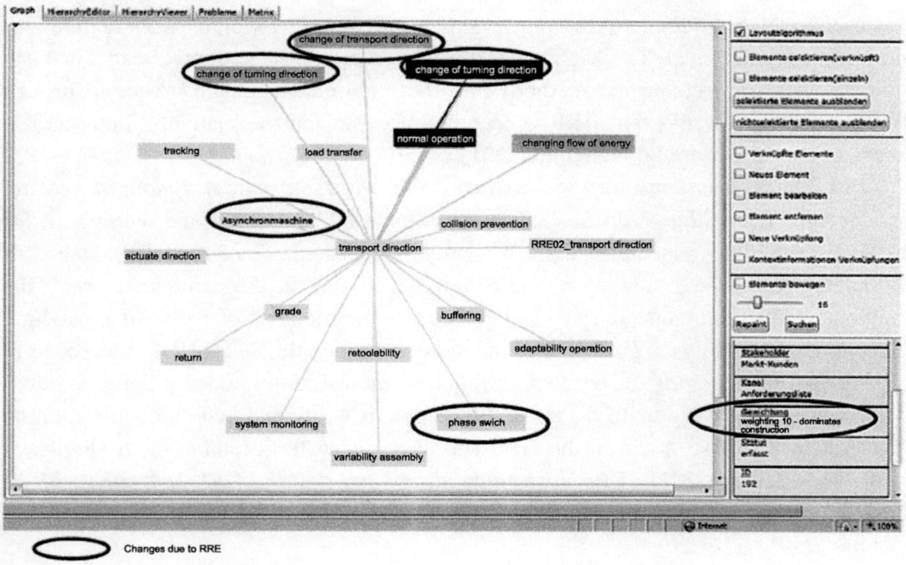

Fig. 6.9 The specified requirement "transport with direction change" in relation to other views of the GSE thinking model using the DeCoDe tools (see Schlund 2011, p. 134)

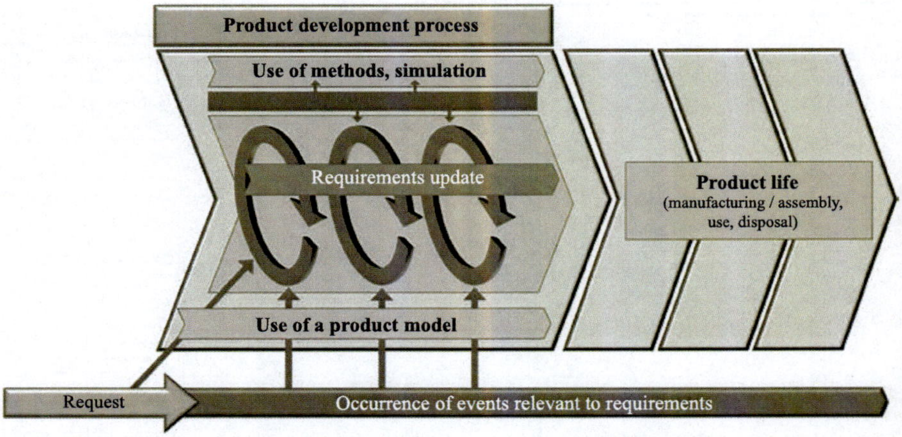

Fig. 6.10 Method and simulation use, controlled via the GSE procedural concept and their classification in the overall concept of the RRE methodology (see Schlund 2011, p. 134)

Through the RRE method (see Fig. 6.10), essentially two basic principles of systematic thinking and action were applied, i.e., ***the basic principle of information encapsulation*** (requirement-relevant events) and ***the basic principle of minimal models,*** i.e., it was asked which requirement-relevant events have effects on which views or relations of the conceptual model. The RRE method was used by transdisciplinary teams that can integrate their subject-specific methods into the requirement update process. Through it, requirement-relevant events (RRE) become traceable and transparently representable (see Künne and Richard 2009; Schlund 2011).

In summary, it is important to emphasize that the requirement update in product development is possible with the GSE approach. SCHLUND used and modified it for the RRE methodology (see Fig. 6.10). By applying the standardized modules of the GSE approach, i.e., the GSE thinking model with its views such as the requirement view, the component, function, and process view, as well as the modules of the GSE procedural concept, i.e., the analysis, goal formation, and design module, SCHLUND succeeded in systematically integrating the problem of requirement updating into the product development process. He was able to demonstrate the practicability and benefit of this solution approach, which was based on the GSE approach, using the example of the logistical system (see Schlund 2011). ***This case study also shows that the GSE approach with its modules can be modified problem-solving oriented and supplemented by subject-specific methods.***

The following example will now examine how it is possible to integrate the GSE approach into existing procedural concepts of product development—specifically the standardized procedure for the development of mechatronic products.

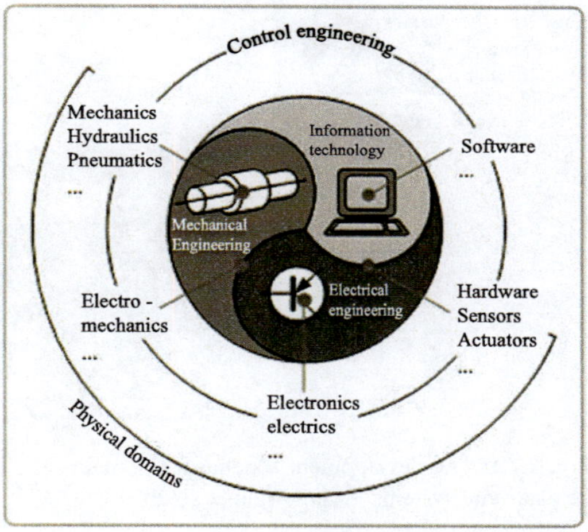

Fig. 6.11 The characteristics of mechatronic systems. (After Lippold 2001)

6.2 Development of Mechatronic Systems

While the example presented in Sect. 6.1 of the application of the GSE approach represents a possible response to the trend of product individualization, the same should apply to the trend of miniaturization tendency. Mechatronic systems are a result of this trend. They are systems with merging system boundaries, i.e., the domains of information technology, electrical engineering, and mechanical engineering are difficult to delineate. The same applies when characterizing the domains from a physical or information technology perspective, as illustrated by LIPPOLD in Fig. 6.11.

The development of mechatronic systems must be cross-domain. However, practice clearly shows that although the system specification for the mechatronic system as a whole is carried out via the black-box model, the development steps thereafter are carried out separately according to their specific field. Thus, a mechatronic system is developed in parallel according to the specific procedural concepts of mechanical engineering, electrical engineering, computer science, and control engineering. Only in the process of "the marriage", i.e., when the mechatronic system is integrated into a complete system, such as a car, can it be tested to see if it meets the requirements. Only at this stage of the process is it possible to verify whether these field-specific parallel developments lead to a functional mechatronic system. This basic development scheme is shown in Fig. 6.12.

As proven in Sect. 2.4.2 and detailed by OTT (see Ott 2009), computer science, electrical engineering, control engineering, and mechanical engineering use different procedural concepts. Parallel to this, due to the variety of procedural concepts that are domain-specific, special procedural models have been developed for mechatronic systems special procedural models (see Ott 2009). The best known is the V-model according to VDI 2206 or the phase model for the development of mechatronic systems. The latter, unlike the V-model, already has iteration loops for specifying requirements, but also

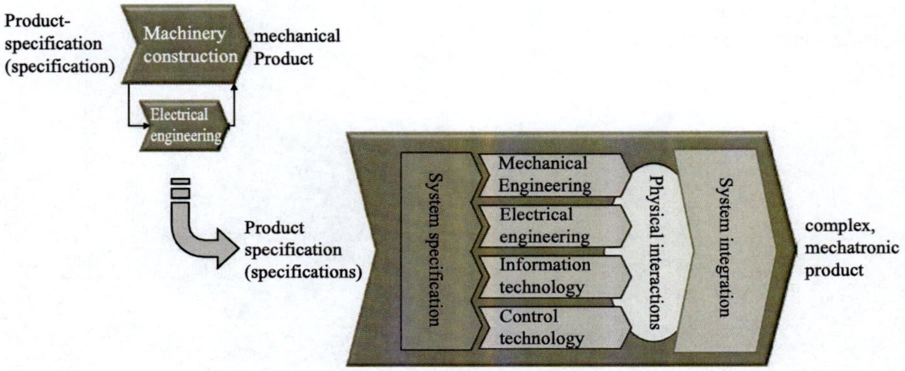

Fig. 6.12 The development scheme for domain- and system-integrated product development for mechatronic systems. (After Welp et al. 2001)

relies on domain-specific models. OTT demonstrates through the comparison of various procedural models that they do not use a uniform thinking model, i.e., a meta-model (see Ott 2009). He proves that developers of mechatronic systems work with specific conceptual models in their domain, i.e., the control engineer uses his conceptual model of the mechatronic system from the perspective of control engineering and the electrical engineer and mechanical engineer also each have their own conceptual model. The same applies to the computer scientist. These field-specific conceptual models of the same mechatronic system only partially coincide.

If mechatronic systems are to be developed cross-domain, as OTT demands, a common conceptual model is needed. This can be created by applying the GSE approach using the DeCoDe tools. At the same time, he demands a common procedural concept that applies across domains. Based on the evaluation of various procedural concepts and using the GSE approach in a problem-specific, modified coupling of the GSE thinking model with the GSE procedural concept, OTT develops the so-called double-cycle model, shown in Fig. 6.13.

OTT clearly points out that the design of mechatronic systems based on the GSE thinking model takes place in several iteration loops and not cyclically sequentially, as previously assumed. This corresponds to a GSE procedural concept modified for mechatronic systems. However, each phase of the double-cycle model must always be coupled with the GSE thinking model, as the basic principle in Fig. 6.14 illustrates.

Each of these steps of the GSE procedural concept modified for mechatronic systems serves to update the GSE thinking model, in which the information flows into the GSE thinking model in a standardized manner. This must be done continuously, i.e., both at each individual step of the problem-solving specific GSE procedural concept and at the cyclic repetition of the entire modified GSE approach, as Fig. 6.15 shows.

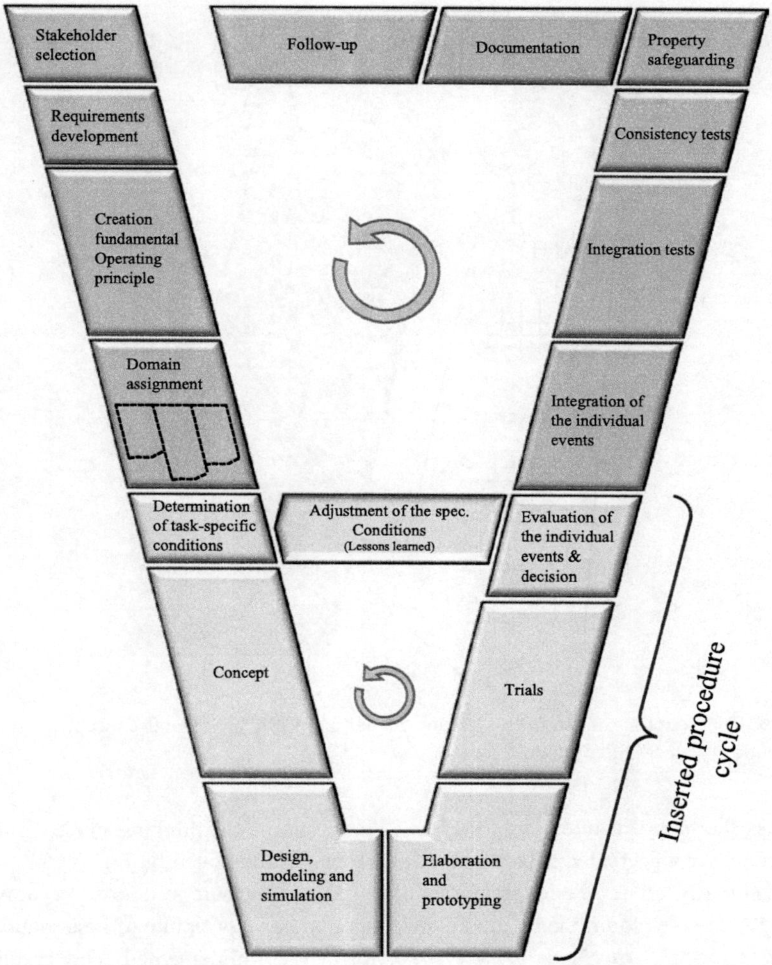

Fig. 6.13 Double-cycle model for the requirement-oriented development of mechatronic systems, developed as a result of the application of the GSE approach. (After Ott 2009)

In the following steps, OTT demonstrates using the example of the development of a mechatronic system how the interaction can occur at individual steps of the problem-specific GSE procedural concept (double-cycle model) with the GSE thinking model. The modified GSE procedural concept for the development of mechatronic systems, for example, provides for the requirements to be determined for each stakeholder in the phase of requirement development, to be included in requirement catalogs, to be compared and prioritized.

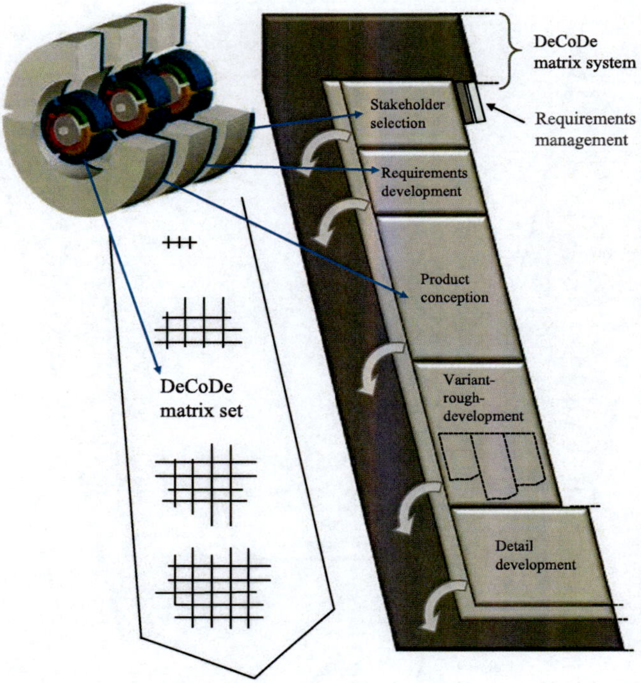

Fig. 6.14 The coupling of the GSE thinking model and GSE procedural concept, modified for mechatronic systems. (After Ott 2009)

This is the basis for integrating these results into the GSE thinking model. Thus, the requirement view is created using the DeCoDe tools, as illustrated by Fig. 6.16.

Subsequently, solution-neutral functions of the mechatronic system to be developed are assigned to the requirements. In doing so, another basic principle of systematic thinking and action, i.e., ***the basic principleof neutrality,*** is implemented. This results in a consistent separation of the realization of requirements by systematically and structured answering of core questions, i.e., from "what" (functions) to "with what" (components) and "how" (process).

Returning to the phase of projecting the functions of the mechatronic system, these need to be structured. For example, the safety function is assigned the reliability function and the reset function. In this function structure, the interactions of the functions with each other must be further considered.

In doing so, the ***basic principle of information encapsulation*** should be observed as another basic principle of systematic thinking and action. It allows for a clear, function-oriented structure of the mechatronic system, which provides a good basis for identifying and eliminating potential failures. Subsequently, the function view is created as another view of the image of the mechatronic system to be developed. It is to be connected with

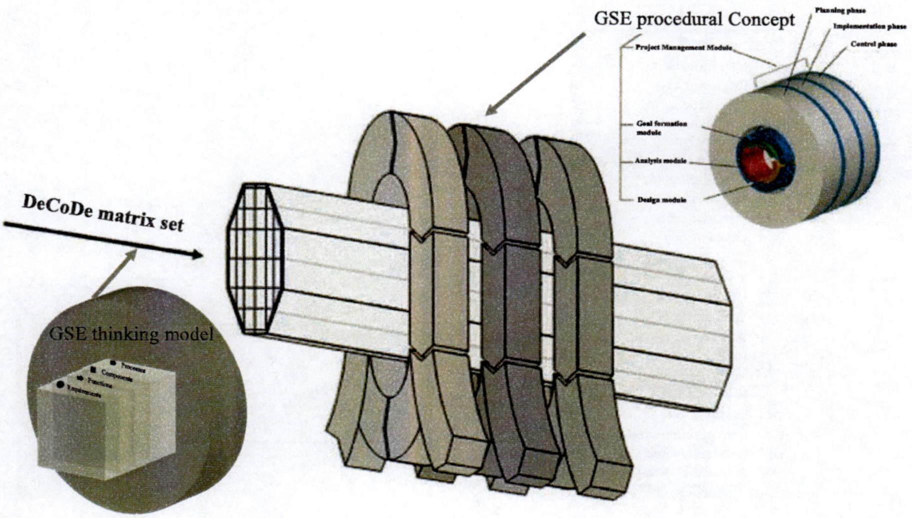

Fig. 6.15 The standardized exchange of information between the developed GSE thinking model and the modified GSE procedural concept for mechatronic systems. (After Ott 2009, p. 179)

the requirement view of the modified GSE thinking model for the mechatronic system according to the double-cycle model of Fig. 6.17.

At the same time, this step shown in overview in Fig. 6.17 also illustrates the coupling with project management, because the development of mechatronic products in companies is considered a project and is controlled via project management.

Using the example of the phase of creating the conception of mechatronic systems, it becomes clear (see Fig. 6.18) how the combination of GSE thinking model and GSE procedural concept occurs in the specific case. This is particularly important when developing solution variants. Each of them leads to a changed conceptual model, which should be documented using the GSE approach to keep the solution transparent. By standardizing the depiction of solution variants, their comparative consideration is also simplified, thus making the solution-finding process more effective.

In summary, the procedure outlined by OTT for developing mechatronic systems corresponds to an application-oriented specification of the GSE approach. He creates the GSE thinking model with the DeCoDe tools and thus represents the mechatronic system via the requirement, component, function, and process view.

From the modules of the GSE procedural concept, i.e., from the goal formation, analysis, and design module as well as the phases of project management, OTT develops the double-cycle model for the development of mechatronic systems. In the individual phases, it was proven how the GSE thinking model and GSE procedural concept are synergistically coupled with each other and how both analysis and prioritization (goal formation) and design are carried out in each phase.

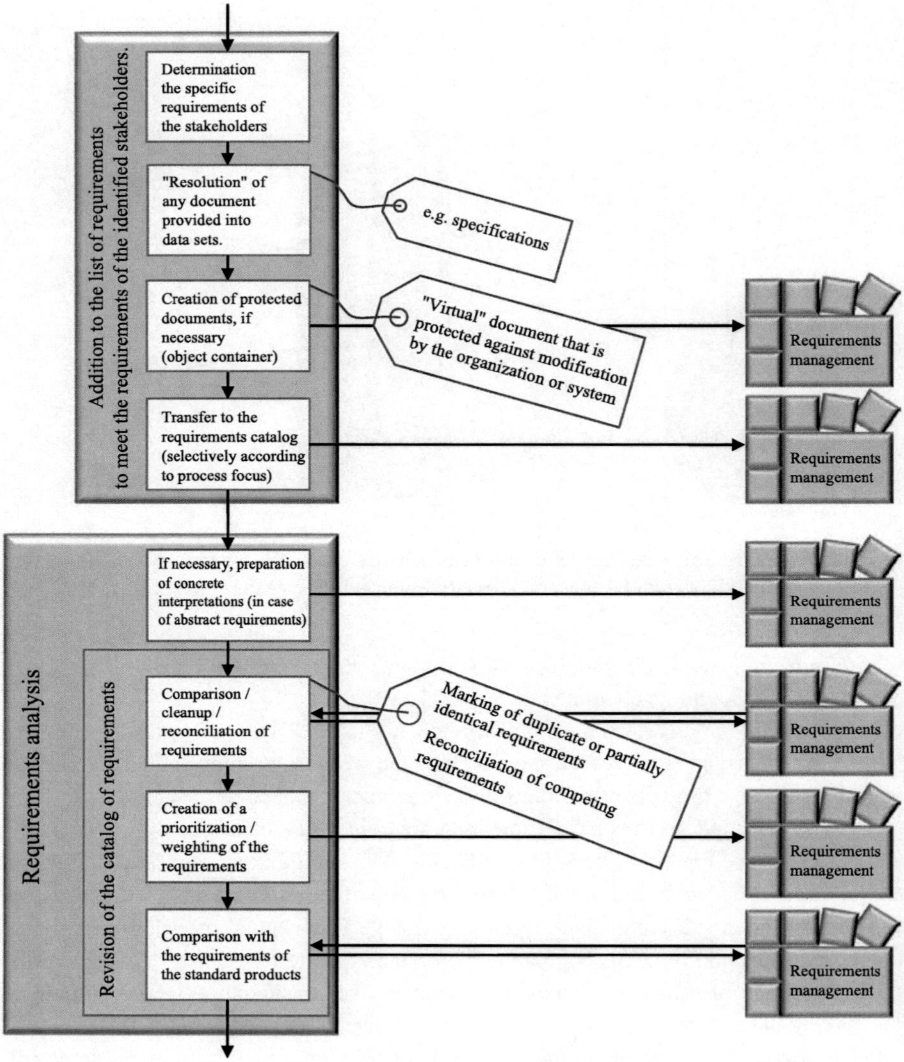

Fig. 6.16 Creating the requirement view, as part of the GSE thinking model for mechatronic systems using the DeCoDe tools. (After Ott 2009, p. 184)

Through the application of a common thinking model and procedural concept, it was proven that mechatronic systems could be better developed in practice (see Ott 2009). Although the development of a common thinking model for mechatronic systems takes more time because the different understanding of components, functions, and

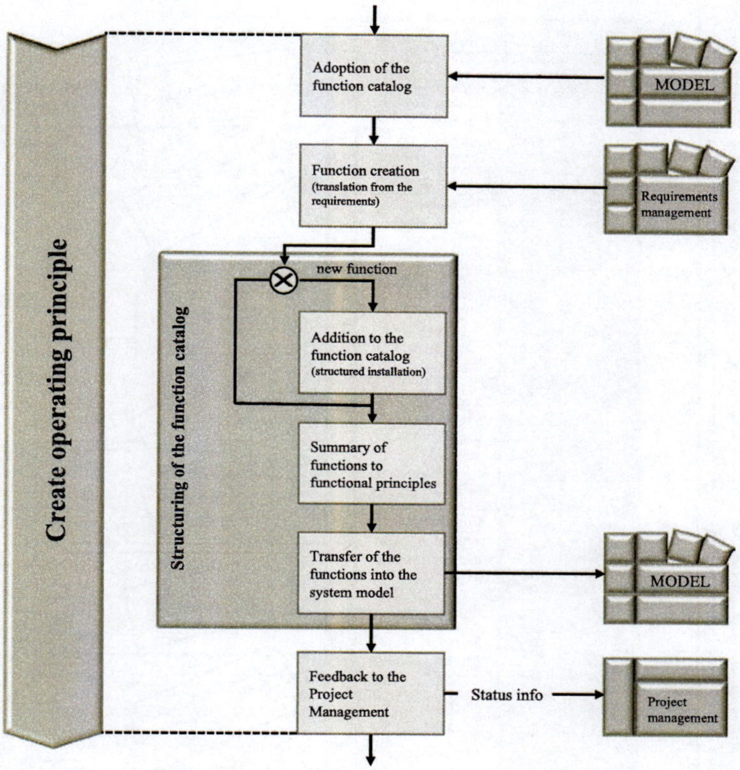

Fig. 6.17 Structuring of functions of a mechatronic system depending on the fixed requirements (see Ott 2009, p. 187)

requirements of the respective specialists (computer scientists, control engineers, electrical engineers, and mechatronics engineers) had to be transferred into a transdisciplinary concept of terms or system model. Once this has been successfully achieved, the phases of the double-cycle model can be completed more quickly by the transdisciplinary team. The jointly developed model for mechatronic systems is an expression of the basic understanding of the various disciplines for each other. On this basis, the transdisciplinary design process for mechatronic systems is better manageable. The testing of the double-cycle model also shows that it can contribute to managing complexity.

While OTT has proven how the GSE approach can be modified for the development of mechatronic systems, the following example will examine whether this approach is also suitable for the reliability consideration of mechatronic systems over their product life cycle.

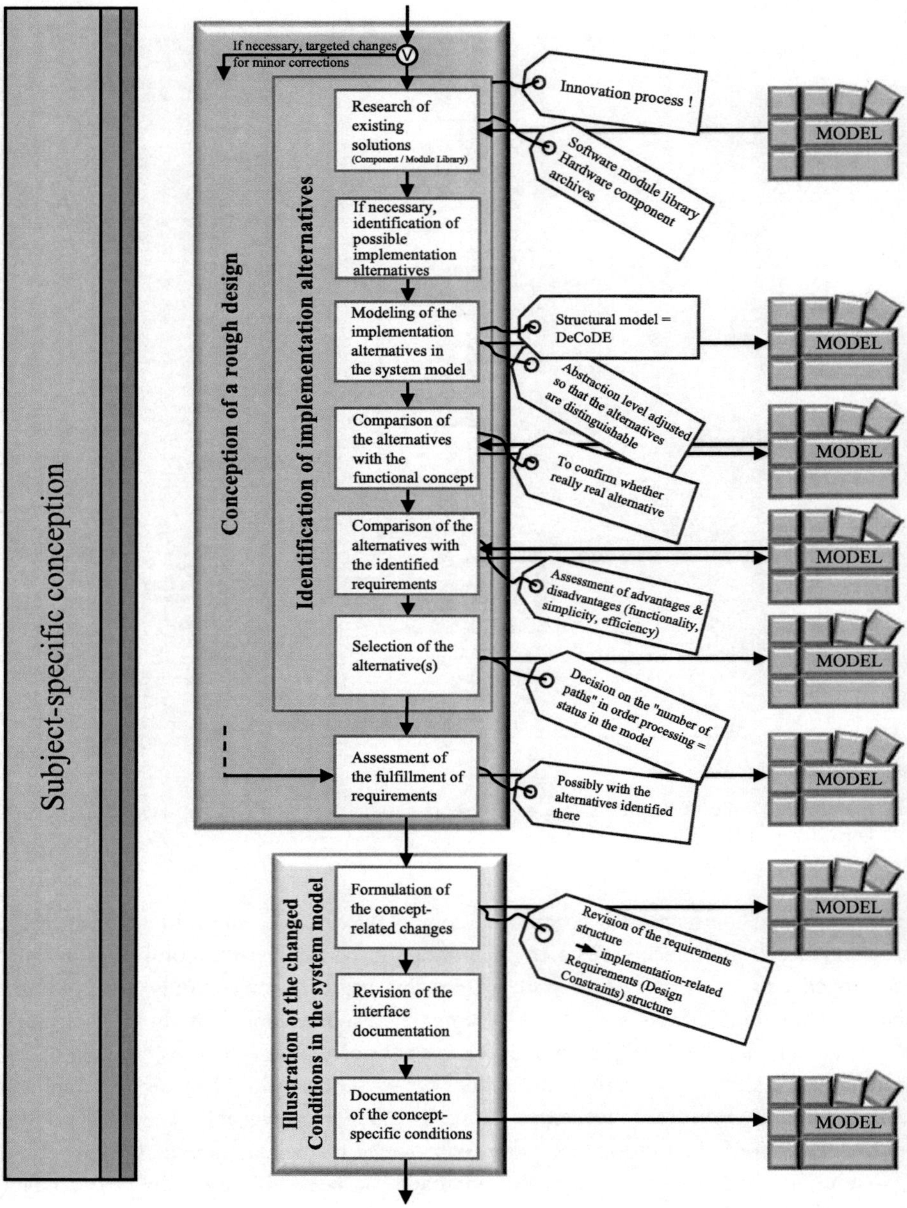

Fig. 6.18 The combination of GSE thinking model and GSE procedural concept in the phase of conception of mechatronic systems. (After Ott 2009, p. 193)

6.3 Reliability Considerations of Mechatronic Systems Over the Product Life Cycle

In Sect. 6.2, it was shown that mechatronic systems can in principle be represented using the standardized modules of the GSE thinking model. It was also demonstrated that their requirement-based adaptation is ensured via the double cycle model, which represents a modified procedural concept of the GSE approach. But what about the reliability considerations over the product life cycle of a mechatronic system? Can the GSE approach also contribute to this and thus help to cope with the third dimension of complexity, i.e., the trend towards globalization? This is reflected, among other things, in the increasing division of labor and specialization, so that a large number of globally active companies are involved in the life cycle of a product. All actors involved in this process worldwide must make their specific contribution to ensuring the reliability of a product. The possibilities for this via the GSE approach are presented below.

Reliability forecasts are based in principle on data and information, preferably from tests (test data) and/or from the field (field data). If a reliability forecast for mechatronic systems over the product life cycle is to be created, appropriate data, i.e., characteristics and characteristic values, on the behavior of mechatronic systems from the individual phases of its product life are required in order to derive forecasts for future behavior from this. Consequently, these characteristics and characteristic values (data), which can arise during the product life of a mechatronic system, must be clearly assignable to it. Because this was not possible in the past, a corresponding BMBF project was launched (Winzer 2012).

The aim of this was to develop a standardized model for mechatronic systems and a procedure for systematically controlling their reliability over the product life cycle, as shown in Fig. 6.19.

It becomes clear that the process of ensuring reliability over the product life cycle can be systematically controlled through the standardized representation of mechatronic systems. A reliability forecast can, as outlined in Fig. 6.19, require a design of the product system (reliability control). However, their results must also be represented in the model. Appropriate tests or simulations can demonstrate that the newly designed system is more reliable. Consequently, through the interaction of model and approach, the data and information that arise over the product life cycle of a mechatronic system can be assigned to its model and the control cycle of ensuring reliability can be activated. This is the basis which on the one hand serves the model precision through the target-actual comparison and on the other hand enables better reliability forecasts because more extensive data material from the life of a product can be used. The task of making this

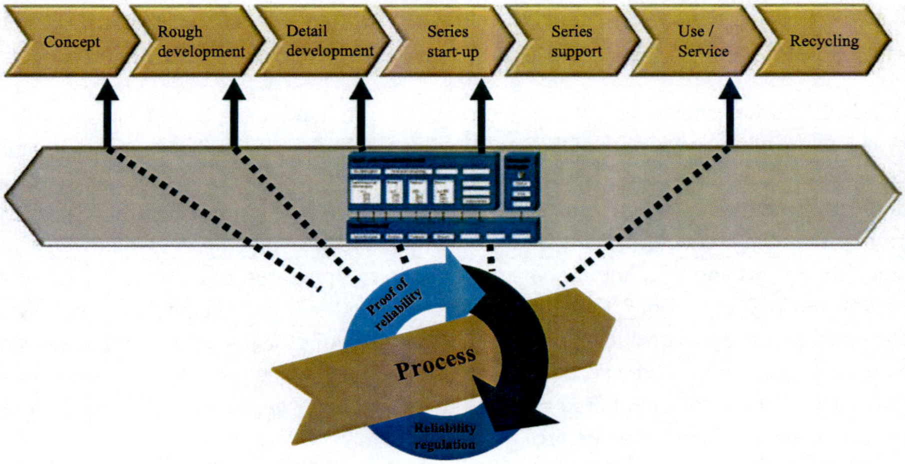

Fig. 6.19 Basic solution approach for ensuring the reliability of mechatronic systems (Müller and Winzer 2009)

usable for practice was the task of the BMBF project PromeSys[2] (Winzer 2012). Thus, the reliability forecast, which was created in the early phases of product development, should be refined with the real data. How the control cycle of the permanent life cycle phase can be repeated methodically is illustrated as a basic principle in Fig. 6.20.

During the course of the project, it had to be determined that the product life cycles of mechatronic systems, which were standardized uniformly according to VDI 4003 (see Fig. 6.21), proceed differently in operational practice, as shown in Fig. 6.22.

Consequently, the processes of the product life cycle of mechatronic systems had to be typified across the various sectors in the project. Representatives of the automotive supplier industry and the capital goods industry were involved in this process. Based on the standardized life cycle for mechatronic systems, keypoints could now be fixed according to the control loop for ensuring reliability, at which a reliability check must necessarily take place.

For this purpose, suitable methods and procedures had to be selected that are already in use in the companies involved (see Fig. 6.23). In corresponding workshops, a lively discussion arose with the partners about which methods and procedures are best suited across companies to derive statements about the reliability of the mechatronic systems.

[2]The BMBF research project PromeSys (**Pro**cess chain-oriented control loop model for a sustainable robust design **me**chatronic **Sys**tems) was funded by the Federal Ministry of Education and Research (BMBF) within the framework concept "Research for the production of tomorrow" and supervised by the project sponsor Research Center Karlsruhe (PTKA), Department of Production and Manufacturing Technologies (PTKA-PFT). Funding code 02PG1323.

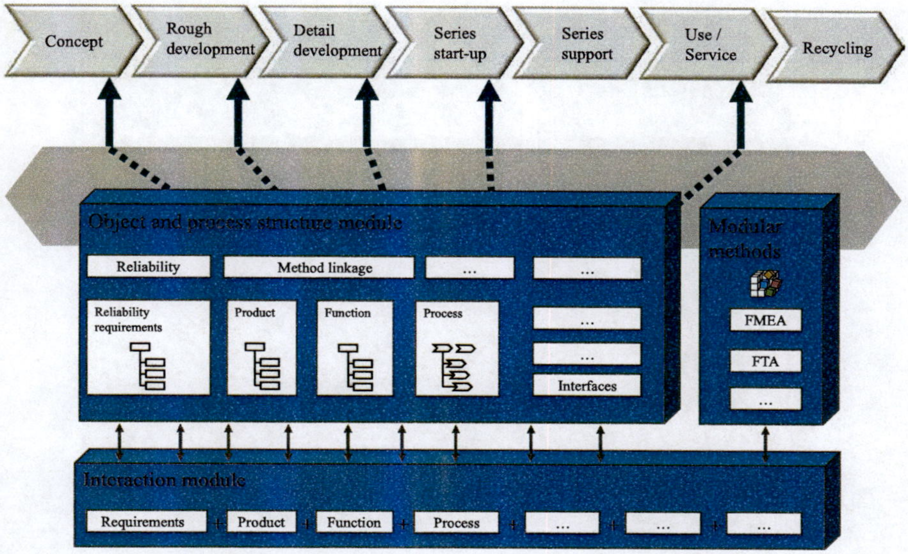

Fig. 6.20 Conceptual model and procedural concept for ensuring the reliability of mechatronic systems over the product life cycle. (After Müller and Winzer 2007)

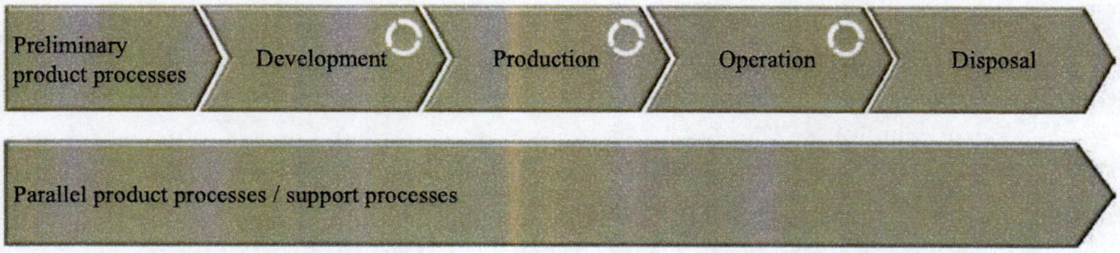

Fig. 6.21 The project-specific product life cycle according to VDI 4003 (see Winzer 2012, p. 23)

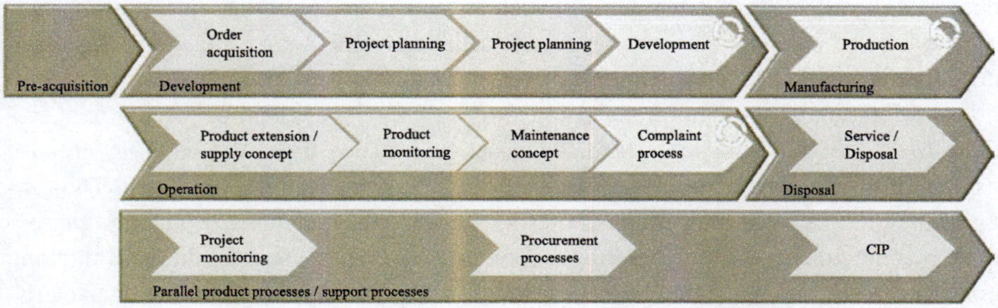

Fig. 6.22 Company-specific product life cycle for a mechatronic system (see Winzer 2012, p. 24)

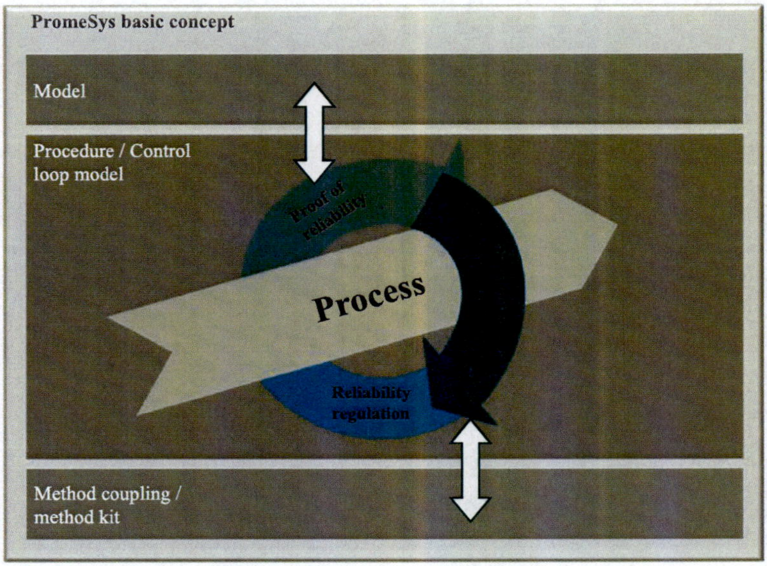

Fig. 6.23 The cross-process control loop model for ensuring the reliability of mechatronic systems (see Winzer 2012, p. 33)

The results of the ***linking of GSE conceptual model(white boxes) and GSE procedural concept (grey arrows) in conjunction with the methods and procedures for the reliability consideration of mechatronic systems over the product life cycle*** are summarized as a basic principle in Fig. 6.24.

Within the framework of the project, the GSE thinking model was supplemented by the free attribution and the problem-solving approach, and the GSE procedural concept was developed by the control loop for reliability control of mechatronic systems in connection with suitable methods, as summarized in Fig. 6.25.

The research activity on the reliability consideration of mechatronic systems over the product life cycle was based on another thesis. If it is possible to break down mechatronic systems into subsystems in such a standardized way that data on these subsystems can be collected across sectors, the reliability forecast for this could be based on a broader data basis and thus become more accurate. This thesis was derived from the consideration that, for example, distance sensors as a subsystem of mechatronic systems are used in many sectors. The distance sensors in the car are known. They are also installed in locking systems or used in overhead current collectors in O-buses. The testing of the truth of the aforementioned thesis required two prerequisites. The first prerequisite was the precision of the thinking model, i.e., the system views of the GSE thinking model had to be supplemented by free attribution. Only in this way can the characteristics and characteristic values, which are recorded over the product life cycle, be clearly

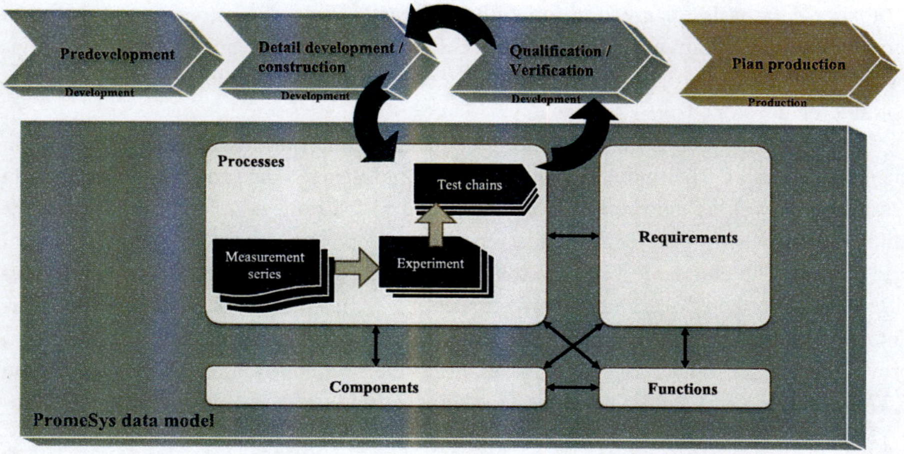

Fig. 6.24 Basic principle of the integration of methods and procedures in the product life cycle-related reliability forecast of mechatronic systems (see Winzer 2012, p. 53)

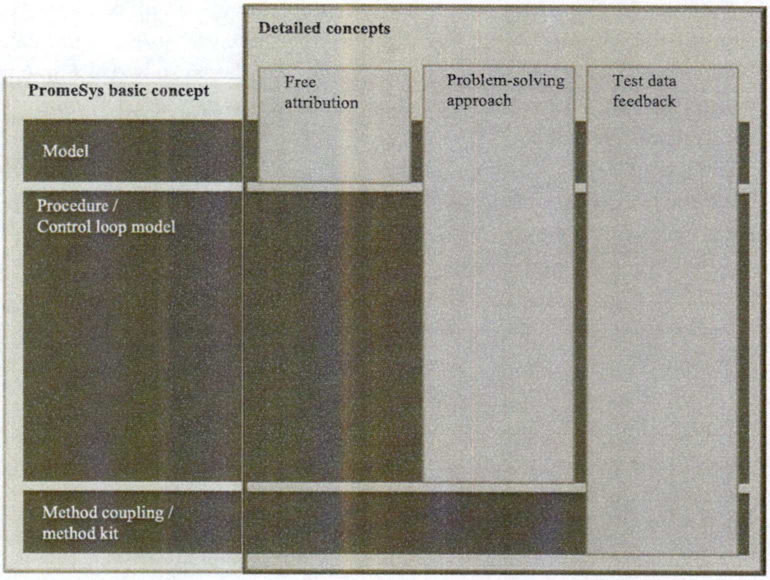

Fig. 6.25 The PromeSys solution approach (see Winzer 2012, p. 25)

assigned to the attributes of the respective system elements (see Fig. 6.25). The second prerequisite was the computer support of the procedural concept and the thinking model via a corresponding internet-capable portal. This portal should enable the cross-sector collection of data and information, e.g., of identical drives and sensors, in order to have a broader data basis for creating reliability forecasts. For this, the connection between the modified GSE thinking model and the GSE aprocedural concept as a temporal logical sequence (see Fig. 6.26) had to be projected. This was the template for the PromeSys portal to be developed.

The portal can be used for the reliability consideration of mechatronic systems both in one's own company, per industry, but also across industries. Thus, company-specific data protection as well as the protection rights agreed in the project per industry and across industries had to be ensured. For this reason, a graded security concept for the portal was created. The companies that use the portal can thus be sure that their company-specific data is protected from unauthorized access by third parties. Nevertheless, the participating companies can use data, for example for the already mentioned motion sensor, provided by other industries and integrate it into their reliability forecast models. Figure 6.27 outlines the domain model of the PromeSys portal.

This leads to the basic structure of the PromeSys portal shown in Fig. 6.28.

The PromeSys portal is a tool to provide the GSE thinking model updated and quickly accessible on the one hand, and to systematically store real data system-related on the other hand, in order to be able to derive more precise reliability forecasts on this basis.

The GSE aprocedural concept adapted to the problems to be solved is supported in a comprehensible way by the portal. For this purpose, the portal uses several databases, which were connected using SharePoint according to the schema shown in Fig. 6.29.

This can make the interrelationships between the views of the image of the mechatronic system transparent.

Figure 6.30 shows this for the component and the requirement view of a mechatronic system. The same is also possible for the function and process view.

The networking of the system views of the modified GSE thinking model can be represented via the PromeSys portal using graphs. This form of representation of Fig. 6.30 enabled portal users to recognize various relationships of the mechatronic system. It could be checked which requirements are met by which functions and components within the framework of the fixed processes. It also became clear which functions are particularly relevant and must be secured by a robust design of the mechatronic system.

Many further detailed questions could be clarified by the graph model. This graphical illustration of the networking of the views of the GSE thinking model modified for mechatronic systems using the portal made it easier for the companies involved in the project to analyze and evaluate their solution concepts (Fig. 6.31).

In addition, context information about the individual elements, which Fig. 6.32 represents, can be read immediately. This user-friendly solution saves the search for related information.

In summary, this example of the reliability consideration of mechatronic systems over the product life cycle also illustrates how the GSE approach, with its standardized modules, applied in a problem-oriented manner, leads to very efficient solutions. A collection of data on a specific reliability problem of a mechatronic system or its subsystems is possible on the basis of the GSE approach. Which subsystems of mechatronic systems are

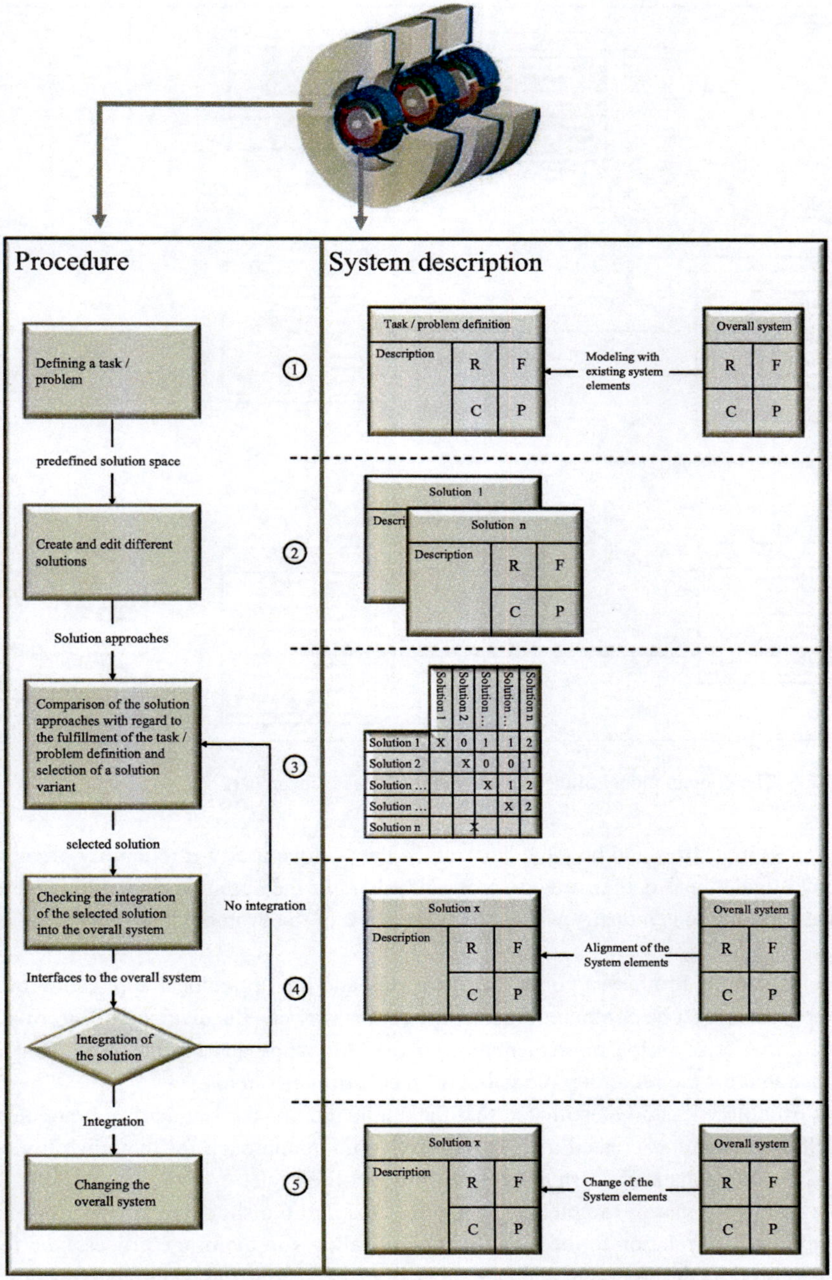

Fig. 6.26 The connection of the GSE thinking model with the modified GSE procedural concept as the basis for the PromSys portal (see Winzer 2012, p. 45)

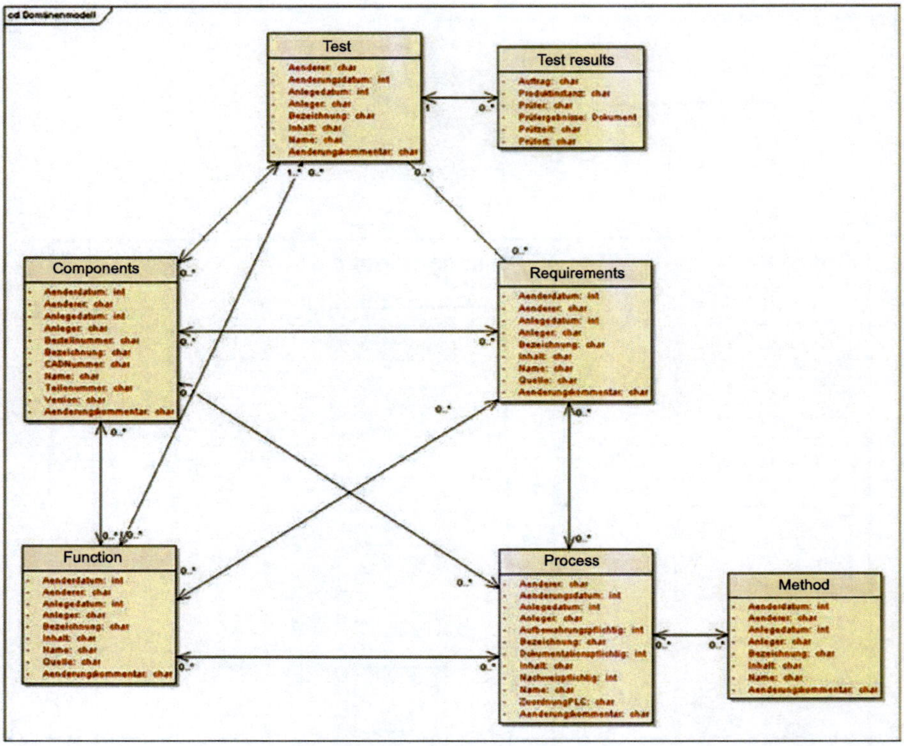

Fig. 6.27 The domain model of the PromeSys portal (see Winzer 2012, p. 56)

used across industries and how the determined data can be used for reliability prediction was investigated and demonstrated across industries in the automotive supplier industry and the capital goods industry within the framework of the PromeSys project (cf. Winzer 2012).

The developed PromeSys portal meets the demand that a practical application of the GSE approach must be computer-aided. Only in this way can the diversity of information and data that arise in the implementation of the GSE approach be managed. This was demonstrated for the reliability prediction of mechatronic systems.

But even now it can be estimated that the application of the standardized modules of the GSE approach has created a cross-industry solution approach for the reliability considerations of mechatronic systems over the product life cycle.

The following three examples of Sects. 6.4, 6.5 and 6.6 show the application of the GSE approach in relation to the identification of failures in the usage process, the feedback into product development and how an error cause search and solution algorithm can be designed according to GSE principles.

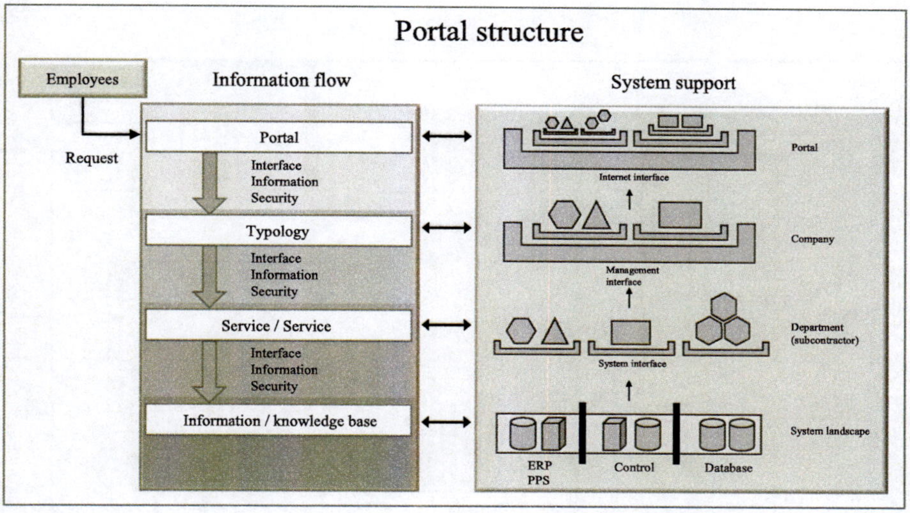

Fig. 6.28 The structure of the PromeSys portal (see Winzer 2012, p. 57)

6.4 Failure Identification in Critical Usage Processes by BIELEFELD

An essential part of the GSE procedural concept is the GSE analysis module, which uses various analysis methods. The analysis module is applied to systematically analyze the interaction between the system elements and the interaction between the system and its environment. In this process, the analysis module interacts with the GSE thinking model in the form of mutual information exchange.

In the following example, the DeCoDe model and the analysis module are used as the basis for the development of a methodology for a model-based and holistic failure analysis (MemogaFa) to identify failures in the interaction of the product with its environment in the early phases of product development.

A lack of consideration of the interaction between a technical product and its environment in the usage phase can lead to devastating and even fatal consequences, as the accident with the Tesla car in 2018 has shown (Web news 2018). The autopilot system was not able to distinguish between the bright sky as a background and the white side of the truck. As a result, the car crashed into the white truck at a speed of over 100 km/h and the driver died. This dramatic example specifically shows the negative consequences that can arise from the interaction between a technical system (or product system) and its environment (system environment). Figure 6.33 sketches the mutual relationship and interaction between the system and its environment.

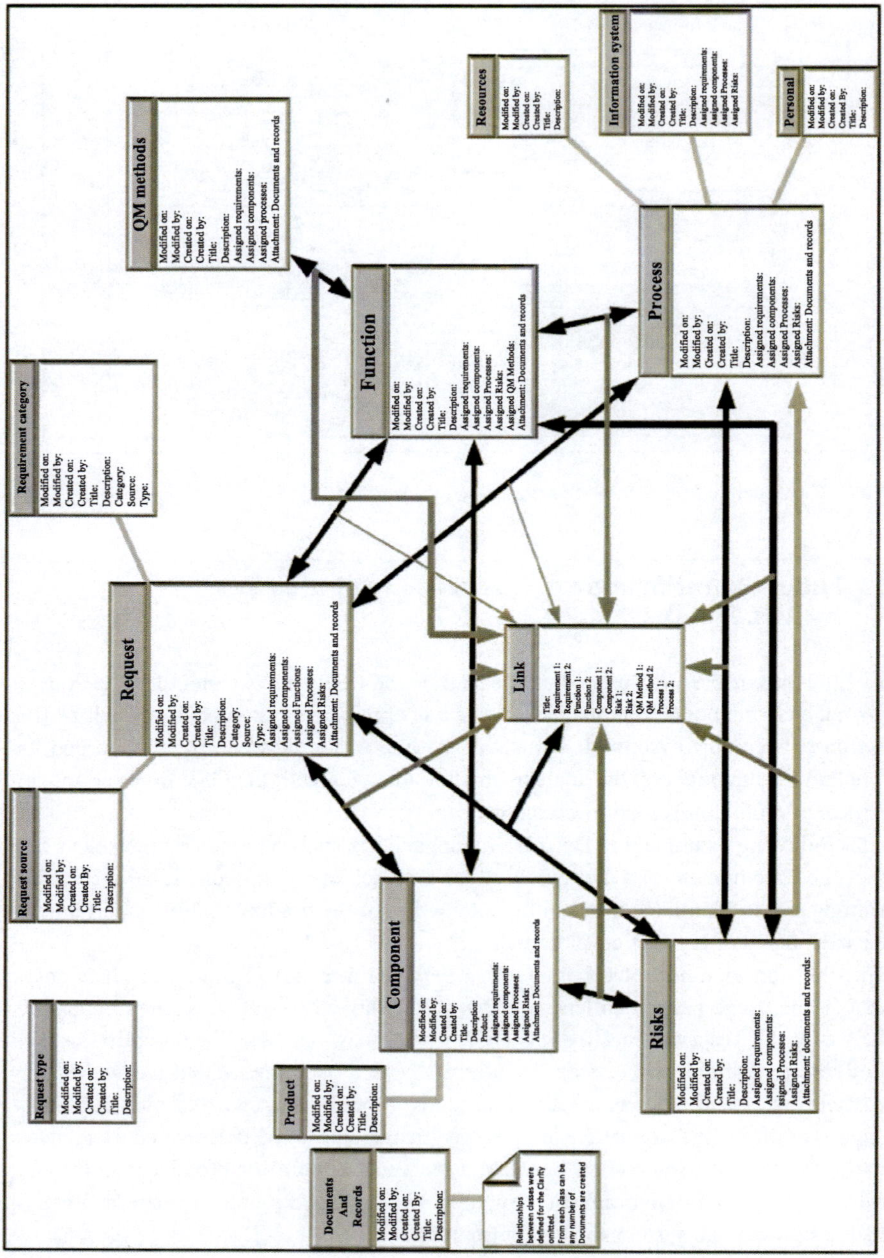

Fig. 6.29 The database schema of the PromeSys portal (see Winzer 2012, p. 63)

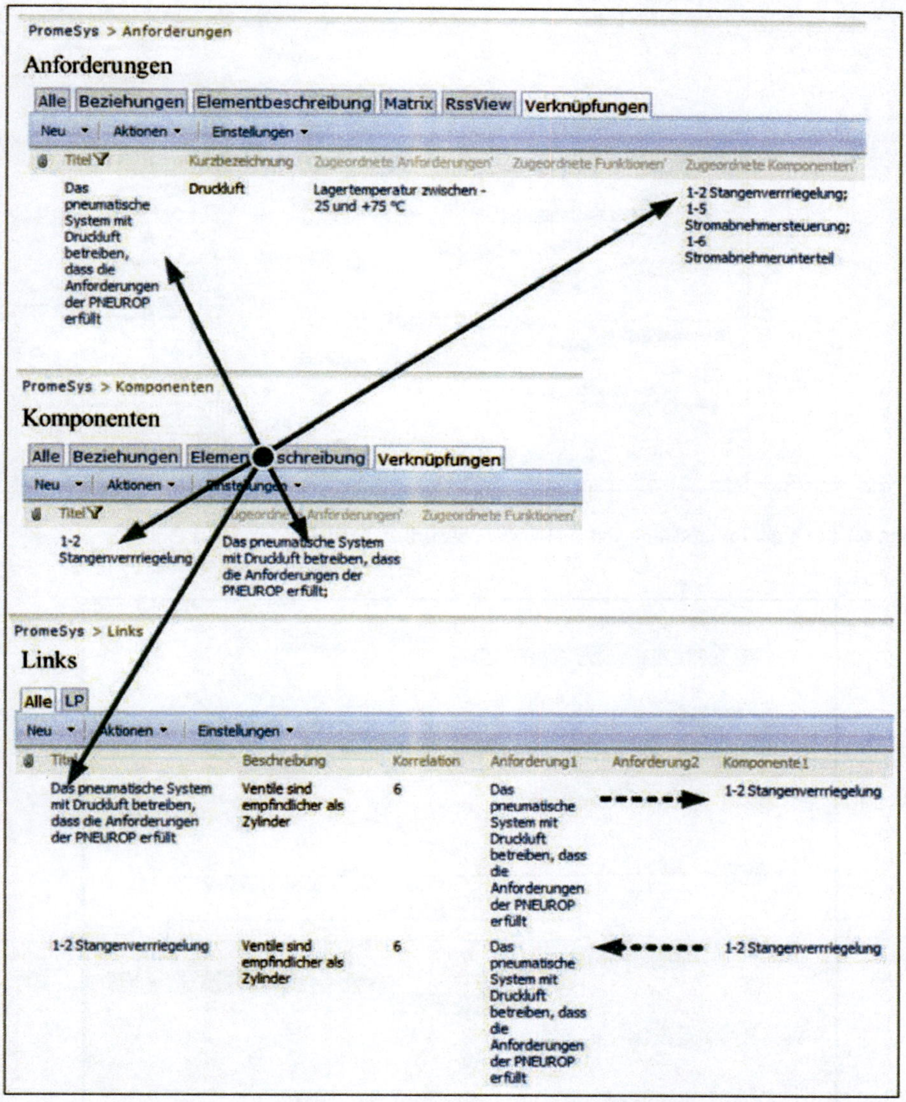

Fig. 6.30 An example of linking system views of the GSE thinking model using the PromeSys portal (cf. Winzer 2012, p. 65)

Therefore, it is important to prospectively analyze this interaction and relationship and identify potential failures in the early phases of product development. For this purpose, the MemogaFa was developed using GSE. the MemogaFa is a approach, a holistic methodology with which potential failures from the interaction between the system and its environment can be systematically recorded and analyzed.

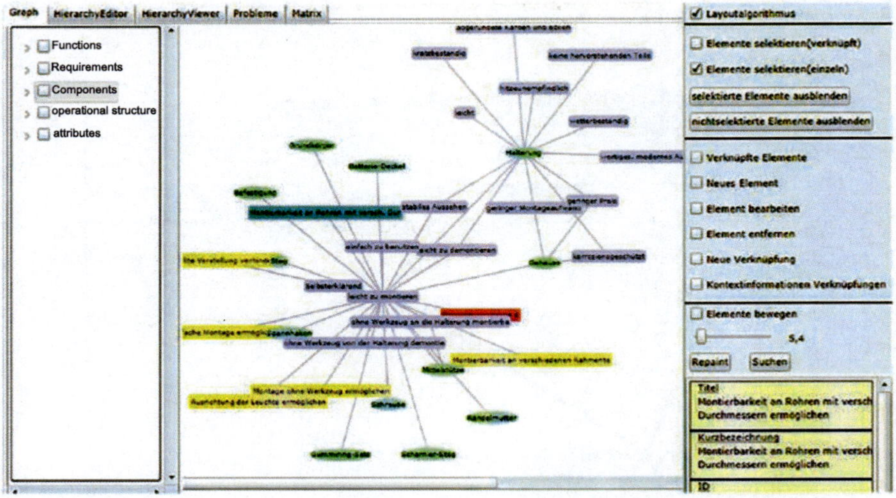

Fig. 6.31 Graph for modeling and analysis of networked data (cf. Winzer 2012, p. 68)

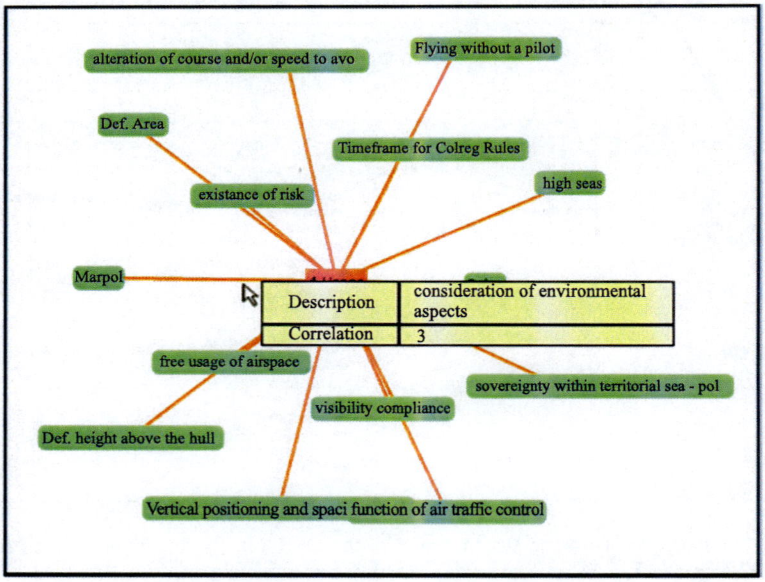

Fig. 6.32 Retrieval of context information on elements of the graph (cf. Winzer 2012, p. 69)

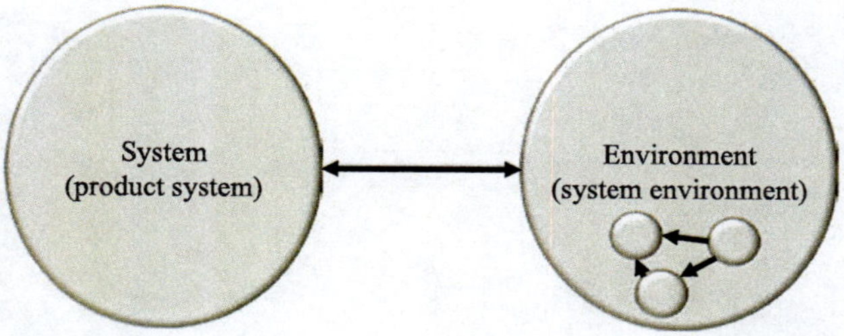

Fig. 6.33 Interrelationship between system and environment according to Hitchins (Hitchins 2007, p. 71)

The next Sect. (6.4.1) explains the methodology and validates it using a case study (Sect. 6.4.2). This is followed by a discussion of the practicality of the methodology as well as its advantages and disadvantages. Finally, an outlook on further research work based on the methodology is given (Sect. 6.4.3).

6.4.1 MemogaFa—Methodology for a Model-Based and Holistic Failure Analysis

The focus of MemogaFa is on the analysis and identification of relevant factors, which can lead to potential failures in the usage phase of a technical product from the interaction between the product and its environment.

To represent this interaction holistically, the methodology uses the DeCoDe model as a basis. Holistic in this context means that all relevant factors should be considered individually or in combination in the failure analysis. Relevant factors can be both components and functions of a technical product system as well as influencing factors (temperature, humidity, etc.).

MemogaFa consists of two phases and six steps (A to F), which also form the individual steps for explaining the methodology using the case study in the next subchapter (cf. Sect. 6.4.2 Validation of the methodology). Each individual step interacts with the extended system model, which is at the center of the methodology. The methodology can be repeated as often as desired, hence the cyclical process. For this, the knowledge from the extended system can be used specifically for each individual step, while individual steps provide information (as output) for a continuous extension of the central system model. It shows how the GSE initially forms the goal and then individual methods of the GSE analysis module as well as the GSE modeling are used to systematically and comprehensibly exclude potential product failures through interaction with the environment (Fig. 6.34).

In the first phase, all information necessary for carrying out the methodology is recorded and made available. In addition, this information is structured and transparently

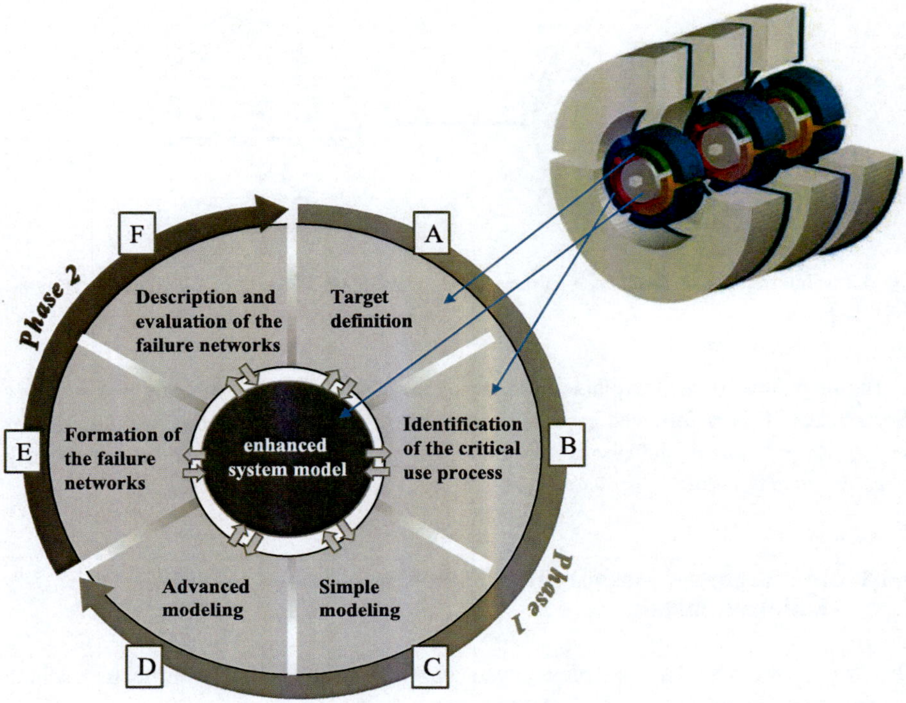

Fig. 6.34 New methodology for a model-based and holistic failure analysis (Bielefeld et al. 2021)

presented using the model structured, so that relationships and interactions between the product system and its environment can be better identified. Because only on the basis of a system model can failure causes and effects be recorded holistically on several system levels (Winzer 2016; Schnellbach 2016). For this reason, the first phase is called "Model-based provision of information for failures analysis and failure description".

The phase consists of four consecutive steps, starting with the definition of objectives (A). The goal of the methodology should be defined in the first step by specifically determining "what" is to be achieved with the methodology and "for what" the methodology is to be used. This corresponds to the GSE goal formation module, which is supposed to ensure that the goal, the target value, and the benefit are established for each activity.

The following criteria serve as orientation:

- By applying the methodology, it should be ensured that the main requirement for the product is met and
- with the help of the methodology, potential failures should be identified that could possibly lead to personal injury.

In addition, in step A of the methodology, the usage processes in the usage phase are determined. From the usage processes, the critical usage process can be identified in the second step (B). Expert knowledge is required for this, as the selection of the critical usage process is based on the assumption that most "critical" failures can occur in this process. Expert knowledge on the subject can be generated in the context of expert workshops or through targeted questioning.

The critical usage process forms the framework for the next step—the "simple" modeling (C). The modeling is titled "simple" because initially "only" the components and functions involved in the critical usage process are modeled.

In order to enable the mentioned "holism", in the fourth step (D) of the methodology, the extended system model is generated using an extended modeling. For this, all aspects important for the failure analysis, which can cause potential failures, are taken into account:

- the events (e.g., car accident)
- effects (e.g., physical) and
- environmental factors (e.g., weather conditions).

The importance of these factors in the context of a sound failure analysis has already been postulated in several scientific papers (cf. (Bielefeld 2020; Mamrot 2014; Schlund 2011). After all necessary information is available in a model-based manner, the failure analysis begins in the second phase of MemogaFa. For the holistic failure analysis, a tool was developed to transparently present information so that interactions and combinations of error causes from the interaction between the product system (components and functions) and its environment (environmental factors, as well as effects and events) can be better identified (E). The so-called "failure network" is the result of this step. An "failure network" is the structure of an failure that is formed from several failure causes, effects, failure effects or impacts.

After the formation of the failure networks, the description and evaluation of the failures or failures networks follows in the next step (F). This is done according to a certain system and structure as well as using a template, which is presented below and shows how analysis methods and DeCoDe modeling harmonize. The application of the methodology ends with the description and documentation of failures or failure networks. The development of concrete measures to avoid failures and their implementation is no longer part of MemogaFa. At this point, it is then necessary to switch to the GSE design module with its methods. This can be, for example, the FMEA.

The six steps of MemogaFa are shown in the following using the example of a linear drive.

6.4.2 Validation of the Methodology

To better understand, the methodology is explained using a case study from a research project named Q-ELF ("Quality-oriented method workflow for the product development of a linear drive in conveyor technology"). The goal of the project was to ensure that quality and reliability requirements are met in the development of linear drives for logistical systems. In a linear drive, electrical energy is converted into translational energy, based on the principle of linear machines (Riekhof et al. 2012). The operating principle of linear machines or linear drives is analogous to rotating machines. The two types of machines differ only in the resulting motion: a translational motion in linear machines and a rotational motion in rotating machines (see sketch in Fig. 6.35).

The individual steps of the MemogaFa are then concretized using the case study "Linear drive in the usage phase".

Step A: Goal definition according to the GSE goal formation module
The goal of the methodology is to identify potential failures that can have the following effects:

- Non-fulfillment of the main requirement of the technical product "linear drive" and
- Occurrence of personal injury during the operation of the linear drive.

The main requirement was defined in the Q-ELF research project as follows: "The linear drive must be able to transport a mass of 50 kg over a gradient of 30° at a speed of 2 m/s" (Riekhof et al. 2012). In terms of the usage phase, five usage processes emerge with a focus on energy transfer and material flow:

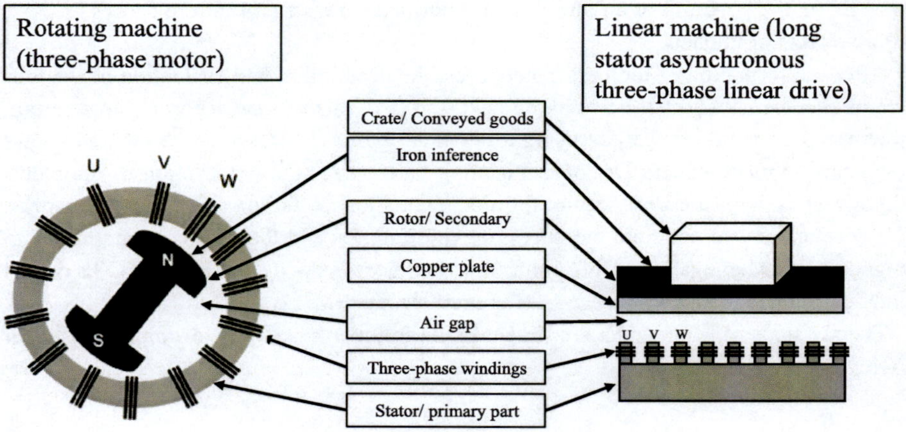

Fig. 6.35 Operating principle of the linear machine (linear drive) compared to the rotating machine. (After Wörner 2013)

- Loading,
- Accelerating,
- Constant conveying,
- Decelerating and
- Removal.

Based on these usage processes, the critical usage process is determined in the next step of the methodology.

Step B: Identification of the critical usage process according to the GSE analysis module

To determine the critical usage process, process mapping can initially be used as an analysis method. Here, the usage processes are recorded in order to then decide which of the usage processes are critical. In this case, the critical usage process can be easily determined from the five usage processes. Only the usage process "constant conveying" fulfills the main requirement ("...speed of 2 m/s to convey") of the technical system. This means that the failure of this usage process results in the non-fulfillment of the main requirement. Therefore, the usage process "constant conveying" is classified as critical and determines the framework for the subsequent, simple modeling.

Step C: Simple modeling in the GSE thinking model

The simple modeling is done according to the principles of the DeCoDe approach. The goal is to get a rough overview of the most important components of the product and their functions. The advantages of using the GSE thinking model DeCoDe in the context of failure analysis compared to other modeling approaches have been demonstrated by numerous publications (see Schlueter et al. 2018; Bielefeld and Schlüter 2019) and will not be further elaborated here.

The first step of modeling is to determine the system boundary, which is done with a focus on the critical usage process from step B. Consequently, the product system to be analyzed includes the entire route of the linear drive and the box (conveyed goods) (see Fig. 6.35). After determining the system boundary, the functions and components of the technical product system are identified. Only the functions and components involved in the critical usage process "constant conveying" are included in the analysis (see Fig. 6.36).

The consideration of functions and components is important because a failure of components and/or functions can also cause a failure of the critical usage process.

In addition, potential failures can occur due to negative influences from external factors. The systematic determination of these external factors is done in the next step by extending the simple model to include these factors. On the foundation of the extended model, the comprehensive failure analysis can then be carried out.

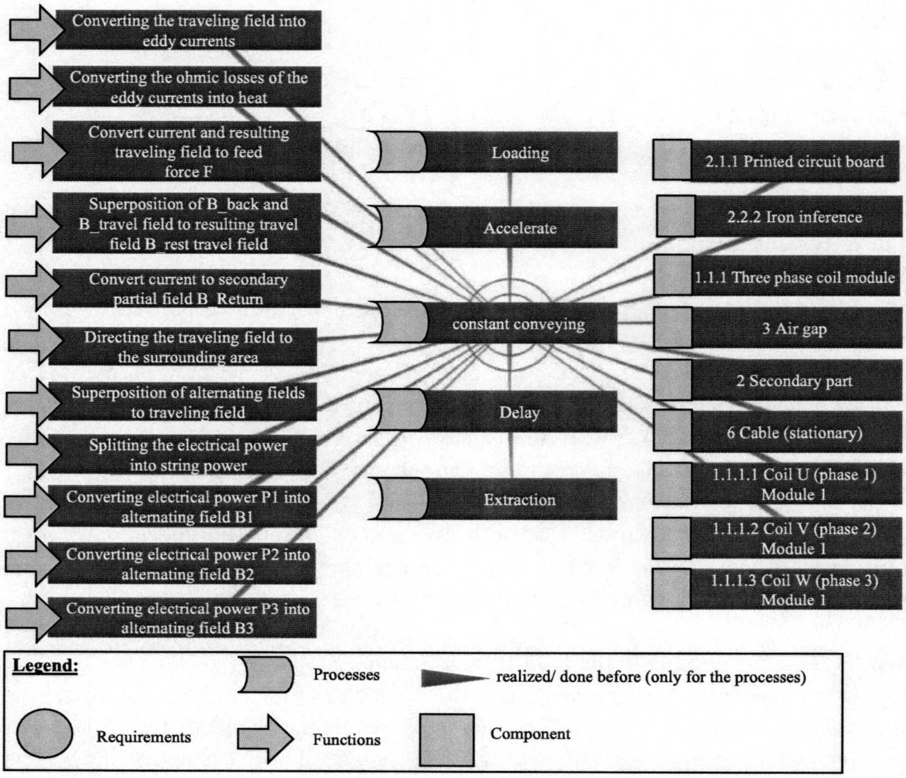

Fig. 6.36 Functions and components of the linear drive with a focus on the usage process "Constant conveying". (After Bielefeld et al. 2021)

Step D: Extended modeling in the GSE thinking model

For the purpose of a comprehensive failure analysis, the simple model is supplemented in this step with events, effects, and environmental factors in the GSE thinking model. To make interactions and combinations of failure causes transparent and structured for the analysis, a new tool has been developed. Using the tool, the four aspects (product system, events, effects, and environmental factors) can be juxtaposed in a "field of tension". This field of tension is subsequently referred to as the "quadrants of interactions". The approach to the quadrants is presented in Fig. 6.37.

Ideally, databases should be available to provide the four fields with the necessary information. The tool consists of a visible area (surface of the tool) and an invisible area (databases with information) (see Fig. 6.37).

The surface of the tool provides the area for conducting the comprehensive failure analysis with a focus on the critical usage process, as here only the information and ele-

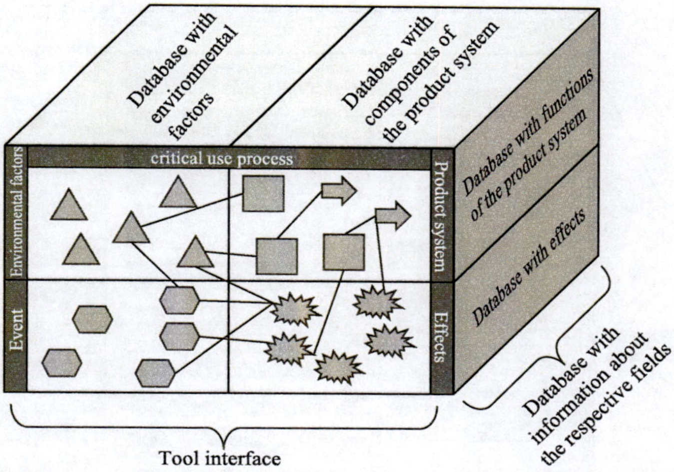

Fig. 6.37 Quadrants of interactions for identifying potential failures from the interaction of product system and environment in the usage phase. (After Bielefeld et al. 2021)

ments from the four fields that realize the critical usage process or have a direct influence on it become "visible".

In summary, the four quadrants consist of the following aspects and elements:

- Product system: Components and functions that are involved in the critical usage process.
- Environmental factors: Factors (e.g., temperature) from the immediate environment that can negatively affect the product system.
- Events: Events (e.g., car accident) can occur in the critical usage process from the interaction of several factors.
- Effects: e.g., physical effects that can arise from the interaction between components and functions and/or from the interaction between these and the environmental factors.

The input for the databases can be generated from checklists and templates from the technical literature. A good example of this is the taxonomy of root causes according to Krishna (see Krishna 2008) which includes all possible failure causes, events, and environmental factors in a structured list. Another example is the TRIZ Effects Database with a list of all possible positive and negative effects (see Oxford Creativity 2020).

The described approach to the four quadrants (Fig. 6.37) is applied to the case study of the linear drive. Since the focus in the case study is on the critical usage process "constant conveying", all events, environmental factors, and effects that can influence this usage process are listed on the surface of the tool.

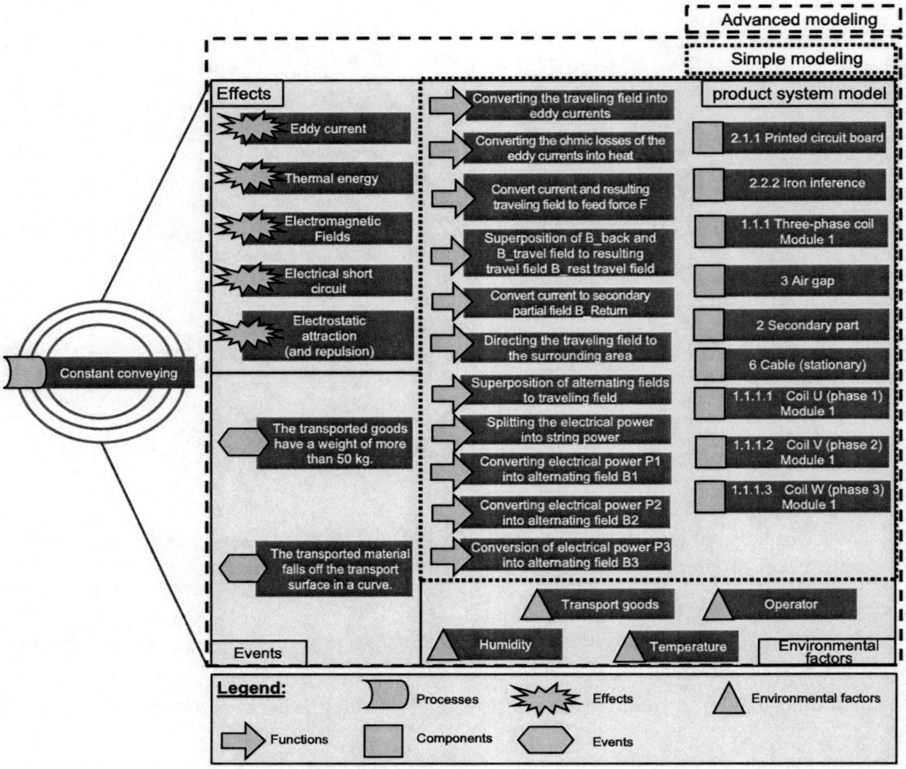

Fig. 6.38 Filled quadrants of interactions with a focus on the usage process "Constant conveying". (After Bielefeld et al. 2021)

The component and functions from the quadrants "Product System Model" were already determined in the previous step and are here supplemented by three more quadrants: Events, Environmental Factors (temperature, humidity, etc.) and Effects (eddy current, thermal energy, etc.) (see Fig. 6.38).

Due to the procedure, it should be mentioned at this point that the quadrants with the effects and events are not complete at this time, as further effects and events can arise from the formation of failure networks from the interactions of different factors. The formation of the failure networks takes place in the next step.

Step E: Formation of Failure Networks within the GSE Analysis Module

After the expansion of the GSE thinking model, the current state of information is provided for the GSE analysis module and the appropriate analysis method for potential failure determination is selected. In the situation described here, the formation of failure networks is pursued as a procedure. The quadrants of interactions, which open up a new perspective for carrying out a holistic and model-based failure analysis, serve as the basis for this.

Another aid for the formation of various failure networks for the critical usage process are the so-called scenarios. A scenario is considered “a concrete manifestation of a use case directed far into the future depending on the discipline, which considers causal relationships” (according to Gausemeier and Fink 1999; Weilkiens 2015). These causal relationships are at the center of the model-based and holistic failure analysis and are formed using scenarios and based on the quadrants of interactions. The sketch below explains the relationship between the usage process, which represents the current point in time of the analysis (e.g., in product development), and different scenarios, which are considered as projection surfaces in the future (see Fig. 6.39).

Based on the quadrants of interactions and the scenario, different failure networks can be identified. To ensure that the failure networks are valid and consistent, the failure analysis should be conducted in the form of workshops with experts and development teams from product development. In this way, the expertise and specialist knowledge of experienced employees can be used to recognize different failure networks based on the quadrants of interactions.

The case study of the linear actuator (Q-ELF) illustrates an failure network using the scenario “Electrical Short Circuit” in Figure 8. The failure network shows a possible scenario (use case in the future) in the usage phase of the technical product system linear actuator.

The individual elements in the failure network are numbered to express the temporal sequence. The potential failure network begins with the cause of the failure (1) and ends with the failure effect (6). The directed relations (e.g., influences, provides input,

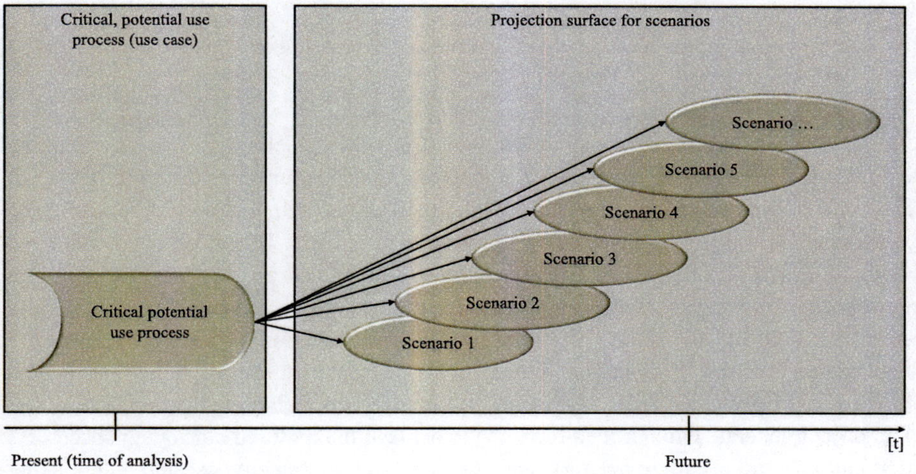

Fig. 6.39 Relationship between the critical usage process and possible scenarios. (Own illustration after Gausemeier and Fink 1999)

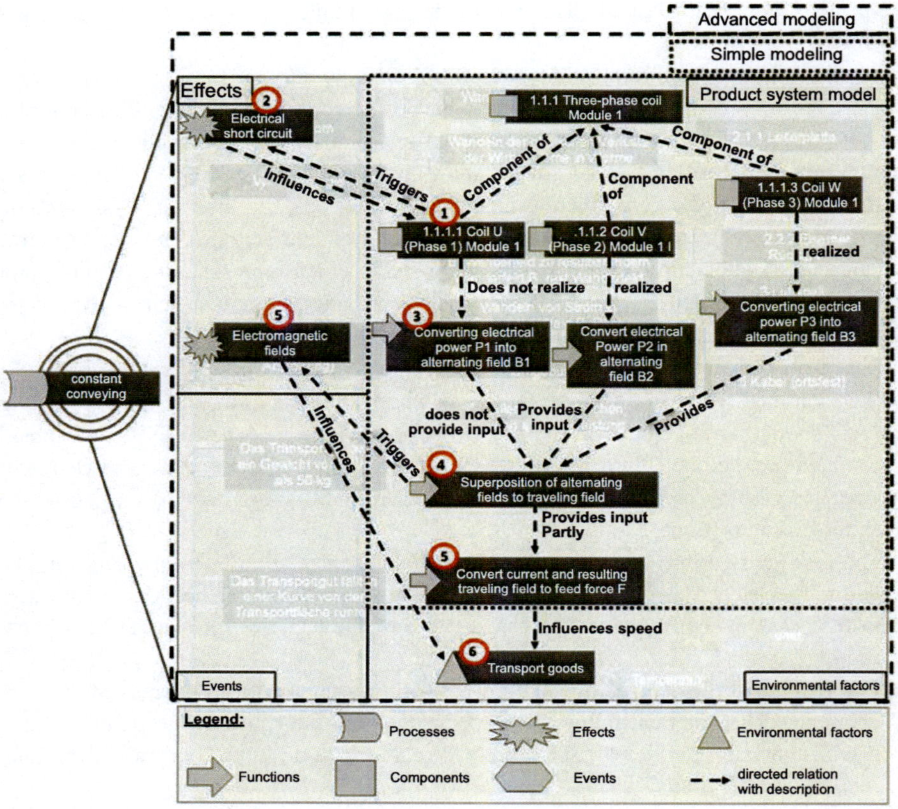

Fig. 6.40 Potential failure network using the scenario "Electrical Short Circuit". (After Bielefeld et al. 2021)

etc.) are formulated according to the rules of DeCoDe modeling, so that relationships and networks between the individual elements and their relationships are well described. The context from the failure network in Fig. 6.40 is to be understood as follows. The components at coil U (1) trigger an electrical short circuit (2). This effect can result from a "poor" insulation of the coil. Therefore, the coil is defined as the cause of the failure network (1). Due to the failure at coil U, the function "Convert electrical power P1 into alternating field B1" (3) is not realized. As a result, this function does not provide further input to the function "Overlay alternating fields to create a traveling field" (4), so that this can only be realized "partially". Consequently, the current strength and the traveling field do not generate enough feed force (5) to achieve the required conveying speed of 2 m/s. The scenario or failure network leads to the main requirement (see step A) not being realized.

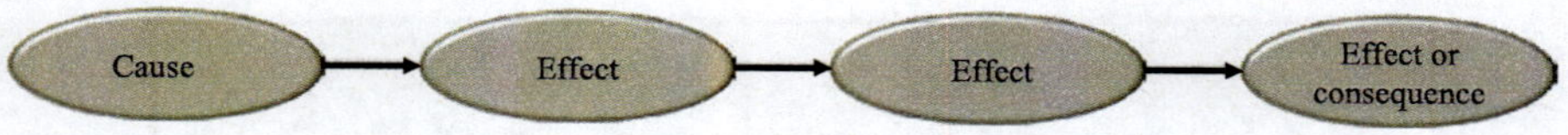

Fig. 6.41 Holistic failure description. (Own illustration after Zingel 2013, p. 50 and Westkämper 1997)

From Fig. 6.40 it becomes apparent that the quadrants of interaction are a good tool for a transparent representation of the interaction between the product system model and the environment, but the representation quickly becomes unstructured and confusing.

Especially with more complex failure networks, there is a lack of a clear structure to consistently describe errors. Therefore, in the next step, another new tool is introduced that allows a detailed description and evaluation of the failure network from the quadrants of interactions.

Step F: Description and Evaluation of Failure Networks

To realize a cross-disciplinary, consistent failure description, a comprehensive analysis of technical literature and standards is necessary. Basically, an failure consists of a cause of failure, failure effect, and failure impact (see DIN EN ISO 9000 2015; ISO 26262 2011; Westkämper 1997). By adding the effect, which arises from failure causes (Zingel 2013), a holistic picture for a comprehensive and detailed failure description results (see Fig. 6.41).

This comprehensive description serves as a "template" to describe failure networks and to structure them better. Individual elements from the quadrants of interactions (see Fig. 6.38) are assigned to the four components of the failure description. An Excel-based tool can be used for this detailed description with subsequent evaluation of the failure network. The content of the tool for failure description is based on extensive literature analysis of technical literature and standards. The structural framework for the tool is provided by the holistic failure description from Fig. 6.41, which also forms the basis for the tool. Using the tool, the failures or failure networks are uniformly described and documented. The tool was programmed in Excel—Visual Basic for Applications (VBA) and consists of five windows or steps (see Fig. 6.42):

1. General information,
2. Failure causes,
3. Effects of the failure cause,
4. Failure impacts and
5. Failure risk.

In the first step, general information (e.g., version, scenario ID, user, date, etc.) is entered for feedback and traceability (1). The transfer of information from the quadrants of interactions begins from the second step with the description of the failure cause. The failure causes usually represent the beginning of the failure networks. Failure causes can be, for

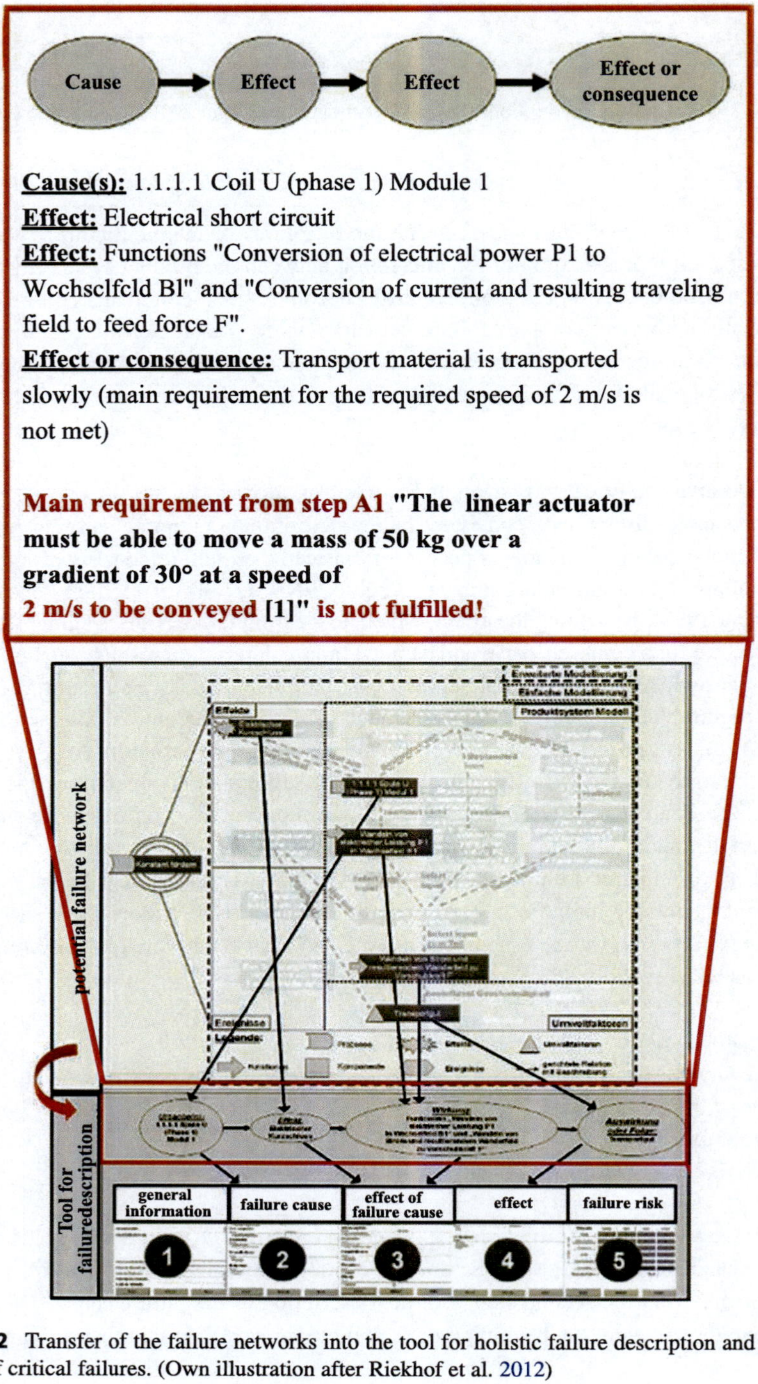

Fig. 6.42 Transfer of the failure networks into the tool for holistic failure description and prioritization of critical failures. (Own illustration after Riekhof et al. 2012)

example, function and component failure. In the case study, the failure of the components at coil U (see Fig. 6.38) is the failure cause, in other cases failures can have multiple causes (2). In the third step, the effects of the failure causes, including the effects, are described (3). Failures can affect different product system levels (e.g., subsystems or the overall system) as well as the environment of the technical product system in its usage phase. These relationships are documented in the fourth step (4).

Subsequently, failures are evaluated and prioritized in the last step using a so-called risk graph, a helper tool from risk management and product development for the evaluation of complex systems (Cox 2009). The risk of potential failures from the quadrants of interactions is defined as the product of damage severity and probability of occurrence (Cox 2009). The transfer of the potential failure network "Electrical Short Circuit" from the quadrants of interactions (see Fig. 6.37) into the Excel-based tool for failure description is shown in Fig. 6.42. The individual elements of the failure network are structured into the tool for failure description using the template of the holistic failure description. The assignment of the elements of the failure network to the four components of the failure description (see Fig. 6.42) is as follows:

Cause(s): 1.1.1.1 Coil U (Phase 1) Module 1

Effect: Electrical Short Circuit

Impact: Functions "Convert electrical power P1 into alternating field B1" and "Convert current and resulting traveling field to feed force F"

Consequence or result: Transported goods are conveyed slowly → **Main requirement** for the conveyed speed of 2 m/s is **not met**

In this way, the failure network can be described uniformly and structured. In addition, based on the evaluation from the risk graph, a prioritization of failures is possible if MemogaFa is executed multiple times. Based on the results from the methodology, measures and strategies for preventive failure avoidance, similar to an FMEA, can be developed. Furthermore, further failure analyses can be carried out based on information and knowledge from the expanded system model, for example for further usage processes (e.g., loading). Due to its cyclical approach, MemogaFa can be applied result-oriented multiple times to achieve continuous improvement in terms of quality, safety, and reliability of technical products already in the early stages of product development.

6.4.3 Conclusion and Outlook

The aim of this contribution was to present a model-based and holistic failure analysis (MemogaFa), which follows the principles and modeling rules of the GSE, using a case study. With the help of this methodology, potential failures from the interaction and interaction between a technical product and its environment in the usage phase should be systematically identified. The basis for the methodology were the GSE thinking model, the GSE analysis module, and various newly developed tools. The case study of the linear actuator showed how potential failures from the interaction between product and environment in the form of failure networks can be identified and transparently represented using the newly developed quadrants of interaction. For this, structured information from an expanded system model was provided for a holistic failure analysis. With the help of the methodology, it was possible to systematically identify failure cause(s) and failure networks from the product-environment interaction based on scenarios (see Fig. 6.40) and to describe and document them uniformly and holistically using the Excel-based tool (see Fig. 6.42).

Compared to other methods of failure analysis (e.g., Ishikawa or FMEA), which only consider individual failure causes, the application of the methodology is advantageous because potential failures from cause combinations can be identified and uniformly described using MemogaFa. However, not only advantages but also some disadvantages result from the application of the methodology. These are presented summarized in the table below (see Table 6.1).

From the disadvantages, promising research points emerge that can significantly increase the efficiency of the methodology.

The high time expenditure of the methodology can be greatly reduced by a suitable software solution in connection with a meta-model. Different databases (e.g., TRIZ Effects Database) can be used to generate information and to put them in context, for example, through the use of intelligent algorithms. Furthermore, an automatic or semi-automatic identification of potential failure networks on the surface of the tool can be realized through the use of algorithms.

6.5 Model-based Field Data Feedback into Product Development of MAMROT

What motivated the development of the approach to model-based field data feedback into product development?

Quality costs continue to account for a significant portion of a company's expenses. In many areas, such as the automotive industry, there is a growing number of safety-related recalls (KBA 2012; NHTSA 2012). To ensure competitiveness, it is therefore necessary to deal sustainably with the resolution of the quality problems that have arisen (Ehrlenspiel 2009). A systematic approach, like the developed approach to model-based field data feedback into product development, is particularly used in complex technical products (Mamrot 2014). These are characterized primarily by an increased number of components, requirements, and functions of various types and disciplines, interacting dynamically with each other. That this complexity is also a cause of the quality problems is often found in relevant literature (Zhang and Luo 2007; Automotive Industry 2011; Schlummer 2012; Bracke and Haller 2012). Thus, the innovation and scientific novelty of the approach to model-based field data feedback lie in the synergistic coupling of a

Table 6.1 Advantages, disadvantages, and potentials of MemogaFa

Advantages	Use of expert knowledge and experience to determine critical, potential utilization processes
	Systematic recording of relevant factors (effects, events and environmental factors), which can cause potential failures in a possible utilization process.
	Generation of scenarios with potential failure networks using the expert knowledge and the quadrants of interactions as tool
	Error descriptions on different levels (from rough to detailed) and use of a new tool for failure description
	Documentation and prioritization of potential failures
	Holistic failure analysis of interactions and interactions between product system and environment in the use phase
Disadvantages	Large time investment
	With a high number of elements in the quadrants of the interactions, the tool becomes confusing and thus the application very difficult
	The failure description is not in the form of an automatically generated and complete sentence with a regulated syntax
	Troubleshooting and system design activities as well as the impact(s) after troubleshooting on the levels of the overall system were not considered within the scope of the methodology

procedure for problem-solving based on field data and the systemic representation of the product system in the system model for dealing with complexity. This principle is based on the GSE.

6.5.1 Application of GSE Using the Example of Model-Based Field Data Feedback

The following explanation illustrates how this universal problem-solving approach (cf. Mamrot 2014) is applied to practical problem situations. The basic prerequisite for the application is the representation of the product in the system model, i.e., the require-

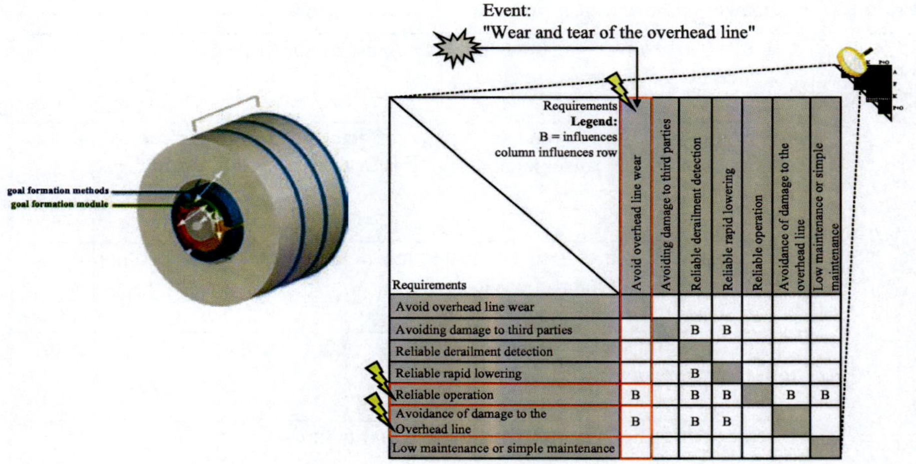

Fig. 6.43 Assignment of the occurred event to the system model with subsequent goal setting for problem-solving (D2) (Mamrot 2014)

ments, functions, processes, and components, as well as the relationships between these elements. All new insights about the product must be stored in this model, and an 8D report[3] is used for the history of problem-solving documentation.

As an example, a pantograph (OSA) was chosen, which was implemented in different variants in various trolleybuses and is used by several transport companies, i.e., cities and routes. During this use, it has been shown that customers repeatedly complained about wear and tear on the overhead line, possibly caused by the OSA.

To investigate the problem, a process was initiated and a processing team was determined (D1). The next step in the 8D systematics is the problem description (D2), for this, however, the problem or event that occurred had to be analyzed in its effects. This requires an assignment to the requirements (how the OSA was modeled is explained in Sect. 5.1).

As this example clearly shows, by assigning occurred problems, here "wear of the overhead line", these can only be recognized as such. The requirement-requirement matrix shown in Fig. 6.43 also allows the quick conclusion that long-term damage to the overhead line can occur due to single wear and tear, and thus reliable operation can no longer be guaranteed. Contractual penalties for downtime and failure correction costs are the possible effects of this problem.

[3] The 8D report consists of the following eight steps: D1 assemble team; D2 problem description; D3 define immediate measures; D4 search for errors; D5 define improvement measures; D6 implement improvement measures; D7 prevent recurrence; D8 appreciate team leadership (according to Linß 2011).

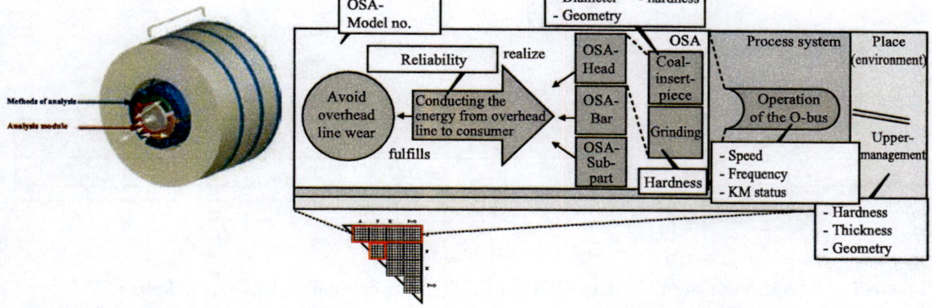

Fig. 6.44 Identification of the effect chain in the system model and derivation of the field data filter (Mamrot 2014)

No immediate measures (D3) were taken, as the causes were unknown and there was no safety relevance that could have led to a shutdown of operations. For the failure and cause search (D4), the system model was initially used, through which elements and element properties could be identified where a correlation with the occurred failure was possible. This analysis corresponds, as illustrated in Fig. 6.44, to the GSE analysis module.

Starting from the problem that occurred and the associated requirements, functions such as "conducting energy from overhead line to consumer" could be recognized, which coupled the components OSA lower part, OSA rod, and OSA head and its subordinate assembly sliding shoe and carbon insert piece. Furthermore, it became clear that these elements had certain properties, such as the diameter of the carbon, and were integrated into the process "operation of the O-bus" and the overhead line, where relevant properties such as the speed of the process or the geometry of the overhead line were also recognized.

In addition to the analysis in the model, an examination of affected components led to the realization that the overhead line had derailed from the groove in the carbon insert piece and thus contact of the overhead line with the hard sliding shoe must have led to the damage. The causes of this derailment were unclear, as it did not occur in all cities and on all routes. Therefore, a further analysis was needed, for which targeted field data should be collected using the field data filter shown in Fig. 6.45.

This field data filter serves for the structured collection of field data or the structuring of already existing data sets for each event "wear of the overhead line". Attributes such as "diameter of overhead line", reliability of the function "conducting energy" at the time of damage, diameter of the carbon insert piece, etc. could thus be recorded in their manifestations, here in the context of complaint and claim management.

Figure 6.46 visualizes the use of the field data filter for structured data collection in connection with the occurred event. A subsequent statistical analysis (GSE analysis module) was able to uncover certain correlations between attributes, which caused the derailment. These were a small curve radius on certain routes, too high speed of the O-buses,

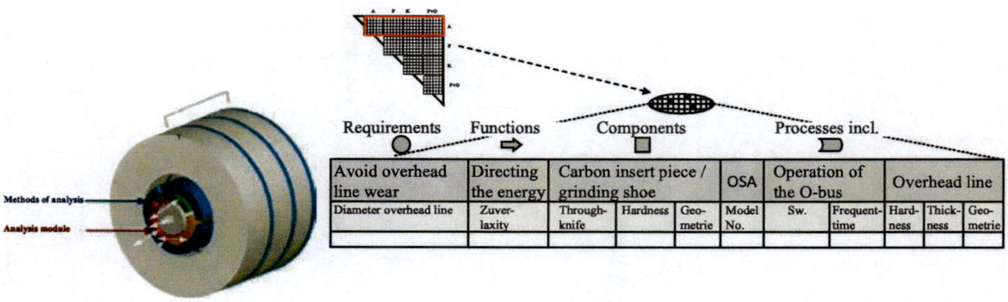

Fig. 6.45 Derivation of the field data filter for problem-oriented data analysis (Mamrot 2014)

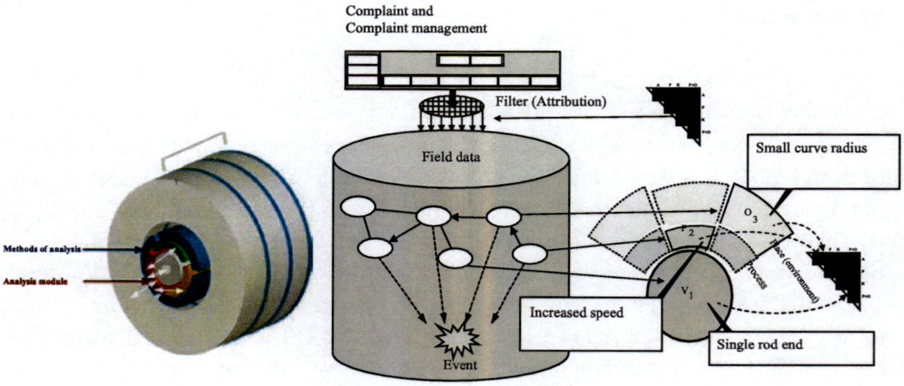

Fig. 6.46 Collection and evaluation of field data (Mamrot 2014)

and a certain OSA head (single-joint head). These findings were documented in the system model, in the form of attributes of the affected elements.

For the analysis of the causes, different techniques or methods are available (cf. Mamrot 2014). For the present problem with the different influencing factors, which only led to a problem in their combination, the cause-effect diagram was chosen for clear visualization and recognition of correlations (cf. Fig. 6.47).

In the application of the cause-effect diagram and in the discussion with experts, it became clear that a combination of too high speed of the O-bus, too small radius of the overhead line during curve driving, and too high mass of the OSA head and the OSA rod must have led to the derailment and thus the damage (cf. Fig. 6.48).

A derivation of the centrifugal force as a physical quantity depending on the various causes is as follows:

$$F = \frac{m * v^2}{r}$$

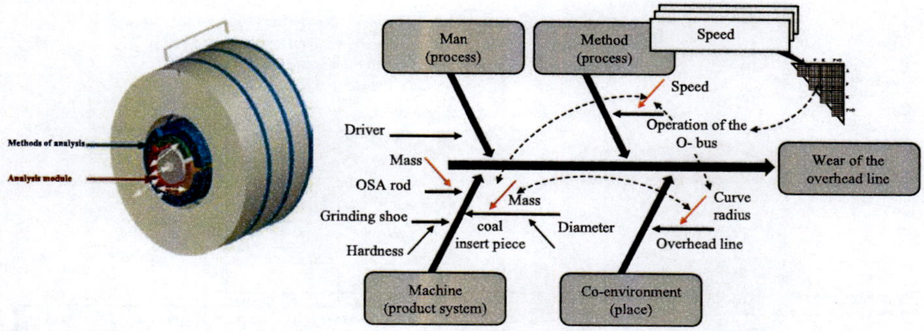

Fig. 6.47 Cause analysis (D4) using cause-effect diagram (Mamrot 2014)

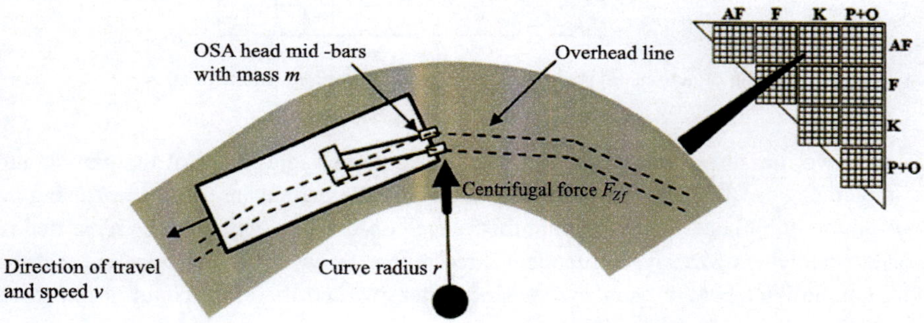

Fig. 6.48 Derailment due to centrifugal force and causing influencing factors (Mamrot 2014)

(for this applies: m = mass of the OSA, v = speed of the O-bus, r = curve radius)

Due to the time-dependent variable sizes, force impulses occur again and again, causing the derailment of the overhead line. Only then can the wear of the overhead line occur and the reliable operation be endangered. This realization was, as shown in Fig. 5.48, again stored in the system model and additionally documented in the context of the 8D report (D4).

To permanently eliminate the recognized problem (GSE design module), possible remedial actions were derived in a further step (D5). For this, as shown below in Fig. 6.49, the system model could be used.

In deriving the actions, those were focused that directly aim at eliminating the causes. These are:

- reducing the OSA mass,
- reducing the O-bus speed, and
- increasing the curve radii.

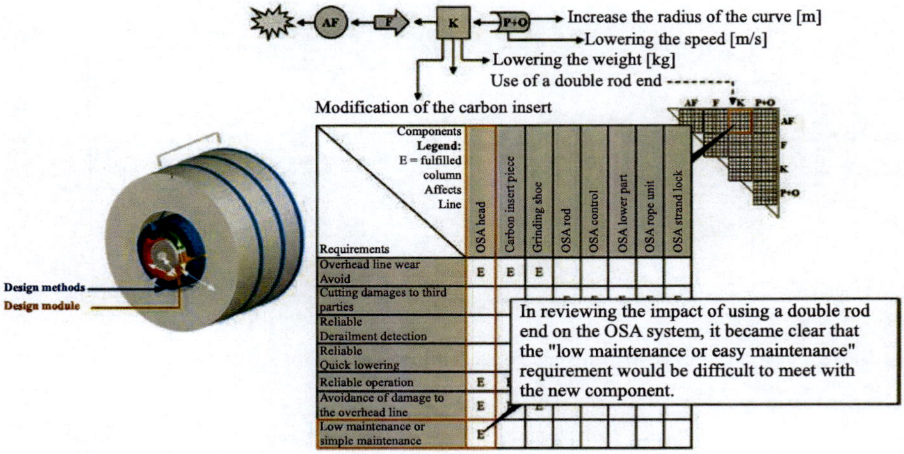

Fig. 6.49 Derivation of actions (D5) to remedy the quality problem (Mamrot 2014)

The last two measures proved to be impractical, only the adjustment of the mass could be implemented. Other measures, such as adjusting the carbon insert piece or using an OSA double-joint head with more mobility, were conceivable, but had to be identified in their effects by using the system model. Through the review, it became apparent that the OSA double-joint head is very maintenance-intensive and therefore could not be used according to the requirements "low maintenance or easy maintenance" (cf. Fig. 6.49).

Following this action derivation (D5), the actions "modification of the carbon insert piece" and "reduction of the OSA mass" were further detailed and implemented (D6). A review of the effectiveness of the actions taken (D7) was realized through continuous field observation, after which the problem-solving process can finally be completed (D8). The insights from this problem-solving process were also stored in the updated OSA system model and are thus also available for further developments, so that a repetition of the failure is not to be expected.

6.5.2 Insights from the Newly Developed Method of Field Data Feedback for the GSE

The application of the GSE for model-based field data feedback, among other things, using the example of the pantograph, has shown that the GSE and the associated modeling can be used profitably for problem-solving. Thus, relationships can be made transparent using the system model and newly gained insights can be continuously documented in it. The GSE modules (goal formation, analysis, and design) were applied in different orders and transferred into the specific procedure of field data feedback.

There is potential for optimization in the information technology implementation of the GSE, especially in modeling. Thus, existing modeling solutions (e.g., LOOMEO or Excel) should be networked with requirement management tools, with PDM/PLM solutions (component bill of materials), or with quality tools for FMEA (components, functions, and their networking) via interfaces. An exchange of elements and element relationships could thus be designed much more efficiently and reliably. It would also be conceivable, especially in the case of field data feedback, to connect a CAQ system in order to use the field data filter directly for structuring the field data in the CAQ system and vice versa to be able to feed back the insights gained from it into the modeling. There is further need for action here.

The following chapter outlines how such a partially automated use of data from the usage phase for reducing failures in production using an algorithm is possible.

6.6 Failure Cause Search and Solution Algorithm by HEINRICHSMEYER

In the next example, the eDeCode approach was applied in an algorithm for failure localization and problem-solving. The goal was to make the continuous complexity of constantly evolving product and production systems more manageable. By applying the model approach in the context of complaint management, the handling of complaints and the resulting amount of information should be designed more contemporarily.

For a transparent presentation of the research results, the algorithm itself is first discussed. In a further step, the eDeCoDe integration is examined from both a theoretical and practical perspective, and based on this, the validation in the industry is discussed. Finally, an outlook is given on the potential improvements that modeling with eDeCode offers for this specific use case and how these potentials could possibly be realized through further research projects.

6.6.1 Concept of the Failure Cause Search and Solution Algorithm

The algorithm for determining failure causes in the production system and for subsequent problem-solving was programmed based on a model. The model from Figure 1 is based, among other things, on an extensive analysis of research projects based on high-ranking literature and the evaluation of 13 different software systems currently used in the industry. The algorithm is divided into four main phases, as shown in Fig. 5.50. These include "information probing", "prioritization", "failure cause localization" and "problem-solving" (Heinrichsmeyer et al. 2019a) (Fig. 6.50).

In the first phase (information probing), relevant complaint information is queried from the complaint texts that arise during the usage phase. Based on this, quantitative values are calculated in the second phase (prioritization) to determine a largely objective

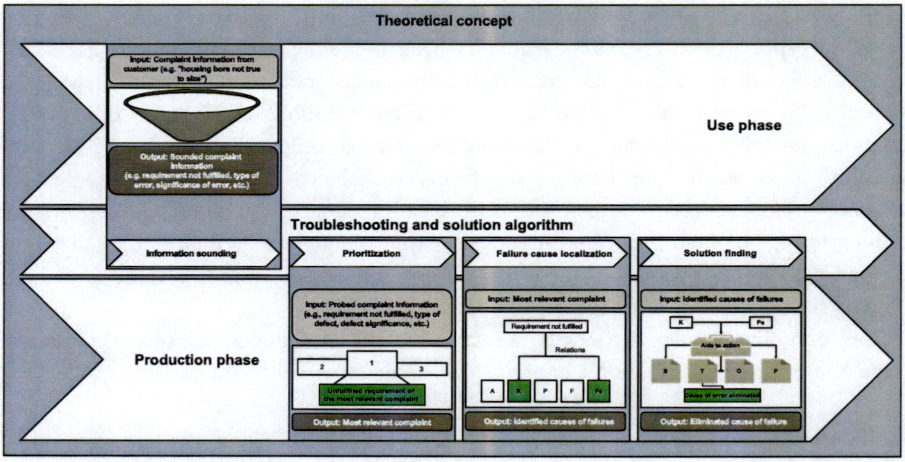

Fig. 6.50 Model of the failure cause search and solution algorithm (FusLa) (Heinrichsmeyer et al. 2019a)

priority of the complaint. Both the complaint information queried in the first phase and information from the systems of an organization are used to calculate the priority. With knowledge of the most relevant complaint, the third phase (error cause localization) is initiated. It serves to locate the existing failure cause within a production system during the production phase. In the last phase (problem-solving), the localized failure cause is finally eliminated (Heinrichsmeyer et al. 2019a).

Of all the phases mentioned, the eDeCoDe approach is primarily integrated in the failure cause localization. For this reason, this phase will be focused on from a theoretical and practical perspective in the following chapter.

6.6.2 Failure Cause Localization

The purpose of failure cause localization is to identify the system element or part of the production system in which the occurrence of the failure cause can most likely be suspected. While current approaches, as described in [Heinrichsmeyer et al. 2019a], rely on the team assigned to handle the complaint to locate the failure cause in the production system, it is essential to automate this process in the face of increasing complexity. The employee assigned to handle the complaint is hindered, despite his experience or expertise, from overseeing the enormous complexity of a production system, e.g., with all its components, processes, or people. This is not necessarily due to the employee, but simply to the variety of system elements and their interaction. Especially with changing production systems or newly hired employees, a subjective search for failure causes is significantly more difficult with increasing complexity. Therefore, the use of a model approach to map the production system and reduce complexity is necessary. By linking probed complaint information and the model of the production system, a more extensive and especially more effective localization of failure causes should be possible (Ansari et al. 2020; Heinrichsmeyer 2020).

6.6.3 Theoretical Implementation

To counteract the described problem, the FusLa was developed based on the so-called Enhanced Demand Compliant Design (eDeCoDe) approach by (Winzer 2016) and (Nicklas 2016) for modeling sociotechnical systems. It uses five standardized views (requirements, components, functions, processes, and people) for classifying system elements. With the help of various matrices, including design structure or domain mapping matrices, the relationships between the individual system elements are recorded and transparently displayed with a notation of 1 (relationship) and 0 (no relationship), as shown in Fig. 6.51 (Heinrichsmeyer 2018).

Figure 6.51 shows that in addition to the five mentioned views, the input (I), the output (P), and the environmental influences (E) of the process are also specified. This is

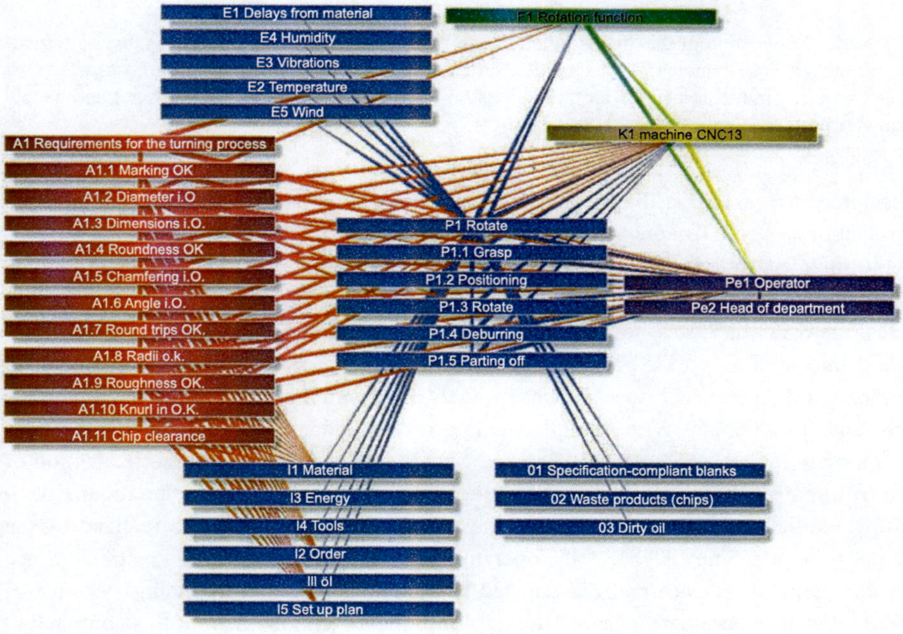

Fig. 6.51 Complex representation of the subsystem manufacturing via requirements (A—Orange), processes (P—Blue), inputs (I—Blue), outputs (O—Blue), external influences (E—Blue), functions (F—Green), components (K—Yellow) and people (Pe—Purple) (Heinrichsmeyer 2018)

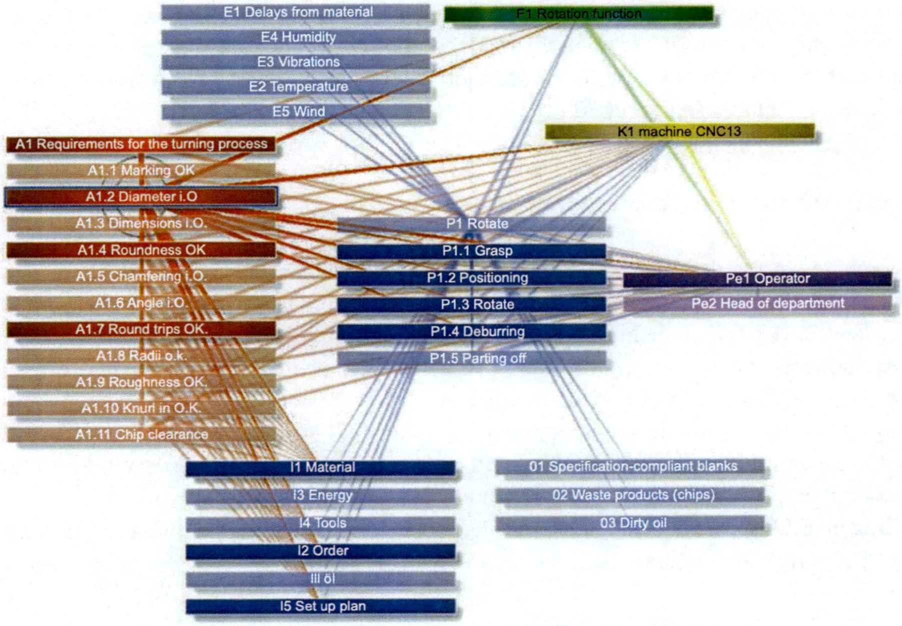

Fig. 6.52 Application of the "focus function" on "Complex representation of the subsystem manufacturing via requirements" (A—Orange), processes (P—Blue), inputs (I—Blue), outputs (O—Blue), external influences (E—Blue), functions (F—Green), components (K—Yellow) and people (Pe—Purple) (Heinrichsmeyer 2018)

absolutely necessary, as these can also be considered as possible failure causes in the production system. To now identify the possible failure causes of an unfulfilled requirement, their interrelationships with other system elements are evaluated. This process is exemplarily shown in Fig. 6.52 (Heinrichsmeyer 2018).

Figure 6.52 illustrates how the process of failure cause localization based on complaint information works. For this, the specification of the unfulfilled requirement, in this case a diameter with the "dimension Ø 24 ± 0.1 mm", is used as the basis for failure cause localization. With this information, it can be demonstrated using the production system model which system elements should have contributed to the realization of the requirement. It can also be traced which system elements influence the requirement. In this explicit case, for example, it is the process "rotating", which is realized by the component "machine CNC 13", the operator, and the function "turning function". Since the requirement was not realized and led to a complaint, a corresponding system element must have necessarily caused the non-fulfillment. With the help of this approach, it should be possible to make even extremely complex production systems more manageable (Heinrichsmeyer 2018; Heinrichsmeyer et al. 2019d).

6.6.4 Practical Implementation

To investigate whether this form of failure cause localization is also feasible in practice, the conceptual model from Fig. 6.50 was implemented in VBA in a next step. This serves primarily to enable a practical examination and validation of the algorithm. Of course, it should be noted at this point that other programming languages from the perspective of software development would certainly be better suited to implement the theoretical concept of the algorithm. However, the reason for choosing VBA-Excel is that many companies still have a lot of information in the form of Excel lists, which simplifies the transfer of information. On the other hand, programming using VBA is only for checking applicability. A complex programming of a complete software tool would be too extensive (Heinrichsmeyer 2020).

The programming was done using Design Structure (DSM) as well as Domain Mapping Matrices (DMM). With the notation of 0 and 1, this interrelationship can be represented. In the case of a notation of 1, the subordination is applicable and in the case of notation 0, this subordination does not exist. Starting from the total of five DSMs for the total of five different eDeCoDe views, the individual elements of the system views were summarized in a DMM, as exemplarily shown in a requirement process matrix in Table 6.2, and related to each other. In this way, an exemplary production system was created, on the basis of which the programming of failure cause localization could be realized (Heinrichsmeyer 2020).

For the relationships of the elements within the eDeCoDe views themselves, the following rules applied in the considered use case (Heinrichsmeyer et al. 2019b):

1. Rule: eDeCoDe element (row) is subordinate to eDeCoDe element (column).
2. Example: Process A1 (row) is subordinate to Requirement D4 (column).

Starting from the programming of the model, an interface was first developed that contains both the information about the unfulfilled requirement and about all system

Table 6.2 Requirement and process view for failure cause localization (Heinrichsmeyer 2020)

Processes realize requirements			Requirements (columns)					
			A	B	C	D	E	F
			1	2	3	4	5	6
Processes (lines)	A	1	1	0	0	1	1	0
	B	2	0	1	1	1	0	1
	C	3	1	0	0	1	1	0
	D	4	1	0	1	1	1	1
Legend	Notation 1 Column stands according to figure 5 in relation to the line				Notation 0 Column stands according to figure 5 not in relation to the line			

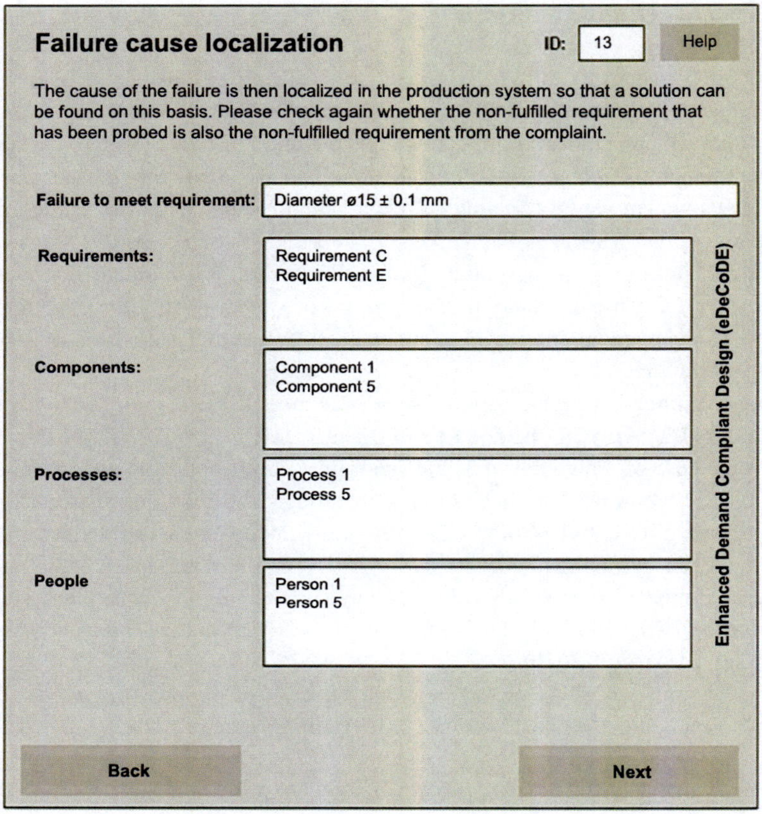

Fig. 6.53 Interface—Failure cause localization according to (Heinrichsmeyer et al. 2019a)

elements related to each other. This interface is shown in Fig. 6.53 (Heinrichsmeyer 2020).

Figure 6.53 shows that the structure of the interface is based on the eDeCoDe approach. Also, the unfulfilled requirement queried from the complaint is listed again, whose relationships must be analyzed in the step of localizing the failure cause. This is achieved by the algorithm examining all relationships of the unfulfilled requirement and transferring those with a relation (value 1) to the interface. In this way, it is ensured that all system elements that could be the cause of the failure are displayed to the user. Due to the complexity of a production system, this would not be conceivable in a manual evaluation by the user within a few seconds (Heinrichsmeyer et al. 2019b).

A special feature of the user interface is that it does not address the functional view. This is due to the fact that according to the eDeCoDe approach, a realization of the functions is done via the components. This means that in the event of a failure of a function, the component must always be identified as the failurecause. For this reason, relationships of the unfulfilled requirements to functions are also recorded via the component view, so that a listing of the function view is not necessary. In the case that a function is realized through the interaction of several components, all these components are also considered as possible failure causes and thus taken over into the interface. With the knowledge of all possible causes, the user can proceed with problem-solving in the next step (Heinrichsmeyer et al. 2019b).

After successful implementation in VBA, a validation in the industry is carried out based on this.

6.6.5 Validation of Implementation

The aim of validation in the industry is to determine whether the algorithm provides meaningful information and evaluations and how the varying quality of complaint information affects the results of the algorithm. To maintain the anonymity of the company and protect internal know-how, all company-related information was deliberately obscured. The validation was carried out using a company in the field of precision machining and cold forming as an example. This sector produces, among other things, cold-formed and cold-hardened parts, e.g. shafts or spindles, which are mainly manufactured for the automotive industry. This industrial example is noteworthy because the complaint processing is subject to the high demands of the automotive industry. This shows that the algorithm can meet these high demands. In order for the algorithm to provide usable results for troubleshooting and solution finding, a corresponding information base had to be created first. This means that all necessary customer, product or order information had to be sifted through and then a model of the production system had to be prepared. In a first step, all the company's information systems were examined for existing information on customers, products and orders. Since the company used very different systems for the respective information, the required information was prepared by the algorithm in Excel tables and compiled for evaluation. This made it unnecessary to program the interfaces for each individual information system. However, it should be noted at this point that for the practical implementation of the algorithm in the industry, such an interface to the existing information systems of the respective company must be programmed and set up by software developers. After the successful mapping of the information systems, a model of the socio-technical production system was developed in a second step using the eDeCoDe approach. In addition to the use of existing documents (e.g. technical drawings, test plans) and their discussion with the process managers of the company and their examination, this industrial example offered the opportunity to systematically go through all processes for the claimed product with the production manager. This allowed requirements, components, functions and people to be linked together. This not only contributed to a better understanding of the relationships within the company, but also showed that the company was very interested in the implementation.

The relationships between the elements were mapped using design structure and domain mapping matrices. The result of the collaboration was a production system that included 69 requirements for a requested product, 21 functions, 22 processes (25 inputs/11 outputs) as well as 11 components and 9 involved persons. After the two steps to prepare the validation were completed, the actual validation of the algorithm could take place. The basis for the validation was a very detailed customer complaint in connection with a requirement not met for the SGW product. The SGW product is usually installed in passenger cars and is a critical component for safety. In this case, the complaint text was available in digital form, so that the complaint text could be transferred to the intended interface, as shown in Fig. 6.54, within a few seconds (Heinrichsmeyer et al. 2019a, c).

It can be seen that the algorithm was able to assign various causes of failure. Among other things, four different components (machines/tools) and three processes were identified. In the case of the KSGD product complaint, requirement A9.2 (KSGD): Flatness 0.35 was recognized as an unfulfilled requirement. With this knowledge, the algorithm examined the production system of the KSGD and listed very different, but possible causes of failure in the arrangement of the eDeCoDe approach. It is shown that the algorithm suspects the cause of the failure among other things in four components, including K1: Ef (machine), K2: Ef St (machine), K4: Ea Ee (machine) or K5: Wg KSGD (tool). It also lists two processes, P5.1 (KSGD): Fr In + Vn (manufacturing) and P5.2: 100 % Ke (testing), in which the requirement should have been met. The algorithm suspects further causes of failure in two people, both the Pe5: Br 'Sn (operator) and the Pe8: Mr' 100 % Ke (tester). With regard to the requirements, the algorithm only shows the requirements that are superior to the flatness. Therefore, this cannot be identified as a cause of failure (Heinrichsmeyer et al. 2019a, c).

6.6.6 Discussion

In order to demonstrate to what extent the algorithm provides meaningful results in the context of failure cause localization, both complaint management employees and production managers were interviewed. In this conversation, it was determined that the indicated causes of failure could have led to the non-fulfillment of the flatness requirement, but they do not represent the actual cause of failure. The background is that the algorithm only identifies causes of failure that contribute to the realization of a requirement at this point. However, this is problematic when the causes of failure are very case-specific and have no direct connection with the requirement. To address this problem, the company's employees suggested also considering the system elements that can cause a change in the requirement. This change could influence the requirement both prospectively (e.g. through the heat treatment process) and retrospectively (e.g. through the post-processing process). It would also be helpful if the algorithm could be linked with existing findings, e.g. from already conducted risk assessments or FMEAs (Heinrichsmeyer et al. 2019a, c).

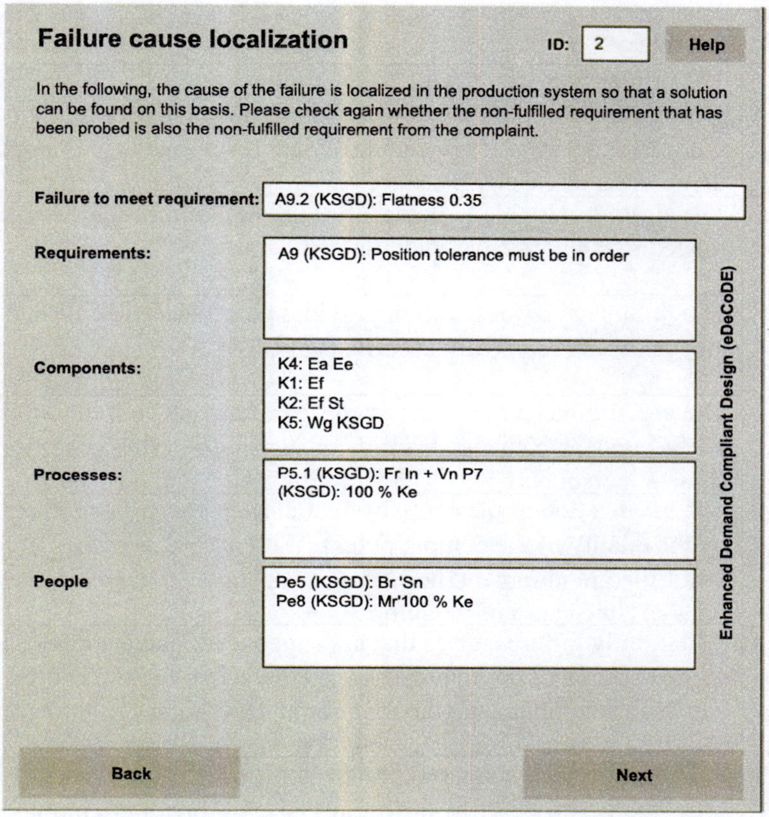

Fig. 6.54 Localization of the causes of complaints of the KSGD product according to (Heinrichsmeyer et al. 2019c)

The algorithm should use these insights to question the existing relationships within the production system and to check them for completeness. Nevertheless, the evaluation has shown that the algorithm already has great potential to master the complexity of a production system. Assuming that the algorithm already had this ability at the time of validation, the algorithm would certainly have been able to identify the correct cause of failure and communicate it to the user. This assumption was confirmed by repeating the failure cause localization after adjusting the interactions with the detection of the failure cause P6 (KSGD): 'Gn (post-processing) and Pe6 (KSGD): Br' Gn (operator) (Heinrichsmeyer et al. 2019a, c).

Overall, based on the validation, the following advantages and disadvantages as well as potential for improvement for model-based failure cause localization emerged (Table 6.3):

Table 6.3 Advantages, disadvantages and potential for improvement of failure cause localization (Heinrichsmeyer 2020)

Advantages	The influence on the quality of the production system can be greatly minimized by the developed interface for checking the production system. However, a residual risk remains with regard to incorrect entries by the user
	The algorithm is able to detect all causes of failures that are directly related to the unfulfilled requirement.
	The failure cause localization provides comprehensible results that also reflect the company-specific evaluation.
	The algorithm can evaluate the entire production system within a few seconds, which also leads to an enormous saving of resources (time & personnel).
Disadvantages	The localization of the cause of the failure is massively influenced by the quality of the complaint text. Without the recording of the unfulfilled requirement, no further evaluation can take place with regard to the cause of the failure
	So far, only failure causes that are supposed to have contributed to the realization of the requirement are detected. Failure causes that can influence or change the requirement are not taken into account
	An interface to existing risk assessments or existing FMEAs, in order to possibly narrow down the cause of the failure more precisely, is currently not possible.
Potentials	Requirements should be specified according to the drawing or the specification in a reclamation text
	System elements that influence requirements both prospectively and retrospectively should be taken into account
	Creation of a possibility for risk estimation by coupling the algorithm with information from data of already performed analyses (e.g. FMEA)

6.6.7 Outlook

The advancing industrial change, whether through the design of intelligent production systems or the development of interactive product systems, suggests that the complexity of production will continue to increase in the future. It therefore seems all the more important to check early on whether current approaches and methods are still sufficient to cope with the advancing complexity. With the described implementation of model-based failure cause search, a first step was taken towards automating complaint management in production. However, the full potential has not yet been exhausted. To achieve this, further research projects are necessary that address the current critical research areas (Heinrichsmeyer 2020).

A decisive factor that plays a role in terms of saving relevant resources is the modeling of a production system. In the course of the research project, this step was carried out manually and in cooperation with the experts of the respective companies. However, this cannot be the goal if we assume that the production systems are becoming increasingly complex. Therefore, it seems sensible to deal with the question of how production systems can be automatically recorded without a huge effort. The use of artificial intelligence approaches that recognize the system elements of a production system according to the eDeCoDe approach and automatically relate them would certainly be helpful. A study by "bitkom" shows that the use of artificial intelligence certainly has its advantages, even if not explicitly for the automated detection of production systems. Thus, 47% (555 industrial companies with 100 and more employees) of the respondents see a "productivity increase" (Berg 2019, p. 5) through the use of artificial intelligence. But also the "optimization of production and manufacturing processes" (Berg 2019, p. 5) could be achieved according to 39% of the respondents, especially with regard to the recording of the production system through the iterative review of artificial intelligence.

6.7 Early Stages of Customer-Integrated Product Development of Industrial Plants

In general, product development requires problem-solving approaches to consider and implement the requirements of various stakeholders as well as different customers. If customer groups are integrated here, it is a customer-integrated product development, which increases the hurdles to communication during the development processes, as the customer does not speak in the language of development engineers. If it is also a product that requires a development team with experts from different engineering disciplines, the development and use of a uniform language becomes indispensable. This should be found and implemented right from the start, i.e. in the early stages of product development. This way, problems can be clearly defined, models can be created that are understandable for all participants, and comprehensible work divisions for the development team can be made, while communication and documentation errors are reduced.

But how can a common language be found? How can the choice of a model and the collaboration of the customer with the development team be designed in such a way that effective and targeted work is enabled?

The following example of the product development of an escape release for a locking system of industrial plants is intended to show how GSE can provide answers to these questions.

6.7.1 Application of the GSE Approach in the Customer-Integrated Development of an Emergency Release for Industrial Plants

A large number of customers complained about defective emergency releases in their industrial plants. A common complaint from customers was that the rods for locking doors were so severely bent that the emergency release was blocked. After examination, the elimination of this problem required a product development of this emergency release for industrial plants.

The task was therefore to find a way to work out and test a solution together with the customer regarding the assembly and use of the emergency release in industrial plants of different sectors (Schlüter and Winzer 2014). The development team used GSE because a common understanding of the problem, a product model usable by the transdisciplinary team, and a basis for communication with the customer were needed. The product development should be sustainable through a standardized system model and a concrete approach that was understandable and acceptable to all involved. How the use of GSE meets these demands and how exactly the customer-integrated product development proceeded is explained below.

1. *Step: System Definition and Delimitation*
Starting from the defined problem of the bending rod of the emergency release, according to GSE, the system (emergency release), the system boundaries (e.g., the interface "mounting for the rod on the door frame"), and the system environment (automated production plant with safety grilles and doors embedded for maintenance personnel) could be determined (Mamrot et al. 2014).

The product under consideration was represented using the Demand Compliant Design (DeCoDe) because it relies on standardized views and thus enables a uniform understanding of terms for customer-integrated and transdisciplinary development teams (Huber et al. 2014).

To avoid additional costs for small and medium-sized enterprises in the project for the acquisition of special modeling software, DeCoDe tools Excel sheets were used for the preparation of the modeling results (Huber et al. 2014).

After modeling the processes, functions, and components of the already existing emergency release, existing as well as new (possibly also weighted) requirements regarding the system "emergency release" itself as well as the environment and the processes "assembly" and "use" were integrated in consultation with the company developers and the customers (Schlüter and Winzer 2014). Furthermore, as also required by GSE, it was decided to continuously update the DeCoDe-based model via the project management module during the specific problem-solving cycle, so that all participants could access the same work status and avoid misunderstandings. Thus, project management not only served to control the development work but also to document all (intermediate) findings sustainably (Huber et al. 2014).

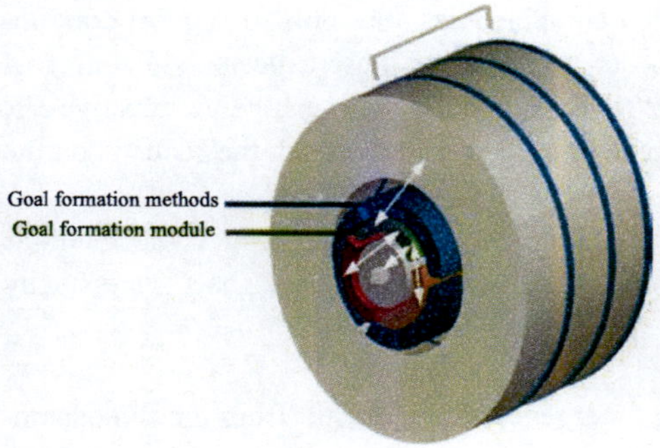

Fig. 6.55 Methods used and results in the goal formation phase according to (Mamrot et al. 2014)

2. *Step: Goal Setting*

To decide the focus of further development together with all participants, the GSE goal formation module was used.

During goal setting, expert discussions between the project team, management, and chief developer, as well as detailed process analyses, were used to identify which problems arose during the installation and use of the emergency release.

This allowed concrete objectives for the VitAmIn project[4] (Huber et al. 2014) and the product development to be derived (see Fig. 6.55).

The main problem regarding the emergency release turned out to be the incorrect dimensioning of the rod. To reduce the risk of incorrect dimensioning during the purchasing phase as well as possible incorrect assembly, the customer, who needs the product for his industrial plant, should be directly integrated into the development process of the emergency release through cooperation with the development team. In addition, the dimensions of safety doors and the point in time when the dimensions of the safety doors for a new plant to be constructed are fixed were queried from various large customers in order to gain insight into the criteria according to which the emergency releases are selected as purchased parts by the customer and which information is already available at this point in time (Schlüter and Winzer 2014). The aim was to find out whether the customer's information is already sufficient to reduce possible "wrong purchases" and complaints.

Furthermore, it was determined regarding the development process that for customer-integrated development in the transdisciplinary team, the development process should be

[4]VitAmIn "Virtual Requirements Management in the Innovation Process"; The IGF project 16716 N of the research association Forschungsgemeinschaft Qualität e. V.—was funded by the AiF within the framework of the program for the promotion of Industrial Collective Research and Development (IGF) by the Federal Ministry for Economic Affairs and Technology on the basis of a decision by the German Bundestag.

as standardized as possible and the required competencies and tools to support communication should be carefully selected in order to minimize these aspects as a source of error or reason for delay (Schlüter et al. 2014). The goal was to enable as effective VR use as possible. These aspects are not further described here. Instead, the focus is on the product and its further development.

As a result of goal setting regarding the emergency release, in which the methods expert interviews and process analysis were used, the following sub-goals were finally defined (Mamrot et al. 2014):

- Sub-goal A: What requirements for the emergency release result from the dimensioning of the safety door?
- Sub-goal B: How should the emergency release be designed so that it can withstand the forces acting on it?
- Sub-goal C: How can the customer's knowledge be used as effectively as possible when selling the emergency release to reduce "wrong purchases" and thus complaints?

The goals were then transferred to the analysis phase to better understand the system, its elements, and their relations to each other as well as with the environment. The following will focus on the sub-goal B "How should the emergency release be designed so that it can withstand the forces acting on it?".

3. *Step: Analysis*
As part of the GSE analysis module, which served to identify, specify, and prioritize the requirement using the methods listed in Fig. 5.56, the focus was, among other things, on the question of which forces act on the emergency release depending on the dimensions of the doors. By using the modeled product, comparing the technical system model, and corresponding calculations on mechanics, variable door heights and the resulting lever effects on the rod of the emergency release were examined. In addition, other variables such as guide rails and their distances were taken into account. The collected data were related to the mechanical forces and calculations were carried out. The insights gained from this were then integrated into the DeCoDe-based product model in the form of requirements (Mamrot et al. 2014).

Subsequently, the newly derived requirements had to be discussed with the customers. To improve the understanding of the topic and thus also the communication among each other, the DeCoDe-based product model was depicted in Virtual Reality (Schlüter and Winzer 2014).

While the development team and customers discussed the product model in Virtual Reality together, the Excel-based DeCoDe variant of the model of the emergency release was displayed and used as a documentation tool at the same time (Huber et al. 2014) (see Fig. 6.56).

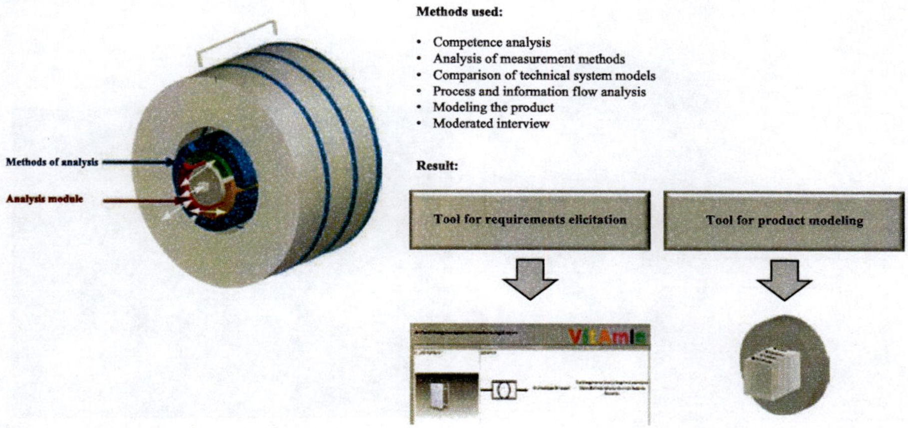

Fig. 6.56 Methods used and results in the analysis phase according to (Mamrot et al. 2014, p. 14)

4. *Step: Design*

The VR session, through the already explained coupling with the product model to be further developed, also served as a link between the analysis and the design module. Based on the preliminary work done in the analysis regarding the mechanical forces and resulting constructive measures, the problem of incorrect dimensioning of the emergency release was illustrated to the development team and the customer. Various solution variants were then discussed based on this (see Fig. 6.57). The Excel-based model of the emergency release was the guide for the moderated interview to derive solutions. This also served as the basis for the feasibility check of the developed solution ideas, as the interactions of new components with the functions, processes, and requirements could be immediately viewed and assessed by the customers and the developers themselves. In the end, guide rails at defined distances were favored. To also consider different designs of safety doors (grating, steel, glass), it was decided to offer the fastenings of the guide rails variably (Mamrot et al. 2014).

The insights regarding the requirements and wishes of the customer regarding the solution alternatives flowed directly into the documentation tool during the VR session, so that an updated DeCoDe system model of the new emergency release was available again after the end of the VR session (Schlüter and Winzer 2014).

The updated product model was finally returned to product development. This allowed the next product development processes to be continued directly without time delay or the risk of loss of information during the transfer of information (Mamrot et al. 2014).

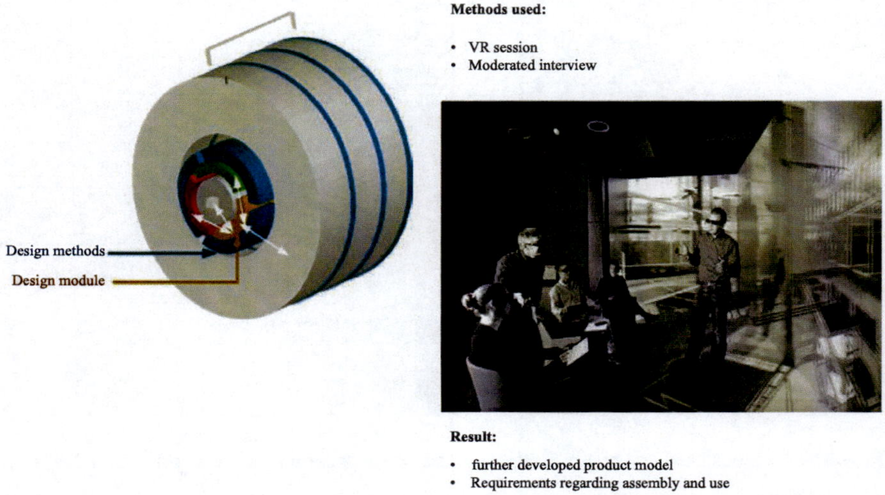

Fig. 6.57 Methods used and results in the design phase according to (Mamrot et al. 2014, p. 15)

Project Management

To manage the work of all three GSE modules (goal, analysis, and design) as well as the continuous updating and provision of the DeCoDe-based GSE system modelto control, the project management module of GSE was used. In the course of the project, product models of different levels of detail were created, the deltas of which illustrated the progress in product development to the project management and the management of the industrial partner. In addition to the development progress, learning effects for future product developments could also be achieved through this (Schlüter and Winzer 2014).

In addition, the use of DeCoDe with its standardized views ensured clear terminology and thus improved communication not only within the company but also in exchange with its customers (Schlüter and Winzer 2014).

6.7.2 Risks and Opportunities of GSE in the Early Stages of Product Development

In the context of the further development of the emergency releases, the problem-solving approach of the GSE was fully completed. It proved to be a challenge to specifically select methods, procedures, and tools that are targeted and efficient for the respective problem case or task. This requires a corresponding level of knowledge about possible methods on the part of the user (Mamrot et al. 2014).

With regard to communication and structured procedure, GSE proved to be a suitable communication platform for a transdisciplinary, customer-integrated product development team. Especially the common language created by the standardization of the system model provided a good basis for effective cooperation. Although the creation and maintenance of current product models proved to be laborious, which also required convincing work by the accompanying research project team. However, these concerns were dispelled towards the end of the project by the fact that this type of work documentation can illustrate the progress of the development process in a simple, understandable way and also generate learning effects for future product developments (Huber et al. 2014).

6.8 Requirement-Appropriate Organizational Development of MISTLER

The REMOt approach presented in Sect. 5.5 (see Fig. 6.58), which is based on the GSE approach for sociotechnical systems, was validated in MISTLER (Mistler 2021) among others at a company from the measurement and control technology industry. This application example, which enables requirement-oriented organizational development, is presented below.

The company is a supplier of products for the use of heating oil and liquefied petroleum gas. It defines itself as a small and medium-sized enterprise (SME). According to a study by the IHK (IHK 2017), such SMEs occupy special niches and are at the beginning of a supply chain. If changes occur in the supply chain, this usually also affects the suppliers. These dynamic changes, which are caused by legislators or the market, often force

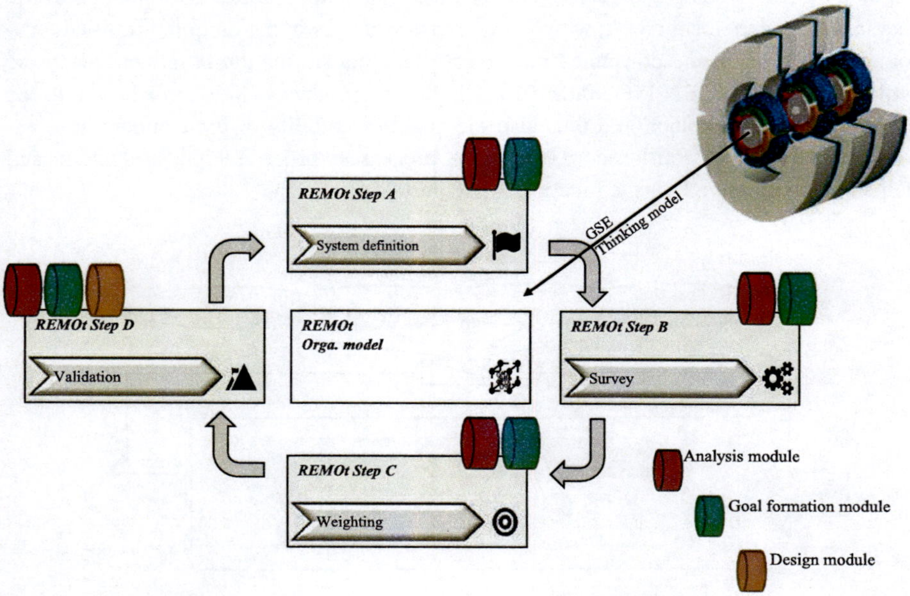

Fig. 6.58 Simplified representation of the REMOt approach according to (Mistler 2021)

companies to be able to serve other industries as well. Keeping up with the speed of these dynamic changes is particularly difficult for SMEs due to a lack of resources (IHK 2017). In addition, rigid or classic organizational management systems, such as quality management systems, complicate the adaptation process (BMWi 2015).

In the following application example of the REMOt approach, it will be demonstrated how agile organizational structures can be built. For this purpose, agile methods and tools have been selected, adapted, and newly developed. This should favor that companies can implement, execute, and make transparent new or changing requirements in the organization in the long term (Mistler 2021).

The REMOt approach begins with the REMOt Step A.

REMOt Step A—Rough Analysis and Goal Setting

The first contact with the measurement and control technology SME begins with the **REMOt Step A** in the **Phase A1.** This phase serves to get to know the company and to narrow down the subject of consideration, i.e., the system, with regard to a problem, as the GSE approach provides. A suitable method for this is the conduct of a **workshop.** For this workshop, the **REMOt project profile** was developed, which is intended to create a common understanding of the organizational system to be considered. With this project profile, the project idea, the project benefit, and the problem can be quickly worked out. The result is recorded with the help of the **REMOt system delimitation approach** (Mistler 2021) (see Fig. 6.59).

It can be noted that the function "manufacturing household regulators" in the organizational system is being considered. For this, the organization needs the input "customer requirements and resources", which are transformed into the output "fulfilled customer requirements and consumed resources". This transformation is influenced by the requirements of DIN EN ISO 9001:2015, GDPR, and product liability, in addition to the customer as a stakeholder, and thus also the quality capability of the company to manufacture the product. Furthermore, the rough processes "order receipt, production, and delivery" are identified for the function (Mistler 2021).

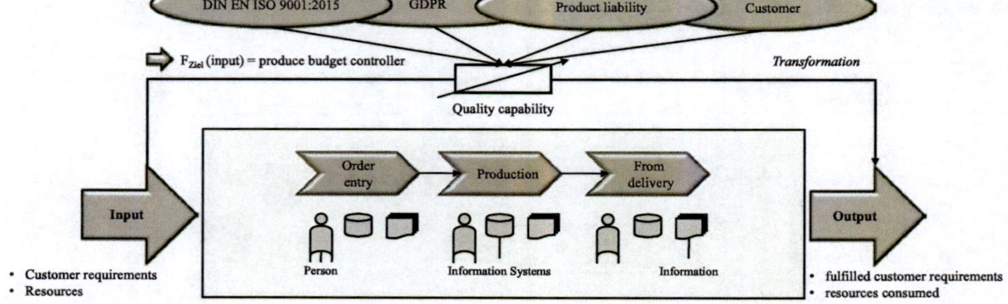

Fig. 6.59 First rough image of the organizational system (Mistler 2021, p. 110)

After the rough understanding of the organizational system and the problem has been worked out, the **Phase A2** begins. In this phase, the organizational system and the problem are further concretized. This phase also is based on the GSE approach, which demands to model the sociotechnical system, in this case, the organization, step by step, focusing on the problem. The developed **REMOt Agenda** serves this purpose, which includes developing a common process understanding of the company's value chain. Furthermore, a schedule for the project is worked out, the type and manner of information use are clarified, and it is agreed on how to proceed after the project. Various tools are used for the REMOt Agenda, such as the **REMOt Value Chain** and the **REMOt Schedule.** The REMOt Value Chain serves as a visualization tool to sketch a rough information flow across the value chain and to create a uniform language for the company's processes. Through this approach, the roles and departments can be identified for the processes to be considered. This information can then help to create the rough REMOt schedule. After the contents in the REMOt Agenda have been clarified, a **production tour** is recommended to gain a further more concrete insight into the delimited organizational system. This approach can further detail the REMOt system delimitation approach (see Fig. 6.60) (Mistler 2021).

Figure 6.60 shows a more detailed image of the organizational system to be considered. This is based on the principles of GSE modeling of sociotechnical systems (see Sect. 3.3). The REMOt system delimitation approach helps to classify the information into the e-DeCoDe modeling (see Sect. 4.3). Thus, the components and persons (roles and departments) can be assigned to the processes. Furthermore, the inputs and outputs

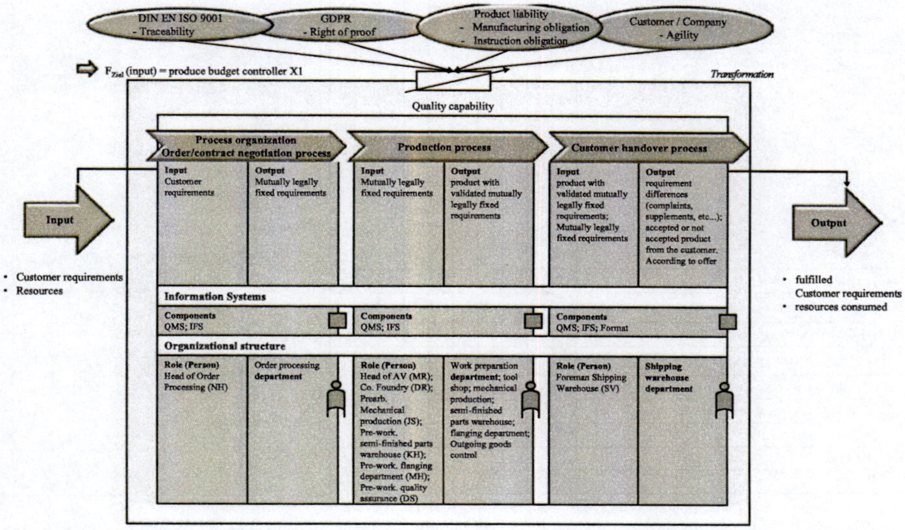

Fig. 6.60 Second rough image of the organizational system (Mistler 2021, p. 111)

of the processes can be specified. This is necessary to systematically delimit the processes from each other. By specifying the system model, it is also possible to concretize the requirements for the organizational system and thus the problem. This means that the GDPR should be considered with regard to the right of proof, the product liability in relation to the manufacturing and instruction obligation, and the DIN EN ISO 9001 with regard to the traceability of the mentioned problems to the customer. Furthermore, it is included as a requirement that the design of the organizational system should also enable agility in the long term through an IT-supported organizational system management tool. Thus, the requirements determine the rough goal of the project (Mistler 2021).

After the rough goal has been worked out, the **Phase A3** begins. For this, the results are documented in the IT tool iQUAVIS in the REMOt organizational model (Mistler 2021). The suitable application of iQUAVIS results from the extensive software analysis, which was described in detail in chapter 3.

To illustrate the modeling with iQUAVIS and the REMOt organizational model in Phase A3, Fig. 6.61 serves as an example. It is based on the modeling with the REMOt organizational model, which was already shown in Sect. 5.5. In Fig. 6.61, it is evident that the inputs and outputs can be assigned to the processes. Furthermore, the roles, departments, and employees are added in the person view. In the component view, the information systems are stored. The individual system elements are connected with each other via the function view. Above the function view, the individual system elements can then be additionally attributed. Thus, it is clear, for example, which information of the

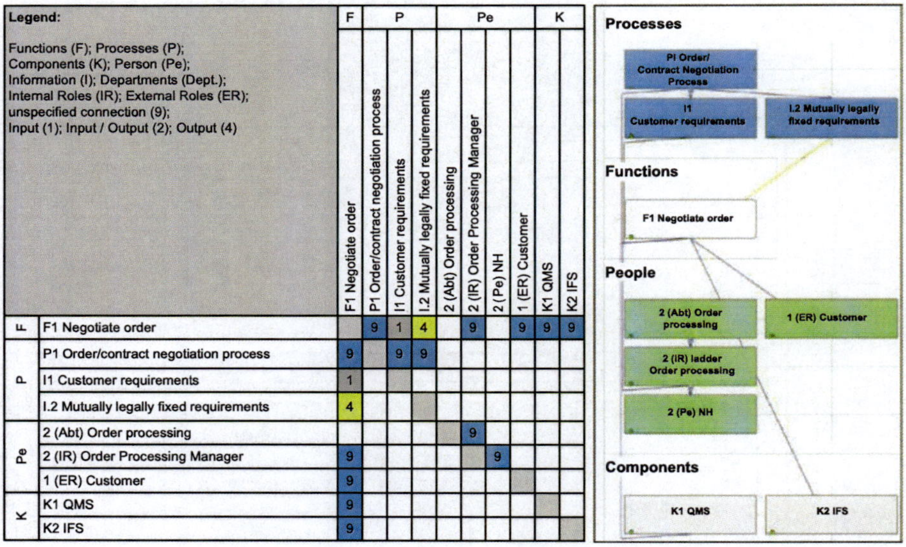

		F	P			Pe				K	
		F1 Negotiate order	P1 Order/contract negotiation process	I1 Customer requirements	I.2 Mutually legally fixed requirements	2 (Abt) Order processing	2 (IR) Order Processing Manager	2 (Pe) NH	1 (ER) Customer	K1 QMS	K2 IFS
F	F1 Negotiate order		9	1	4		9		9	9	9
P	P1 Order/contract negotiation process	9		9	9						
	I1 Customer requirements	1									
	I.2 Mutually legally fixed requirements	4									
Pe	2 (Abt) Order processing						9				
	2 (IR) Order Processing Manager	9						9			
	1 (ER) Customer	9									
K	K1 QMS	9									
	K2 IFS	9									

Fig. 6.61 Graphical and matrix-based modeling of the state t_0 in iQUAVIS using the example of the function "F1 Negotiate Order" (Mistler 2021, p. 112)

processes represents an input or an output for the function "F1 Negotiate Order" (Mistler 2021).

If all the information collected in REMOt Step A is combined, the state t_0 of the model results, as provided by the GSE approach. This again makes it clear that REMOt is based on GSE. This state t_0 of the model represents the starting situation of the project and is secured with the documentation of a system image in iQUAVIS according to the principle of minimal models of systemic thinking and action (see Fig. 6.62).

As Fig. 6.62 shows, the modeling results in a rough process and organizational structure as well as a function and requirement view of the sociotechnical system, i.e., the organization. To document only the essential requirements regarding the problem for the delimited system, the *REMOt requirement filter* was used. This represents an extension of the requirement filter by THIELE (Thiele 2007). Thus, the requirement structure represents the rough goal of the project, which can also be read out in tabular or matrix-based form with iQUAVIS (Mistler 2021).

REMOt Step B—Detailed Analysis and Goal Setting

Following REMOt Step A, *REMOt Step B* begins with the *Phase B1*. In this phase, the information flow analysis is prepared using the *REMOt Checklist* and the *REMOt Interview Guide*. These tools are an extension of the tools by BRAUNHOLZ (Braunholz 2006), which were specifically developed for information flow analysis. First, the REMOt Checklist is used to review the objective. This means determining which attributes should be queried for the system elements in relation to the defined requirements. The result of the REMOt Checklist is the REMOt Interview Guide, which is used to conduct the information flow analysis (see Fig. 6.63).

The information flow analysis is conducted in *Phase B2*. The goal of this phase is a more detailed collection of information about the organizational system. For this purpose, the *REMOt Schedule* (1), the *REMOt Consent Form* (2), the *REMOt Value Chain* (3), the *REMOt Organizational Structure Data Sheet* (4), and the *REMOt Process Organization Data Sheet* (5) are also used (Mistler 2021) (see Fig. 6.64).

The REMOt information flow analysis is conducted through interviews along the value chain. The logical and temporal sequence of the query is recorded with the REMOt Schedule, analogous to the GSE module project management. The REMOt Value Chain serves as a tool to visualize to the interview partner how they have been classified in the value chain. This quickly creates a common understanding of the problem and the system to be considered. Furthermore, the REMOt Consent Form serves as a tool to build trust with the interview partner by making transparent how the information will be further used. The REMOt Organizational Structure Data Sheet and REMOt Process Organization Data Sheet are aids to record the information of the interview partner (Mistler 2021).

The information from the interviews is processed in *Phase B3* and merged in the REMOt Organizational Model. This is illustrated in Fig. 6.65 by the modeling of the

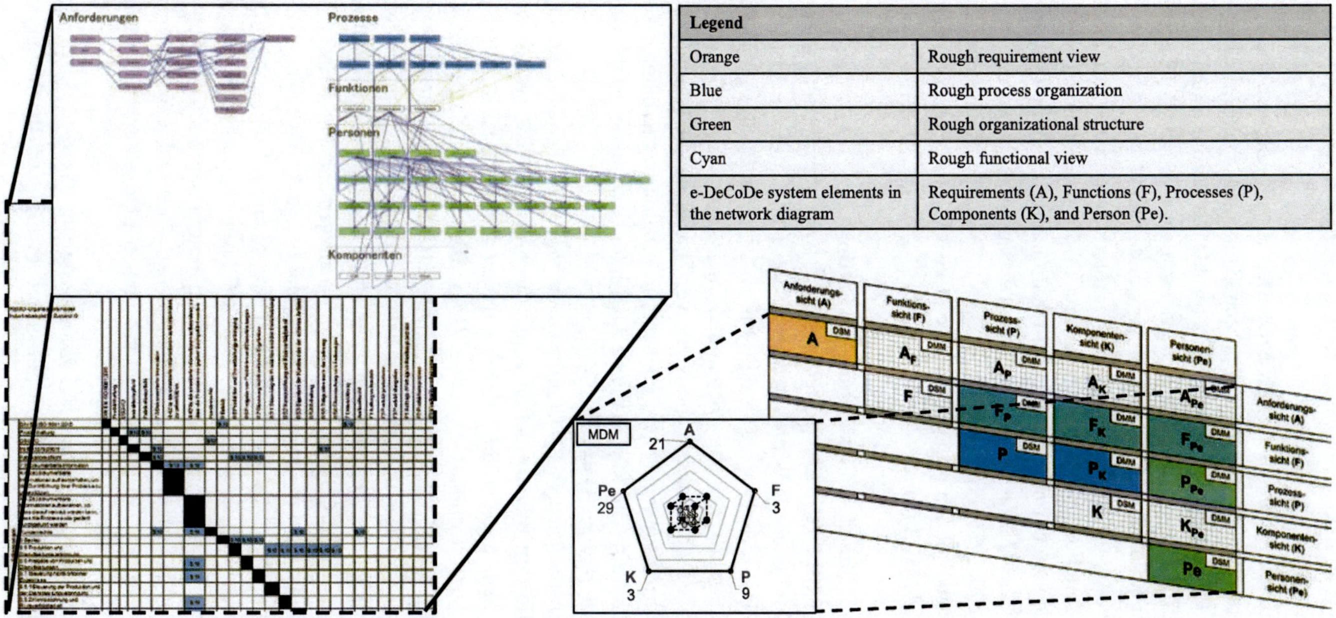

Legend	
Orange	Rough requirement view
Blue	Rough process organization
Green	Rough organizational structure
Cyan	Rough functional view
e-DeCoDe system elements in the network diagram	Requirements (A), Functions (F), Processes (P), Components (K), and Person (Pe).

Fig. 6.62 Principle representation REMOt organizational model state t_0 (Mistler 2021, p. 112)

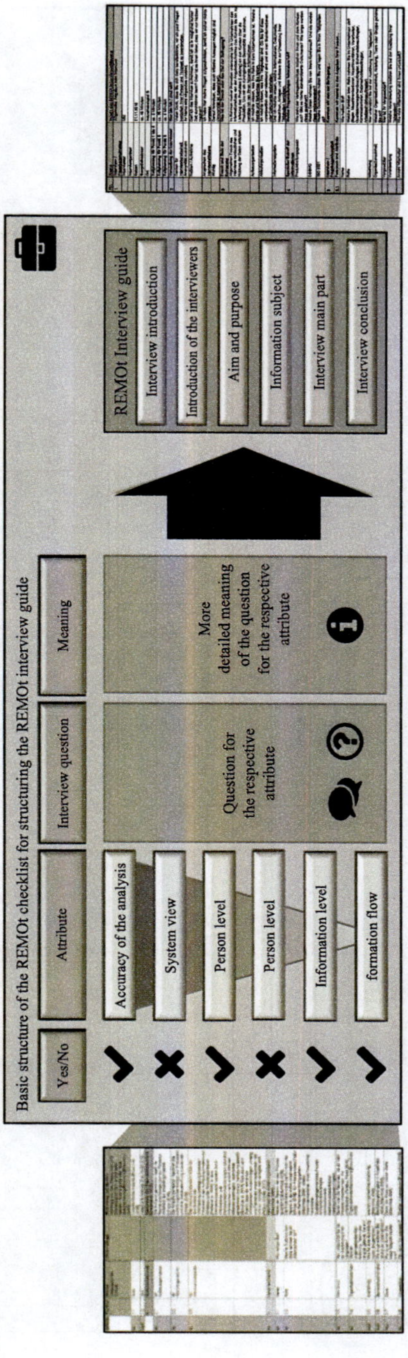

Fig. 6.63 Structure of the REMOt Interview Guide (right) with the REMOt Checklist (left) (Mistler 2021, p. 113)

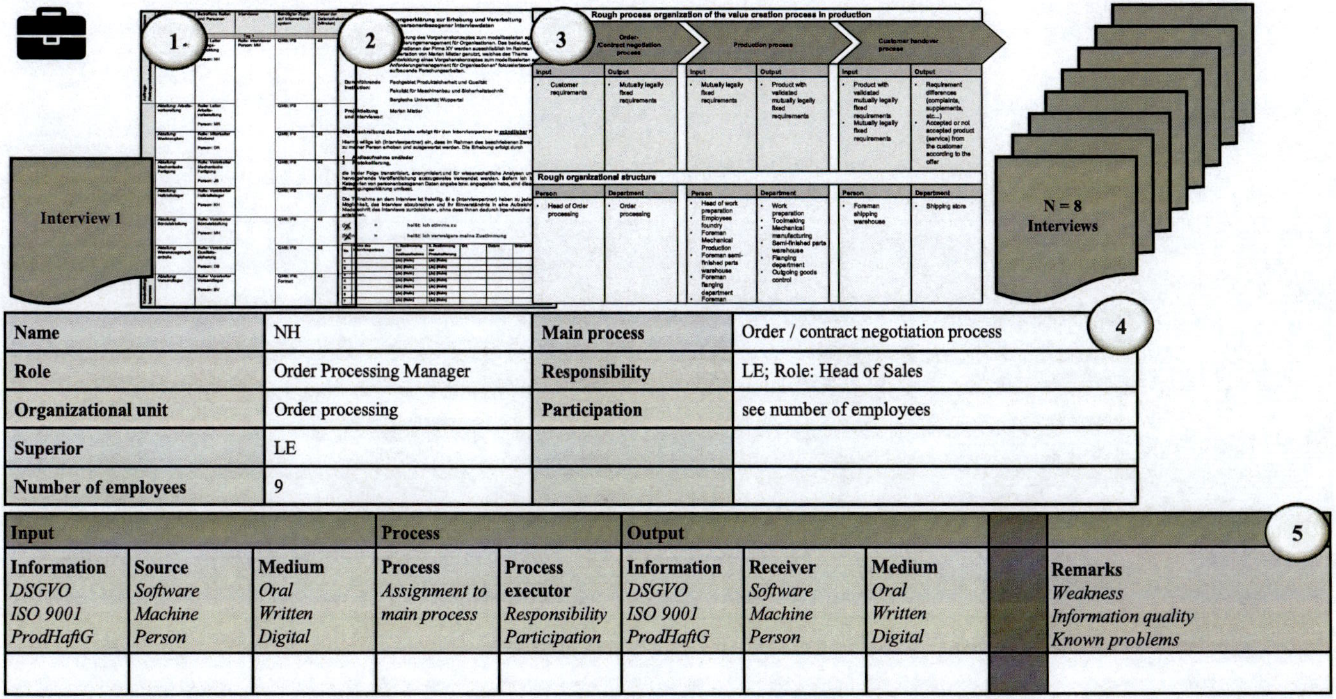

Fig. 6.64 REMOt information flow analysis with the following REMOt tools: REMOt Schedule (1), Consent Form (2), REMOt Value Chain (3), REMOt Organizational Structure Data Sheet (4), and REMOt Process Organization Data Sheet (5) (Mistler 2021, p. 114)

function "F.2.1 Planning Production" and the role "(IR)1.6 Head of Work Preparation" in iQUAVIS with the REMOT Organizational Model (Mistler 2021).

Figure 6.65 visualizes a section of the modeling of the sociotechnical system, i.e., the organization according to the GSE approach. Both the matrix-based and graphical modeling are shown. The classification of the system elements is carried out as presented in Sect. 4.5, but in addition to the function-centered modeling, the relationships between the system elements are also depicted. This means that when focusing on a role, it can be shown, among other things, which components, processes, inputs, and outputs the role uses for which function. Furthermore, it is possible to illustrate which requirements affect the system elements and which function they are supposed to fulfill. The requirement structure can be followed more closely, for example, through the tabular representation of iQUAVIS (Mistler 2021) (see Table 6.4).

The merging of the interviews with the REMOt Organizational Model in iQUAVIS results in the state t_1 (see Fig. 6.66).

As can be seen in Fig. 6.66, the structured modeling with the REMOt Organizational Model and iQUAVIS allows the vast amount of relevant information to be merged. The function view forms the starting point or the module to be able to focus on subsystems of the overall system. This made it possible to determine and document a detailed organizational structure, process organization, requirement structure, and function structure (Mistler 2021). At the same time, a significant contribution was made to make the modeling of sociotechnical systems in GSE practical and to implement it in a problem-focused manner in reality.

REMOt Step C—Focus-oriented Analysis and Goal Setting

The refined REMOt organizational model is the basis for **REMOt Step C.** The step begins with **Phase C1.** In this phase, the information is first structured and discussed. A **workshop** is suitable for this. To make the presentation of the information manageable for each participant, the **REMOt function chain diagram,** also a modified tool from the GSE, is used as the basis for discussion. With this tool, the determined functions can be logically represented and roughly described in iQUAVIS (Mistler 2021) (see Fig. 6.67).

The function chain in Fig. 6.67 serves both as an overview of the entire context of the value chain and as a means to check along the individual functions whether all necessary information about the system has been collected and correctly interpreted. To check the system elements for each function, the detailed representation of a function in table form can help (see Table 6.5). The table breaks down the individual information for each function. For example, it can illustrate which processes, inputs, outputs, requirements, presons, and components are related to the respective function (Mistler 2021).

With the help of the representations, a structured moderation can generate an understanding of the collected organizational system and check for incompleteness or potential for improvement. The analysis can be directly controlled and improved using iQUAVIS (Mistler 2021).

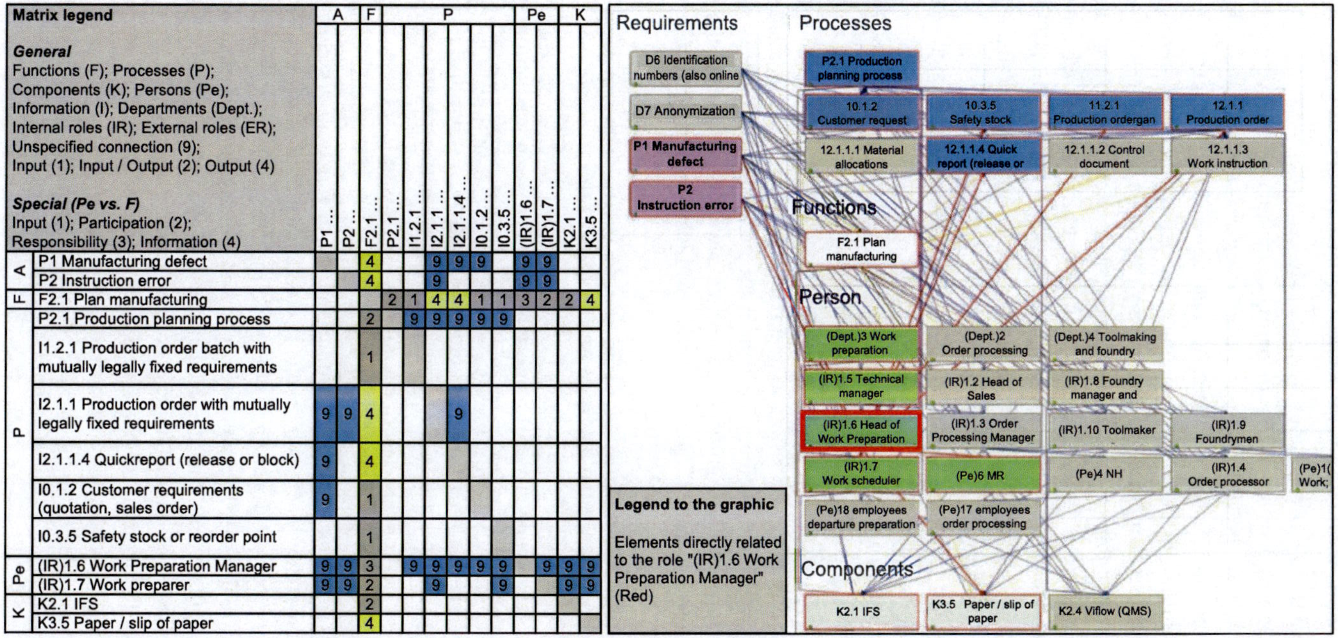

Fig. 6.65 Graph-based modeling of the state t_1 in iQUAVIS using the example of the function "F2.1 Planning Production" and the role "(IR)1.6 Head of Work Preparation" (Mistler 2021, p. 114)

Table 6.4 Excerpt from the REMOt requirement structure with a focus on GDPR requirements (Mistler 2021, p. 115)

<table>
<tr><td rowspan="5">GDPR</td><td rowspan="5">Customer rights</td><td>7.5 Documented information</td><td>…</td></tr>
<tr><td>8.5.3 Ownership by customers or external suppliers</td><td>…</td></tr>
<tr><td rowspan="3">Right to information</td><td>D1 Name</td></tr>
<tr><td>…</td></tr>
<tr><td>D7 Anonymization</td></tr>
</table>

After the analysis in Phase C1 has been completed, *Phase C2* can begin. In this phase, the functions are evaluated with regard to the identified problems (DIN EN ISO 9001, GDPR, product liability) using the *REMOt weighting* (Mistler 2021) (see Fig. 6.68).

The REMOt weighting is done in two stages. In the first stage of the REMOt weighting, each function is evaluated with regard to the defined problems. The scale of the evaluation is four-tiered, as the legend to Fig. 6.68 shows. It was found that DIN EN ISO 9001 has a very high relevance in all functions in relation to the traceability of information. The relevance of the GDPR and product liability issues varies depending on the function considered. The result of the first stage serves to generate a better interpretation with regard to the actual problem. This makes it possible to identify and formulate problem priorities together. The main priorities that could be worked out are the right to proof in the GDPR (1), the obligation to manufacture and instruct in product liability (2), and the joint fulfillment of the GDPR and product liability issue (3). All problems are related to the problem of traceability of information, which is required by DIN EN ISO 9001 (Mistler 2021). This connection is outlined in Fig. 6.69 and can be traced and checked through the requirement structure in iQUAVIS.

In the second stage of the REMOt weighting, the problems are also evaluated (see Fig. 6.69). This serves to highlight the main problem. In this case, it is the joint fulfillment of GDPR and product liability requirements, as on the one hand, the GDPR requires customer data to be anonymized as much as possible and storage of information to be avoided as much as possible. On the other hand, product liability requires that product liability-relevant information must be stored in order to be able to trace relevant information in a product liability case. These accordingly also include the customer's data. Through the various attribution possibilities with iQUAVIS, it is possible to record the evaluations of the REMOt weighting in the REMOt organizational model (Mistler 2021).

After the REMOt weighting was able to work out priorities with regard to organizational system development, a focused analysis of the REMOt organizational model is carried out in *Phase C3* with the help of the use of various attributions in iQUAVIS (Mistler 2021) (see Fig. 6.70). This step is also based on the GSE philosophy.

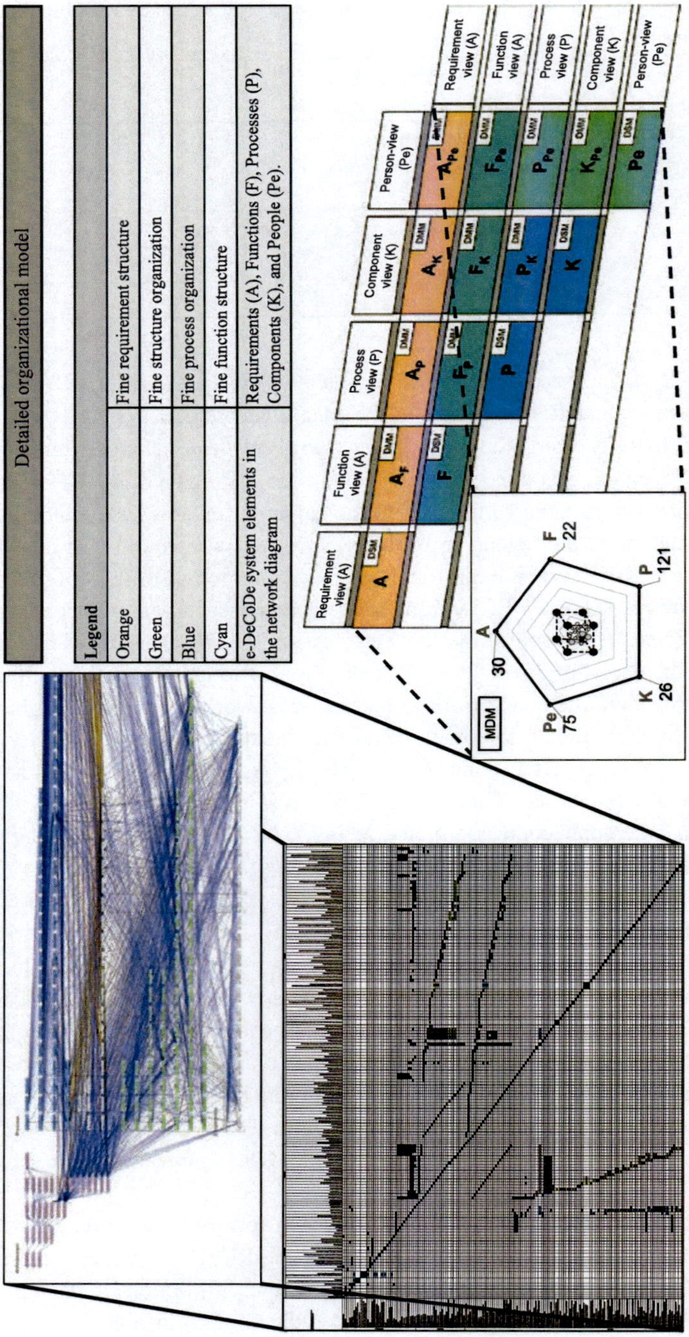

Fig. 6.66 Principle representation REMOt Organizational Model state t_1 (Mistler 2021, p. 115)

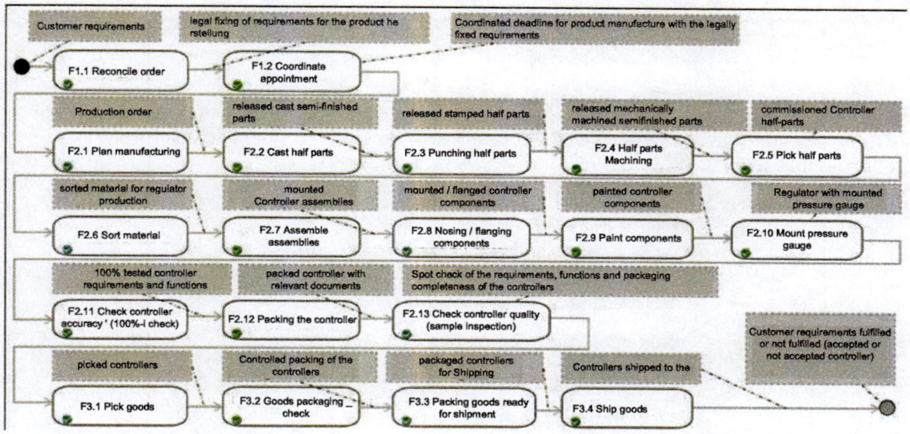

Fig. 6.67 REMOt function chain diagram visualization (Mistler 2021, p. 116)

Table 6.5 REMOt function chain diagram excerpt in table form (Mistler 2021, p. 117)

P	A	Pe	K	F	Pe	Pe	Pe	K	A	P
Input	Input	Input	Input		V	M	I	Output	Output	Output
P1.1 Up-contract stimulat-ion process	D1 Name	(ER)2.1 Customer	K1.1 Phone-annex	**F1.1 Order tee-ments**	(IR)1.3 Manager Order processing	(IR)1.4 Up-TRAnS wROn DER	(IR)1.35 Com-missio-ners	K2.1 IFS	D1 Name	I1.1.9 Delivery-extensive from Ka-talog
I0.3.4 Process descript-ion Order agreement	D2 Adresse		K1.12 Websi-teserver				(IR)1.34 Pre-worker sand loader		D2 dresse	I1.1.8 Require-ments deviating from the catalog
..	..	...	...		...	...	...	...	...	...

As Fig. 6.70 outlines, the interfaces between the system elements can be attributed with iQUAVIS. This is used after the REMOt weighting to specifically mark the relevance of the interfaces with regard to product liability and GDPR. The result of the attributions in the REMOt organizational model with iQUAVIS represents the state t_2. With the help of the attributions, the REMOt organizational system can be systematically examined using various filter functions (Mistler 2021) (see Fig. 6.71).

As can be seen from Fig. 6.71, the various attributions can contribute to systematically reducing the complexity of the REMOt organizational model. This is particularly important in order to be able to examine the system in a targeted manner with regard to

Legend
1 = very low relevance to the problem
2 = low relevance to the problem
3 = high relevance to the problem
4 = very high relevance to the problem

Functions	DIN EN ISO 9001:2015	GDPR	Product liability
F1.1 Coordinate order	4	4	3
F1.2 Coordinate deadline	4	4	2
F2.1 Plan production	4	3	3
F2.2 Cast semi-finished parts	4	1	2
F2.3 Punch semi-finished parts	4	1	2
F2.4 Process semi-finished parts mechanically	4	1	4
F2.5 Pick semi-finished parts	4	1	1
F2.6 Sort material	4	1	1
F2.7 Assemble assemblies	4	1	3
F2.8 Assembling / flanging components	4	1	3
F2.9 Painting components	4	1	1
F2.10 Mount pressure gauge	4	1	4
F2.11 Check controller quality (100% check)	4	1	4
F2.12 Pack controller	4	1	4
F2.13 Check controller quality (sample inspection)	4	1	4
F3.1 Pick goods	4	4	1
F3.2 Check goods packaging	4	4	1
F3.3 Pack goods ready for shipment	4	4	3
F3.4 Ship goods	4	4	3

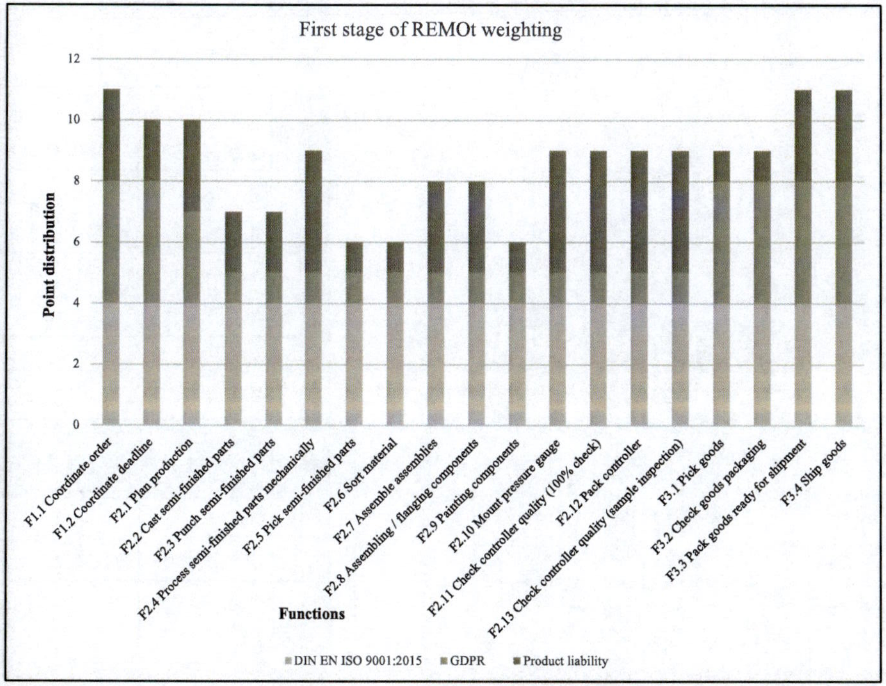

Fig. 6.68 First stage of the REMOt weighting (Mistler 2021, p. 117)

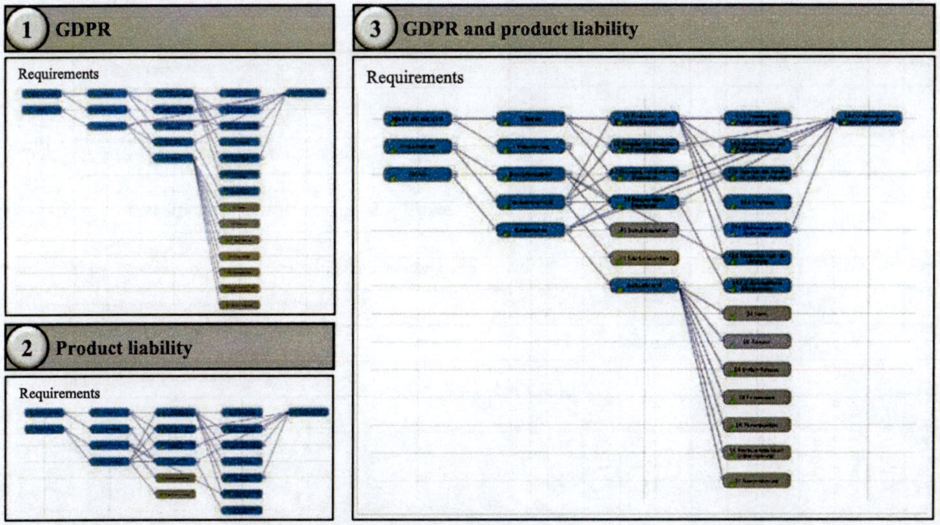

No.	Problem	Scoring
1	GDPR	1
2	Product liability	5
3	DSGVP and product liability	6

Legend	
Blue	Requirements
Gray	Derived requirement attributes

Fig. 6.69 Second stage of the REMOt weighting (Mistler 2021, p. 118)

Matrix legend

General
Functions (F); Processes (P); Components (K);
Person (Pe); Information (I);
Departments (Dept.); Internal Roles (IR);
External Roles (ER); Unspecified Connection (9);
Input (1); Input / Output (2); Output (4)

Special (Pe vs. F)
Input (1); Participation (2);
Responsibility (3); Information (4)

		A		F	P						Pe		K	
		P1 …	P2 …	F2.1 …	P2.1 …	I1.2.1 …	I2.1.1 …	I2.1.1.4 …	I0.1.2 …	I0.3.5 …	(IR)1.6 …	(IR)1.7 …	K2.1 …	K3.5 …
A	P1 Manufacturing defect			4			9	9	9		9	9		
	P2 Instruction error			4			9				9	9		
F	F2.1 Plan manufacturing				2	1	4	4	1	1	3	2	2	4
P	P2.1 Production planning process			2		9	9	9	9	9				
	I1.2.1 Production order infeed with mutual legal fixed requirements			1										
	I2.1.1 Production order with mutually legally fixed requirements	9	9	4				9						
	I2.1.1.4 Quickreport (release or block)	9		4										
	I0.1.2 Customer requirements (quotation, sales order)	9		1										
	I0.3.5 Safety stock or reorder point			1										
Pe	(IR)1.6 Work Preparation Manager	9	9	3		9	9	9	9	9		9	9	9
	(IR)1.7 Work preparer	9	9	2			9			9			9	9
K	K2.1 IFS			2										
	K3.5 Paper / slip of paper			4										

Relationship is relevant to:

☐ Product liability

☐ GDPR

Focus on the role "(IR)1.6 Ladder Work preparation"

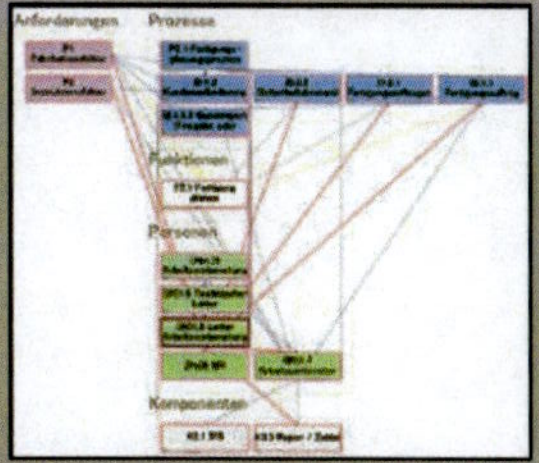

Fig. 6.70 Matrix-based modeling of the state t_2 using the example of the function "F2.1 Planning production" (Mistler 2021, p. 119)

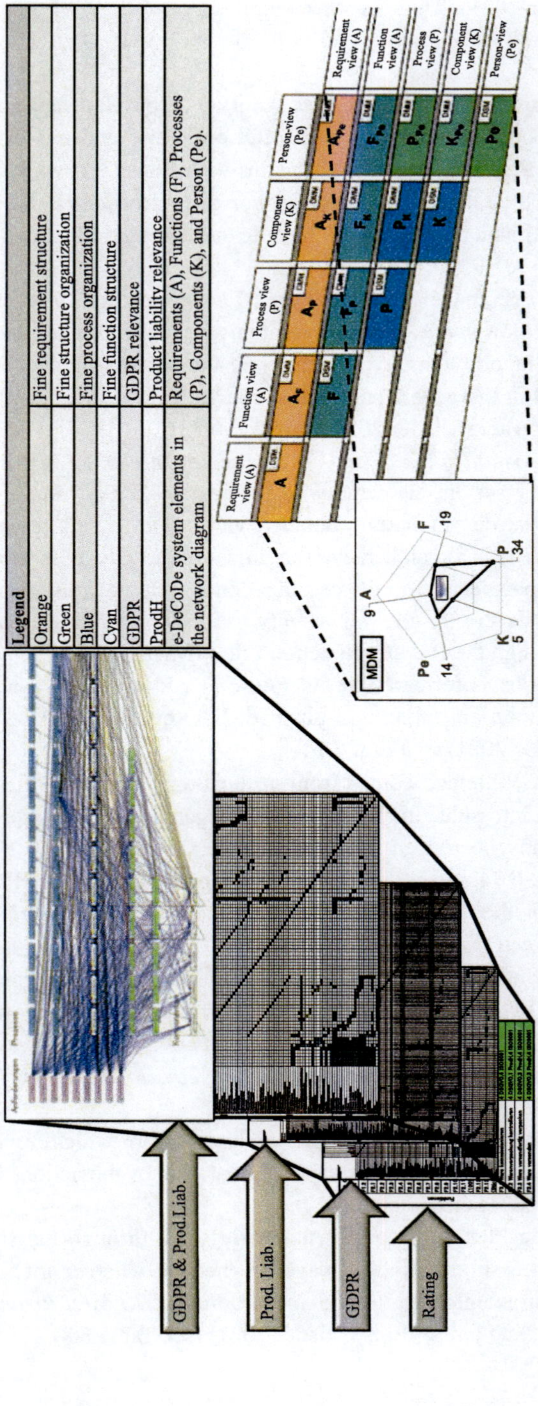

Legend	
Orange	Fine requirement structure
Green	Fine structure organization
Blue	Fine process organization
Cyan	Fine function structure
GDPR	GDPR relevance
ProdH	Product liability relevance
e-DeCoDe system elements in the network diagram	Requirements (A), Functions (F), Processes (P), Components (K), and Person (Pe).

Fig. 6.71 Principle representation REMOt organizational model state t_2 (Mistler 2021, p. 120)

the problems (Mistler 2021). Thus, another requirement of the GSE approach could be practically implemented and demonstrated.

REMOt Step D—Implementation-oriented Analysis, Goal Setting and Design

The systematic attributions stored in the REMOt organizational model with iQUAVIS serve as the basis for **REMOt Step D.** This begins with **Phase D1,** which aims to fix and design, analogous to the GSE design module. For this purpose, the **REMOt Function Filter** was developed, which has various limiting and filtering functions (Mistler 2021) (see Fig. 6.72).

As shown in Fig. 6.72, the application of the REMOt Function Filter can significantly narrow down the solution space, enabling a targeted examination of the organizational system in relation to the problems, even with a vast amount of system elements. To allow detailed analyses within the generated solution space, various tools can be created with the REMOt Function Filter (Mistler 2021) (see Fig. 6.73).

With the *REMOt Visualization Tool* (1), the functions can be highlighted in detail in relation to the other system elements with their respective attributions. The *REMOt Matrices* (2) enable various individual detailed views within and between the system views. The *REMOt Tables* (3) can achieve this in a similar way, with the table structure providing a different perspective on the organizational system. Consequently, the viewer of the organizational system has the opportunity to focus on specific information or add more information through the REMOt Function Filter (Mistler 2021).

The tools form the basis for carrying out *Phase D2*, in which an analysis and development of different solution variants is planned. For this purpose, the *REMOt STOP Method* is used (Mistler 2021) (see Fig. 6.74).

Originally, the STOP method comes from occupational safety to categorize or prioritize solution options into substitutable, technical, organizational, and personal solution options and has already been used in the context of GSE (ArbSchG 2020; Heinrichsmeyer 2020). As Fig. 6.74 shows, the method was modified for the REMOt approach (Mistler 2021). This method helped MISTLER to systematically categorize and prioritize various solution suggestions from different authors (Linß 2018; Eisenberg et al. 2014; Rohrlich 2018; Voigt and Bussche 2018) for the problems GDPR and product liability.

Subsequently, in *Phase D3*, the feasibility of the solution suggestions could be checked with the REMOt tools from the REMOt Function Filter. The result of the validation is summarized by the *REMOt Action Plan Development* (Mistler 2021) (see Table 6.6).

From the validation, the *REMOt Action Plan* is derived, for which responsibilities and deadlines need to be set. The status of the implementation of the actions is continuously documented (Mistler 2021) (see Table 6.7).

The REMOt Action Plan aims to systematically work through the solution suggestions. During the processing, it must always be checked whether the actions actually meet the defined requirements during implementation. The *REMOt Requirement Validation* serves as a supporting tool for this (Mistler 2021) (see Table 6.8).

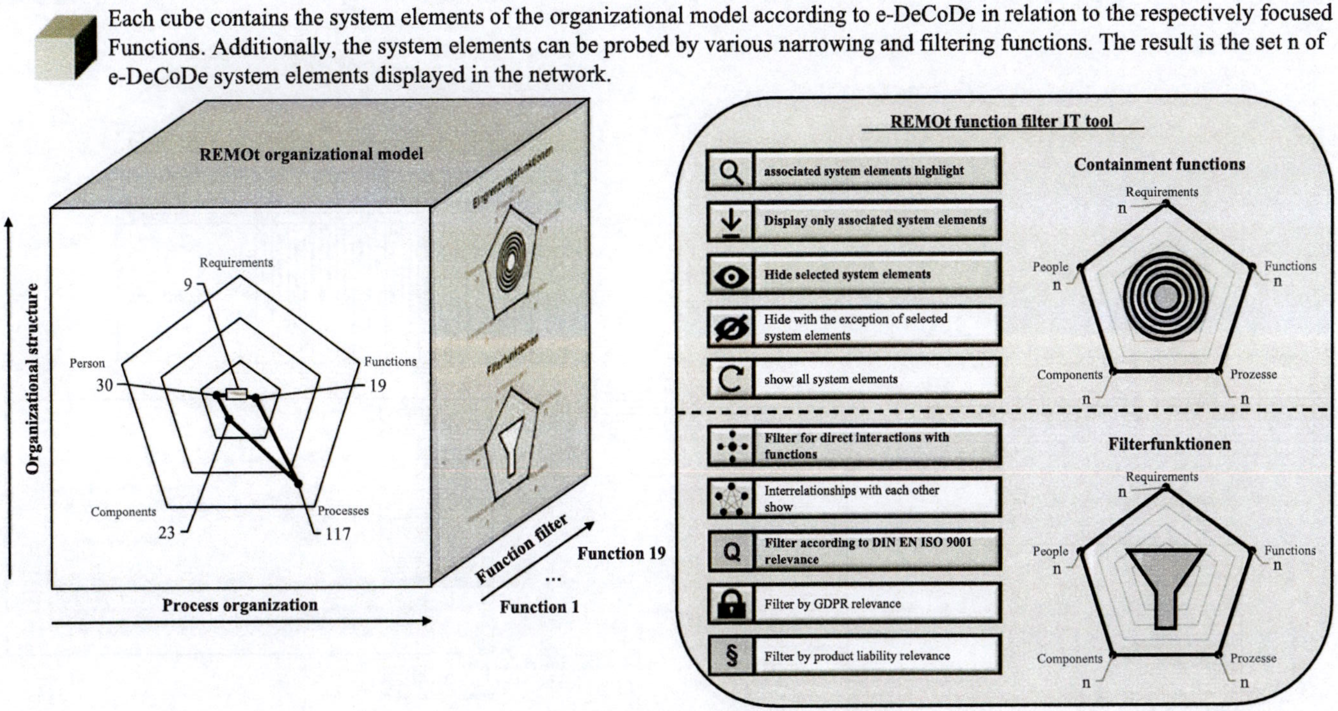

Fig. 6.72 Principle representation of the REMOt Function Filter (Mistler 2021, p. 120)

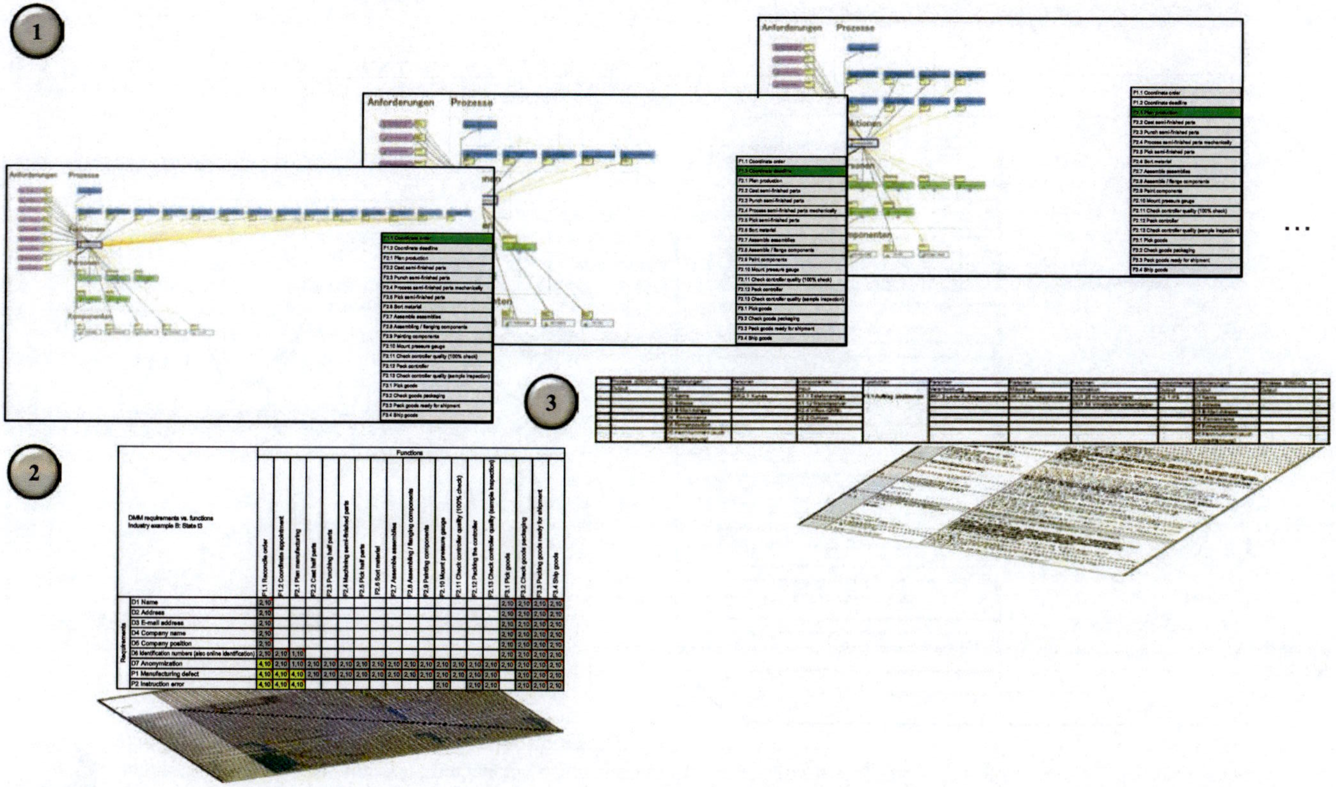

Fig. 6.73 Principle representation of generated REMOt visualization tools (1), REMOt matrices (2) and REMOt tables (3) by the REMOt Function Filter (Mistler 2021, p. 121)

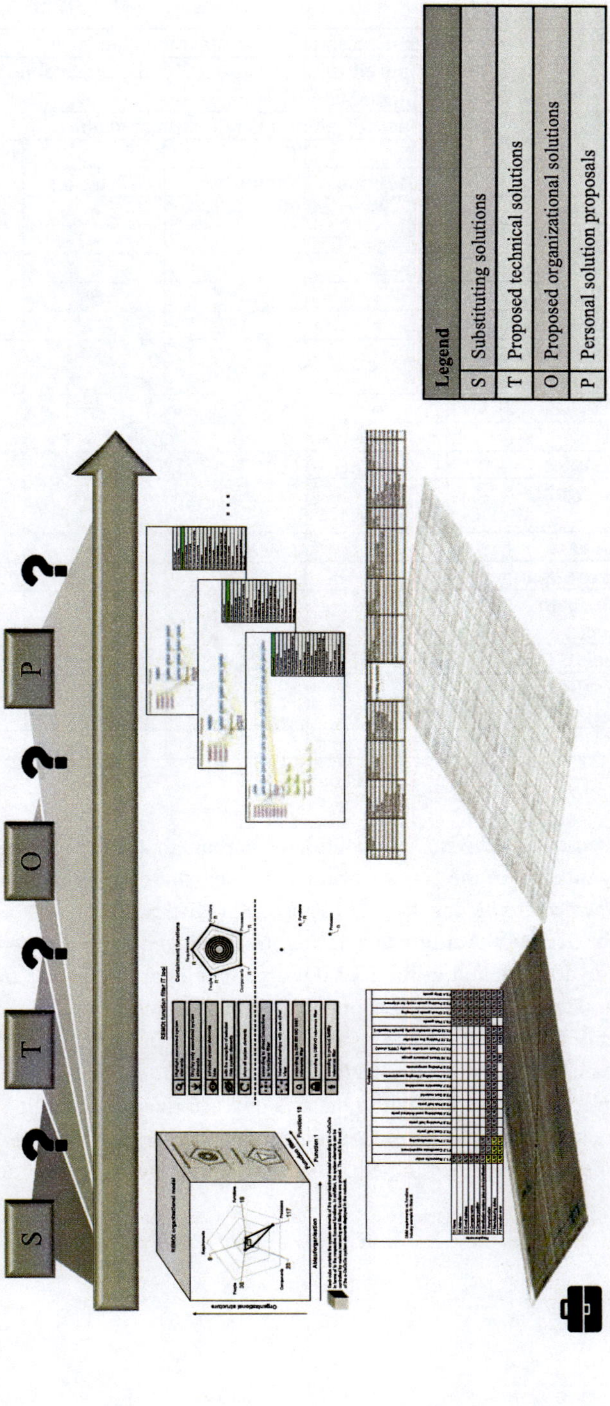

Fig. 6.74 Principle representation of the REMOt STOP Method (Mistler 2021, p. 121)

Table 6.6 Summarized REMOt Action Plan Development (Mistler 2021, p. 122)

●	Proposed solution is implementable			
◑	Proposed solution is partially implementable / the implementability is uncertain			
○	Proposed solution is not implementable			
STOP method	**S**	**T**	**O**	**P**
functions	**Validation**	**Validation**	**Validation**	**Validation**
F1.1 Coordinate order	○	◑	●	●
F1.2 Coordinate deadline	○	●	●	●
F2.1 Plan production	◑	◑	◑	●
F2.2 Cast semi-finished parts	●	◑	◑	●
F2.3 Punch semi-finished parts	●	◑	◑	●
F2.4 Machining semi-finished parts	●	◑	◑	●
F2.5 Pick half parts	●	◑	◑	●
F2.6 Sort material	●	◑	◑	●
F2.7 Assemble modules	●	◑	◑	●
F2.8 Assemble / assemble components	●	◑	◑	●
F2.9 Paint components	●	◑	◑	●
F2.10 Mount pressure gauge	●	◑	◑	●
F2.11 Checking the controller quality (100% check)	●	◑	◑	●
F2.12 Pack controller	●	◑	◑	●
F2.13 Check controller quality (sample inspection)	●	◑	◑	●
F3.1 Pick goods	…	…	…	…

If it is determined during the implementation of actions that the requirements are not met by the implementation of the actions or the action cannot be implemented, then new solution suggestions need to be developed in Phase D2 (Mistler 2021).

Furthermore, the REMOt Action Plan is iteratively linked with *Phase D4*. In this phase, the state t_3 is documented in the REMOt organizational model and transferred to the IT tool Quam. This step is also the practical implementation of the GSE philosophy, which demands a constant update of the meta-model. In the case of the example presented, this corresponds to the *REMOt Sustainability Management* (Mistler 2021). REMOt Sustainability Management with Quam is characterized by enabling sustainable and transparent organizational system management for the company and its employees. This was also extensively described in Sect. 4.3 in the analysis of software tools.

Table 6.7 REMOt KVP Action Plan Excerpt (Mistler 2021, p. 122)

REMOt CIP action plan							
Function	Substitutive actions	Technical actions	Organizational actions	Actions related to persons	Responsibility	Termin	Status
F1.1 Up vote on the contract		Storage of personal data for the traceability of information relevant to product liability for 30 years; …	Preparation of a data protection declaration on the use of personal data for processing purposes; minimization of personal data	Attachment to privacy statement explaining to customers the processes of what data will be used when and for what purpose to complete the order; …	PM (Project manager)	Date	Offen
F1.2 Termin agree	…	…	…	…	…	…	…

The central view in Quam is the REMOt Function Chain Diagram, which has been adapted in principle for REMOt Sustainability Management in Quam (Mistler 2021) (see Fig. 6.75).

Quam offers the possibility to generate various tools. These have been adapted in principle to the REMOt organizational model and serve to be able to store information in a detailed and structured manner. This information can then be transparently traced by all employees in the company (Mistler 2021) (see Fig. 6.76).

Based on the information, employees can see how their role fits into the overall organizational system and what impact their actions could have on the overall system. Furthermore, the information provides the starting point for acquiring new projects. Here, the advantage of reusing the information already obtained when carrying out a new project exists (Mistler 2021).

Table 6.8 Requirement Validation Excerpt (Mistler 2021, p. 123)

<table>
<tr><th colspan="2">REMOt requirements validation</th><th colspan="4" rowspan="2">Requirements</th></tr>
<tr><th>Functions</th><th>Note</th></tr>
<tr><td rowspan="9">F1.1 Reconcile order</td><td rowspan="9">Retention of documented information relevant to product liability for 30 years.</td><td rowspan="7">GDPR</td><td rowspan="7">Customer rights</td><td rowspan="7">Right of information</td><td>D1 Name</td></tr>
<tr><td>D2 Address</td></tr>
<tr><td>D3 E-mail address</td></tr>
<tr><td>D4 Company name</td></tr>
<tr><td>D5 Company position</td></tr>
<tr><td>D6 Identification numbers (also online identification)</td></tr>
<tr><td>D7 Anonymization</td></tr>
<tr><td rowspan="2">Product liability</td><td>Instruction duty</td><td colspan="2">P2 Instruction error</td></tr>
<tr><td>Duty to make</td><td colspan="2">P1 Manufacturing defect</td></tr>
<tr><td rowspan="4">F1.2 Coordinate appointment</td><td rowspan="4">The storage of GDPR and product liability relevant information information is already already by the function "F1.1 Order" function The order can be sent to the user.</td><td rowspan="2">GDPR</td><td rowspan="2">Customer rights</td><td rowspan="2">Out-information-right</td><td>D6 Identification numbers (also online identinization)</td></tr>
<tr><td>D7 Anonymization</td></tr>
<tr><td rowspan="2">Product-liability</td><td>Instruction duty</td><td colspan="2">P2 Instruction error</td></tr>
<tr><td>Duty to make</td><td colspan="2">P1 Manufacturing defect</td></tr>
<tr><td>F2.1 Plan finishing</td><td>...</td><td colspan="4">...</td></tr>
</table>

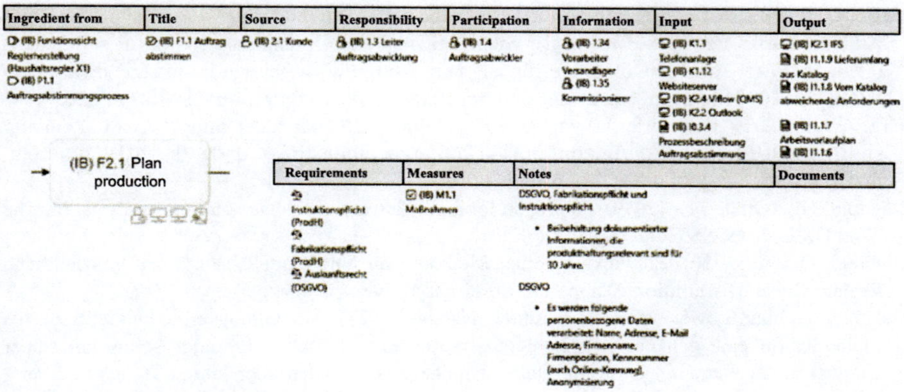

Fig. 6.75 REMOt Sustainability Management Excerpt on the Function Structure for the Function "(IB) F2.1 Plan Production" (Mistler 2021, p. 124)

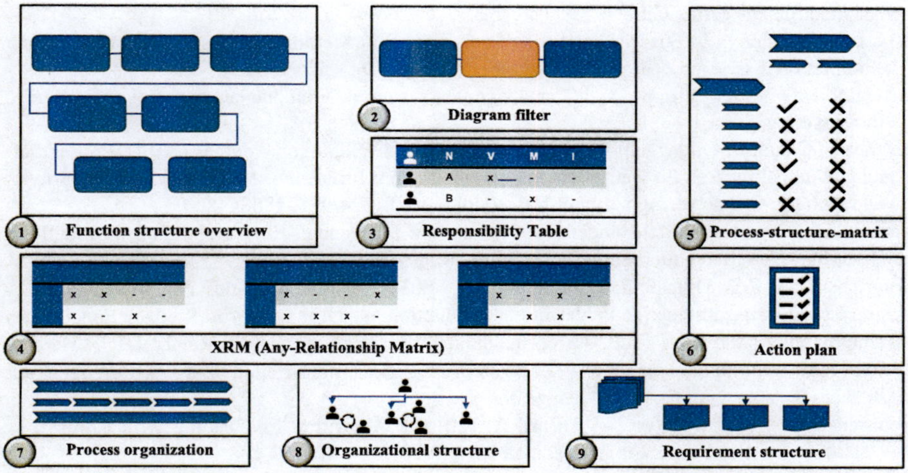

Fig. 6.76 Principle representation of REMOt Sustainability Management with Quam (Mistler 2021, p. 124)

References

Ansari, Amirbabak; Schlüter, Nadine; Heinrichsmeyer, Marius; Löwer, Manuel (2020): Development and Validation of a Failure-Cause-Searching and Solution-Finding Algorithm Based on Complaint Information from the Use Phase. In: 2020 IEEE International Conference on Industrial Engineering and Engineering Management (IEEM). IEEE, S. 1220–1224.

ArbSchG (2020): Arbeitsschutzgesetz. Gesetz über die Durchführung von Maßnahmen des Arbeitsschutzes zur Verbesserung der Sicherheit und des Gesundheitsschutzes der Beschäftigten bei der Arbeit. Online verfügbar unter http://www.gesetze-im-internet.de/arbschg/BJNR124610996.html, zuletzt aktualisiert am 19.06.2020, zuletzt geprüft am 10.11.2020.

Berg, Achim (2019): Industrie 4.0-jetzt mit KI. Online verfügbar unter https://www.bitkom.org/sites/default/files/2019-04/bitkom-pres-sekonferenz_industrie_4.0_01_04_2019_prasentation_0.pdf, zuletzt geprüft am 28.05.2019.

Bielefeld, O.; Schlüter, N. (2019): Approach for a preventive model based failure analysis. In: The 22nd QMOD-ICQSS Conference.

Bielefeld, Ovidiu (2020): Entwicklung einer Methodik für eine modellbasierte und ganzheitliche Fehleranalyse. Dissertation. Wuppertal: Bergische Universität Wuppertal.

Bielefeld, Ovidiu; Löwer, Manuel; Schlüter, Nadine (2021): Vorstellung einer modellbasierten Methodik für eine ganzheitliche Fehleranalyse eines technischen Produktsystems mit seiner Umwelt in der Nutzungsphase. In: Qualitätsmanagement in den 20er Jahren-Trends und Perspektiven: Springer, S. 1–18.

BMWi (2015): Erschließen der Potenziale der Anwendung von 'Industrie 4.0' im Mittelstand. Studie im Auftrag des Bundesministeriums für Wirtschaft und Energie (BMWi): agiplan GmbH.

Braunholz, Helge (2006): Werkzeugentwicklung für informationsflussorientierte Prozessmodelle. Aachen: Shaker Verlag GmbH.

Cox, Louis Anthony (2009): Risk analysis of complex and uncertain systems. New York: Springer (International series in operations research et management science, 129).

DIN EN ISO 9000 (2015): Qualitätsmanagementsysteme—Grundlagen und Begriffe. Berlin: Beuth Verlag.

Eisenberg, Claudius; Gildeggen, Rainer; Reuter, Andreas; Willburger, Andreas (2014): Produkthaftung. Kompaktwissen für Betriebswirte, Ingenieure und Juristen. 2. Aufl. München: De Gruyter Oldenbourg. Online verfügbar unter https://doi.org/10.1524/9783486854800.

Gausemeier, Jürgen; Fink, Alexander (1999): Führung im Wandel. Ein ganzheitliches Modell zur zukunftsorientierten Unternehmensgestaltung. München, Wien: Hanser.

Heinrichsmeyer, M.; Dransfeld, H.; Schlüter, N. (2019a): Validierung des Fehlerursachensuch- und Lösungsalgorithmus für Fehler in der Produktion auf Grundlage von Reklamationen eines Unternehmens für Stanz- und Umformtechnik. In: N. Schlüter und M. Reiche (Hg.): Herausforderungen im Umgang mit Anforderungen in Zeiten des industriellen Wandels. Herzogenrath: Shaker (Berichte zum Generic-Management).

Heinrichsmeyer, M.; Schlüter, N.; Ansari, A. (2019b): Algorithm for Dealing with Complaints Data from the Use Phase. In: Proceedings ICONS 2019b, Bd. 1, S. 1–6.

Heinrichsmeyer, Marius (2018): Algorithmus zum Umgang mit Reklamationsinformationen aus der Nutzungsphase. In: N. Schlüter und M. (Hrsg) Reiche (Hg.): Umgang mit Anforderungen in agilen Organisationen. 1. Aufl. Herzogenrath: Shaker (Berichte zum Generic-Management, 2), S. 113–128.

Heinrichsmeyer, Marius (2020): Entwicklung eines zielgerichteten Fehlerursachensuch- und Lösungsalgorithmus [FusLa]. Dissertation. Wuppertal: Bergische Universität Wuppertal.

Heinrichsmeyer, Marius; Dransfeld, Hendrik; Schlueter, Nadine; Koesling, Fynn (2019c): Validation of a failure cause searching and solution finding algorithm for failures in production. In: *International Journal on Advances in Software Volume 12, Number 3 & 4, 2019c.*

Heinrichsmeyer, Marius; Schlüter, Nadine; Ansari, Amirbabak (2019d): Algorithm-Based Handling of Complaints Data from the Usage Phase. In: 2019d International Conference on Quality, Reliability, Risk, Maintenance, and Safety Engineering (QR2MSE). IEEE, S. 305–312.

Hitchins, Derek K. (2007): Systems engineering. A 21st century systems methodology. Chichester, England, Hoboken, NJ: John Wiley.

Huber, M.; Schlieper, M.; Schlüter, N.; Winzer, P.; Aust, M. (2014): Vitamin für die Produktentwicklung-Virtuelles Anforderungsmanagement im Innovationsprozess. In: *Qualität und Zuverlässigkeit (QZ)* 59, S. 26–29.

IHK (Hg.) (2017): Zulieferer vor der Zerreißprobe. Wie Zulieferer im Automobil- und Maschinenbau den Wandel durch Industrie 4.0 meistern können. Industrie- und Handelskammer (IHK). Stuttgart: Druckerei W. Kohlhammer GmbH + Co. KG. Online verfügbar unter https://www.ipa.fraunhofer.de/content/dam/ipa/de/documents/Publikationen/Studien/Studie_Zulieferer-vor-der-Zerreissprobe_WEB_offen.pdf, zuletzt geprüft am 22.01.2020.

ISO 26262 (2011): Road Verhicles—Functional Safety.

Krishna, B. (2008): Handbook of Performability Engineering. London: Springer Verlag.

Linß, Gerhard (2018): Qualitätsmanagement für Ingenieure. 4., aktualisierte und erweiterte Auflage. München: Carl Hanser Verlag. Online verfügbar unter http://www.hanser-elibrary.com/isbn/9783446439368.

Mamrot, M.; Schlueter, Nadine.; Winzer, P. (2014): Generic Systems Engineering (GSE) in der praktischen Anwendung. In: Petra Winzer (Hg.): Trends zur Handhabung von Komplexität im Qualitätsingenieurwesen. Aachen: Shaker (Berichte zum Generic-Management, 2014, 2), S. 1–18.

Mamrot, Michel (2014): Entwicklung eines Ansatzes zur modellbasierten Felddatenrückführung in die Produktentwicklung. 1. Aufl. Herzogenrath: Shaker (Berichte zum Generic-Management, 2014, 1).

Mistler, Marian (2021): Entwicklung eines Vorgehenskonzeptes zum modellbasierten agilen Anforderungsmanagement (Requirements Engineering und Requirements Management) für Organisationen—REMOt. Dissertation. Wuppertal: Bergische Universität Wuppertal.

Nicklas, Jan-Peter (2016): Ansatz für ein modellbasiertes Anforderungsmanagement für Unternehmensnetzwerke. Dissertation. Aachen: Shaker Verlag (Berichte zum Generic-Management, Band 2016, 2).

Oxford Creativity (2020): Oxford Creativity—TRIZ Effects Database. Hg. v. Oxford Creativity. Online verfügbar unter http://wbam2244.dns-systems.net/EDB, zuletzt aktualisiert am 27.04.2020.

Riekhof, F.; Winzer, P.; Wörner, L.; Kulig, S. (2012): Funktionsorientierte Auslegung eines Linearantriebs. In: Conf. Entwurf komplexer Automatisierungstechnik EKA. Madgeburg, S. 121–138.

Rohrlich, Michael (2018): Die DSGVO verstehen und anwenden. Datenschutzkompetenz für Unternehmen. Frankfurt am Main: entwickler.press.

Schlueter, Nadine; Winzer, Petra; Ansari, Amirbabak; Bielefeld, Ovidiu; Dransfeld, Hendrik; Heinrichsmeyer, Marius (2018): KAUSAL. A New Methodological Approach for Model Based Analysis of Complex Failure Chains by Example of an Electromobility Concept. In: 2018 IEEE International Conference on Systems, Man, and Cybernetics (SMC). Miyazaki, Japan: IEEE, S. 947–952.

Schlund, Sebastian (2011): Anforderungsaktualisierung in der Produktentwicklung. Entwicklung einer Methodik zur Aktualisierung von Anforderungen durch die Einbindung anforderungsrelevanter Ereignisse. Aachen: Shaker.

Schlüter, N.; Huber, M.; Winzer, P. (2014): Mit dem Kunden gemeinsam Produktkonzepte in der Virtuellen Realität entwickeln-ein kompetenz-und prozessorientierter Ansatz. In: *Gröger, Eiselt, Schult (Hrsg): Qualitätsmanagement denken-motivieren-leben,* S. 45–66.

Schlüter, Nadine; Winzer, Petra (2014): Kundenintegrierte virtuelle Produktentwicklung mittels Generic Systems Engineering. In: *Tag des Systems Engineering,* S. 267.

Schnellbach, A. (2016): Komplexität ist ein Feind mit vielen Gesichter. In: *Qualität und Zuverlässigkeit, Bd* 7, S. 11–13.

Thiele, Julia (2007): Entwicklung, Erprobung, Evaluierung und dauerhafte Etablierung eines forderungsgerechten integrierten Managementsystems: Shaker.

Voigt, Paul; dem Bussche, Axel von (2018): EU-Datenschutz-Grundverordnung (DSGVO). Praktikerhandbuch. Berlin, Heidelberg: Springer-Verlag, zuletzt geprüft am 08.06.2018.

Webnachrichten (2018): Tesla Unfall endet tödlich. Autopilot könnte eine Rolle spielen. Online verfügbar unter www.webnachrichten.de/auto/tesla-unfall-endet-toedlich-autopilot-koennte-eine-rolle-spielen-zr-9763945.html, zuletzt geprüft am 14.04.2022.

Weilkiens, T. (2015): SYSMOD—The Systems Modeling Toolbox-Pragmatic MBSE with SysML-Version 4.0.: Tim Weilkiens Verlag (MBSE4U (MBSE4U Booklet Series)).

Westkämper, Engelbert (1997): Null-Fehler-Produktion in Prozessketten. Massnahmen zur Fehlervermeidung und -kompensation. Berlin, Heidelberg, New York, Barcelona, Budapest, Hong Kong, London, Mailand, Paris, Santa Clara, Singapur, Tokio: Springer (Qualitätsmanagement).

Winzer, Petra (2016): Generic Systems Engineering. Ein methodischer Ansatz zur Komplexitätsbewältigung. 2. Aufl. 2016. Berlin: Springer Berlin; Springer Vieweg.

Wörner, L. (2013): Ein heuristisches Verfahren zum automatisierten Eingrenzen des Lösungsraumes in frühen Phasen der Produktentwicklung. TdW. Wuppertal.

Zingel, Johannes Christian (2013): Basisdefinition einer gemeinsamen Sprache der Produktentwicklung im Kontext der Modellbildung technischer Systeme und einer Modellierungstechnik für Zielsystem und Objektsystem technischer Systeme in SysML auf Grundlage des ZHO-Prinzips. Basis definition of a common language of product engineering in the context of modeling of technical systems and a modeling technique for the systems of objectives and objects of technical systems on the basis of the ZHO-principle. Karlsruhe: IPEK Institut für Produktentwicklung (Forschungsberichte IPEK, 70).

7 The New Guise of SE—GSE as a Solution Variant

The frequently cited **increase in complexity** is in its nature, as ***in Chap. 2 proven*** (see also Lindemann 2016; Haberfellner et al. 2012, 2019; Weilkiens 2019; Sitte and Winzer 2011), ***actually only another form of complexity,*** with which we have to deal in the present and in the future. Complex tasks have been solved since the existence of humans. The seven wonders of the world, the pyramids of Cheops and many other historical witnesses prove this. The mastering of the complexity of the present is, as worked out in Chap. 2, characterized by five essential tendencies:

1. ***the globalization tendency,*** which expresses itself in:
   – the increase in stakeholders,
   – the growing diversity of requirements,
   – the increase in division of labor and specialization,
   – the increased networking and globally distributed value creation networks, and
   – the location independence of the individual phases of the product life cycle
   is found.
2. ***the individualization tendency*** of products and services, which includes:
   – shortens their innovation cycles,
   – increases customer expectations and needs,
   – changes the temporal and content aspects of the product life cycle phases, and
   – increases the variety of functions and components.
3. ***the miniaturization tendency,*** which includes:
   – merges system boundaries,
   – requires transdisciplinarity in all phases of the innovation and product life cycle, and
   – conserves resources.

N. Schlüter, *Generic Systems Engineering*, https://doi.org/10.1007/978-3-662-67994-4_7

4. ***the dynamization tendency,*** which requires:
   - increases agility or the necessity of flexibility,
   - integrates digitalization through all areas of social life, for example in the form of artificial intelligence,
   - networks interacting, intelligent technical systems, and
   - increases the dissolution of traditional corporate structures and industry boundaries.
5. ***the sustainability tendency,*** which includes:
   - promotes resource conservation,
   - reduces environmentally harmful emissions,
   - uses environmentally friendly and reusable materials,
   - increases the application of green technologies,
   - strives for social equality, and
   - practices responsible social action on a global level

Current research has to clarify ***how these new dimensions of complexity can be managed. A possible answer could be provided by SE.*** Its nature is to break down complex issues into individual aspects and to network them, as well as to solve details without losing sight of the whole. This means that the mutual dependency of the individual and its entirety are the subject of SE. The ***SE*** was and is a ***simple approach, to handle complexity and to control it*** using simple rules. Consequently, the SE approach would be suitable for managing the mentioned dimensions of complexity. ***In fact, however, the nature of SE mutated.*** This resulted in a ***variety of SE approaches.*** These are primarily used for software development, but also increasingly for product development, factory design, the creation of safety concepts in chemistry, medicine and biology. However, as could also be proven in Chap. 1, the individual disciplines use modified approaches of SE. In part, the procedures are also exclusively discipline-oriented.

This has resulted in the ***universal character of SE and its simplicity being lost.*** Some scientists—representatives for this are BAHILL and GISSNG (see (Bahill and Gissing 1998)—tried to analyze the commonalities of the various variants of SE and to develop modularized building blocks for a universal SE approach based on this. They and many others, including HABERFELLNER (Haberfellner and Daenzer 1999; Haberfellner et al. 2012, 2019; Gausemeier et al. 2013a, b; Dumitrescu et al. 2021) ***demand*** that the ***SE approach must necessarily include a system-theoretical approach and the procedural concept should include the basic principles of systematic thinking and action.***

This means that it must allow a step-by-step exploration of the solution space and a justified derivation of solutions without losing sight of the whole. Through further analyses, the ***GSE approach*** (Sitte and Winzer 2005, 2011) could be developed, ***which is based on SE. The goal*** in its development was to ***restore the universal character of SE,*** thus making a contribution to mastering the new dimensions of complexity. As a result of extensive literature analyses, presented ***in Chap. 3,*** the following ***basic requirements for the GSE approach*** could be derived in summary.

The GSE approach must:

- possess a thinking model, which is based on system thinking and has defined views,
- ensure the synergetic connection of thinking model and procedural concept with targeted use of the basic principles of systematic thinking and action,
- enable the transparent design of the thinking model, and
- provide a standardized, universal, modular procedural concept with the integration of discipline-specific methods and procedures.

This is necessary, as Chap. 3 shows, to guarantee transparency and traceability through the problem-solving process. However, the realization of these basic requirements also ensures that transdisciplinary teams can systematically solve problem-specific tasks over the life cycle of systems. They were further substantiated in Chap. 3 as a result of a comprehensive literature analysis. Thus, the ***GSE thinking model*** must be structured modular and must map causalities between its elements and the dynamics of the system. It must serve as a description, explanation, forecast-, design- and optimization model and implement the principles of systematic thinking and action. By comparing various procedural concepts of SE, it was derived (Sitte and Winzer 2011) that the ***GSE procedural concept*** should consist of universal modules such as the analysis, the goal formation and the design module as well as the project management module. These modules are to be efficiently linked in their temporal logical coupling by the building blocks of the project management module, i.e. the planning, implementation and control phase. Here too, the basic principles of systematic thinking and action are to be implemented. The basic principles of systematic thinking and action thus form the basis for the GSE approach.

In Chap. 4 the ***views of the GSE thinking model,*** the ***creation of the views,*** the creation of ***the relations*** between the views and within the views *for technical systems using DeCoDe* and ***sociotechnical systems using e-DeCoDe*** were presented ***in detail.*** For various examples, it was proven that the ***GSE thinking model serves as a description, explanation, forecast, design- and optimization model.*** Simply by representing the drive of the logistical system via the GSE thinking model using the DeCoDe tools and the DeCoDe method workflow, for example, design approaches such as the change of the rotor rod emerged (see Chap. 4).

For the mechatronic systems, which were represented via the thinking model, their reliability could be predicted with this structured help. The GSE thinking model established for the KitVes system showed that it was possible for a transdisciplinary, international project team to identify highly risk-laden subsystems of the KitVes system, such as the "Kite-Steering Unit". The consideration of the dynamics of the system still poses challenges for the developed GSE thinking model. Because to handle the dynamics and the resulting updates and changes of the information in the technical or sociotechnical system model, a corresponding computer-supported solution is needed. Also, in the GSE thinking model, the relations between the system elements and the system views as well as the description of the system elements in the form of attribution templates need to be further standardized. In doing so, the various variants of the attribution possibilities are to be compared and adapted to both the requirements resulting from the GSE thinking model and those from the Industry 4.0 concept (Müller et al. 2010; Schlund 2011; Dumitrescu et al. 2021; Gausemeier et al.). Currently, findings are available on software of different distribution depth and specialization: Excel, LOOMEO, Quam, Cameo and iQUAVIS. It was explained which IT tools are suitable for implementing the e-DeCoDe approach (Mistler 2021; Bielefeld 2020).

Furthermore, it is to be examined whether it is necessary when creating the GSE thinking model, as HABERFELLNER suggests, to characterize the problem, intervention and solution area (see Haberfellner et al. 2012). The design space of the GSE approach seems sufficient to solve problems in a targeted manner (see Sitte and Winzer 2011).

In Chap. 5 it was demonstrated that the universal ***approach of the GSE procedural concept can be modified problem-specifically or industry-specifically.*** This is made possible by the specific selection and coupling of methods and procedures in each module, i.e., the goal formatino, the analysis, and the design module as well as with the building blocks of both classic and agile project management. Thus, as demonstrated in the examples of risk minimization for the "KiteVes-System" and the feedback of test data from products of the automotive supplier industry into the design and development, the GSE approach can be easily modified in a problem-solving oriented manner. However, it is important that the input information for the methods and procedures as well as the output information are standardized in such a way that they can each be stored in the GSE thinking model. If this is not the case, the results of the method applications cannot be reused during the life cycle of a system.

Consequently, ***the connection between GSE thinking model and GSE procedural concept is absolutely necessary.*** How this can be practically implemented was ***demonstrated in eight examples in Chap. 6 proven,*** i.e.:

- the requirement-based product development,
- the development of mechatronic systems,
- ensuring the reliability over the product life cycle of mechatronic systems,
- the identification of failures in critical usage processes,
- the feedback of field data into the design and development,
- for the failure cause search and solution algorithm,
- the integration of customers into product development and
- the requirement-based organizational development.

The requirement update in the product development process, which should be done permanently to avoid such mishaps as with Airbus and Toyota, can, as demonstrated in Chap. 6, be better achieved using the GSE approach than through requirement management (Schlund 2011). The same was demonstrated by the application of the GSE approach and its modification via the double cycle model in the development of mechatronic systems (see Schlund et al. 2008). The problem of reliability prediction of mechatronic systems over the product life cycle could also be solved using the GSE approach (Winzer 2006, 2015). In the comparison of requirements between the customer and the product developers in the early phases of product creation, the use of GSE is demonstrably helpful (Schlüter and Winzer 2015; Huber et al. 2014). The methodological approach of model-based field data feedback was developed on the basis of GSE and validated using case studies (Mamrot et al. 2014). It represents a way to efficiently transfer field data into the design and development process. With the help of the GSE thinking model and its rules for networking various data lakes, an failure cause search and solution algorithm could be designed that allows the feedback of complaints into production to be partially automated (Heinrichsmeyer 2020). The interactions of the product system with the environment can also be analyzed with GSE in terms of potential failures n critical usage processes (Bielefeld 2020). MISTLER successfully used the GSE approach to better manage the complexity of evolving organizational networks (Mistler 2021).

All eight examples show that through the GSE thinking model and under problem-specifically modified use of the universal modules of the GSE approach in conjunction with the basic principles of systematic thinking and acting (see Fig. 7.1), it was possible to develop solution concepts and implement them efficiently.

Thus, ***the developed GSE approach is a contribution to giving SE a new guise.***

The case studies of Chap. 6 also illustrate that the results achieved in each phase of the implementation of the GSE procedural concept must be continuously stored in the GSE thinking model. This can lead to changes in the GSE thinking model, which should be traceable. This means, ***the synergy between GSE thinking model and GSE procedural concept must be permanently and sustainably ensured.*** This can only be

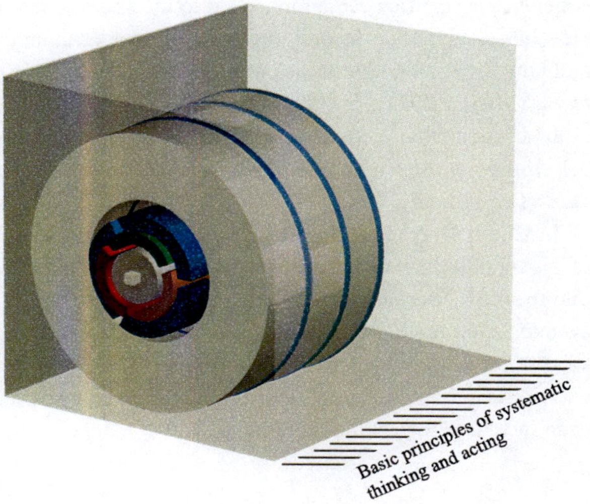

Fig. 7.1 The GSE approach in conjunction with the basic principles of systematic thinking and acting

implemented if this process is computer-supported. The PromeSys portal corresponded to a first solution approach for this (Müller et al. 2010). It is capable of making thinking models of product systems securely usable in the global space while preserving company secrets. However, only static images of the systems can be generated here. These are comparable to snapshots of a system. Similar to computer tomography, images of real systems are taken at certain time intervals. What PromeSys cannot do in contrast is to compare time-delayed models in the GSE approach. This must be done manually in order to then recognize the changes in the system over a certain time interval. Here, software like iQUAVIS has much greater potential. Thus, the statuses of models can be tracked if they are appropriately attributed (Mistler 2021).

The standardized description of the relations between the system elements and the system views as well as the description of the system elements themselves have advanced with the further development of the GSE approach. Thus, in addition to the use and verification of the notation of DeCoDe for technical systems by Bielefeld (Bielefeld 2020) and Heinrichsmeyer (Heinrichsmeyer 2020), the applicability has been further confirmed. Mistler (Mistler 2021) set new milestones in the use of e-DeCoDe for sociotechnical systems. How to proceed efficiently using the basic principle of minimal models has been investigated by all three and corresponding iterative approaches have been derived. The same applies to the creation of the system image according to the basic principle from the whole to the detail.

What remains unclear, however, is the realization of the basic principle of discursive proceeding using inheritance theories. Whether the system properties of a subsystem have effects on the system properties of the overall system when a subsystem has been designed is to be investigated according to SE. However, how this can be done systematically and efficiently in GSE and its modeling is open. Conversely, the question also arises as to how, when a rough solution design for the overall system is available, its conceived properties can be inherited to the subsystems of the overall system. Here, Mistler has sketched corresponding solution proposals and worked out tools for iQUAVIS and Quam with the inheritance of requirements to lower function, process, and personnel levels (Mistler 2021).

In a first conclusion, it can therefore be stated that the ***GSE approach outlined in Fig.*** 7.1 ***fundamentally meets the requirements arising from coping with the new dimension of complexity.***

In Chaps. 4, 5 and 6, product systems as well as sociotechnical systems in the form of organizations were considered. However, since in Chap. 3 the thesis was put forward that the GSE approach is usable for any type of systems, i.e., for biological, for chemical systems, etc., it still needs to be tested for these types of systems. This proof cannot and should not be provided within the scope of this book, as this requires cooperation with various scientists, e.g., biologists, doctors, chemists, and sociologists. Perhaps this book is an incentive to provide this proof.

Furthermore, ***the understanding of the term complexity is currently changing*** in the international SE community, which has various effects on the use of SE and its limits. The International Council of Systems Engineering (INCOSE) opened the discussion with its 2011 White Paper on complexity as to how far the complexity of a system and its environment can be handled and influenced by SE in today's world and which areas cannot be controlled (INCOSE 2021). These areas can only be recognized in their influences, which they have on the considered system, when they occur. Subsequently, it is necessary to react to these influences. A complex system is thus characterized not only by the number of elements and their relations, but also by the fact that subsystems or partial areas are not completely understood and designed by humans. It is therefore consciously accepted in the new definition of complex systems that there are areas within or outside his considered system that cannot be influenced by the system engineer. In view of the generalistic GSE approach, this thought means that when modeling the technical or sociotechnical system and their environment in the (e)-DeCoDe based thinking model, the system engineer not only has to question what goal and benefit the modeling pursues in terms of the types of models description, explanation, forecast, design, and optimization models (see Chap. 4). He must also consider to what extent the granularity or existing information quality is sufficient to reliably design the system within the framework of the approach and to what extent a non-available information quality can still be acceptable for a successful project implementation. Also, it is to be investigated how to deal with black-box subsystems in GSE. Here, a solution could be to set requirements for the black box and the generated output. As long as the requirements are met, no action is possible. If deviations occur, regulators can be integrated through feedback systems (INCOSE 2021). Since the GSE thinking model is based on the fundamental principles of system theory, as demonstrated in Chap. 4, the degree of granularity and the information quality of the system model are already taken into account. The constant interaction between the procedural concept and the thinking model also allows for adaptation with regard to newly occurring influences and their consideration. A feedback system concept is therefore conceivable. The three GSE modules for analysis, design and goal setting also fulfill the new way of thinking demanded by INCOSE, that methods within the framework of the approach should be consciously selected according to their suitability with regard to the task at hand and the existing complexity. In addition, methods for estimating only difficult to predict influences or states from the field of reliability can be integrated. However, it is still to be researched to what extent processes for determining and controlling the necessary information quality as well as decision-making with regard to the VUCA factors volatility, uncertainty, complexity, and ambiguity need to be changed in the area of project management and to what extent these could have effects on GSE approaches and the GSE thinking model.

Also, methods for analyzing and testing AI algorithms in terms of their functionality need to be investigated. To what extent these methods can be integrated into the analysis and design modules of the GSE needs to be investigated.

The ***Systems Engineering Vision 2035*** postulates for the ***SE of the future six global megatrends:***

- Environmental protection and sustainability are of utmost importance.
- Increasing dependencies of systems due to the (communication) networked world.
- Digital transformation changes products and ways of working.
- Industry 4.0 and Society 5.0 change industry and society.
- Exploding system complexity.
- Increase in the number of Smart Systems (INCOSE 2022).

The megatrends are associated with the five tendencies

- Globalization tendency
- Individualization tendency
- Miniaturization tendency,
- Dynamization tendency and
- Sustainability tendency,

which underlie the GSE. However, the SE Vision 2035 is much broader in the field of society, as it targets Society 5.0 and its challenges. With Society 5.0, a future of socio-cyber-physical systems is sought, in which a human-centered society seeks the balance of economic progress with the solution of social problems through a highly integrated system of cyber-space and physical space. Data from sensors in the physical space are accumulated, analyzed by AI, and the results are returned to people in the physical world in various forms (INCOSE 2022).

Here again, the problem of increasing complexity becomes apparent. Technologically speaking, the SE of the future must not only master the complexity in the following disciplines:

- Mechanical engineering and electrical engineering,
- Mechatronics,
- Software,
- Networks,
- Data analysis and AI algorithms as well as
- Human and biotechnology.

The future SE must be able to consider and handle these technologies in combination with AI algorithms. Also, the scope of systems, or the question of where a system boundary should be drawn, must be comprehensively but also dynamically considered by the SE. From individual components, subsystems and application systems to enterprise systems and even system of systems.

All these challenges, which INCOSE wants to tackle by 2035, have been taken into account in the research and development work of the GSE with regard to the basic principles such as inter- and transdisciplinarity in the GSE thinking model as well as in the GSE procedural concept. However, the GSE reaches its limits with regard to software support. The GSE is designed from the perspective of the systems engineer. It is oriented towards the needs and ways of thinking of people. The question of which software is used to implement these ways of thinking and working is currently answered by using complexity management software for cross-disciplinary modeling at the meta level and project management. More concrete goal formation, analysis or design approaches and modeling are carried out in corresponding specific software, which uses the data of the meta-modeling according to DeCoDe rules from the complexity management software as input. The output of the specific software is to be imported back into the complexity management software. This approach is cumbersome and costly. To what extent the MBSE software development and its standards in modeling language and data exchange formats will be compatible or more compatible in the future with the few, but still existential demands of the (e)-DeCoDe modeling, the future will show.

In summary, these explanations illustrate that it is possible to use the ***GSE approach*** for the problem-solving design of technical systems (product systems), but also for the requirement-oriented design of sociotechnical systems (companies and company networks). It is thus ***a general approach*** for various types of systems, which can be adapted generically, i.e. ***specific to the problem.***

By ***ensuring the permanent synergy between the standardized GSE thinking model and GSE procedural concept*** on the basis of simple rules, complex systems can be made transparent and efficiently comprehensible. The latter is particularly supported by the ***targeted use of the basic principles of systematic thinking and action.***

Although there are still numerous research tasks to be solved in this field, it can be estimated that ***the GSE approach*** (see Fig. 7.1) ***meets the demands that were made on the re-design of the SE,*** and which resulted from the comparison and detailed analysis of SE thinking models and SE procedural concepts (see Chaps. 3 to 5).

In conclusion, it must be stated that to ***solve problems, the recognition and implementation of the dialectic between thinking and action is necessary.*** The ***thinking*** is based on a ***thinking model*** and the ***action*** on a procedural ***concept.*** There are a variety of thinking models and procedural concepts (see Chap. 2). Which type of thinking model and procedural concept is used depends on the people who plan and implement the problem-solving process. The basic goal of SE is to solve problems systematically. The ***SE is based*** on ***thinking in systems*** and ***various universal and discipline-specific approaches.*** They result in ***SE system models*** and ***SE procedural concepts.*** Here too, there are ***various SE approaches in the literature*** (see Chap. 2), which make a targeted selection difficult for the potential user. They also do not contribute comprehensively enough to cope with the new dimensions of complexity (see Chaps. 2 and 3). As a result, a ***universal problem-solving approach*** was developed, i.e. the ***GSE approach.*** It is based on ***thinking in standardized system views,*** which was transferred into the ***GSE thinking model*** (see Chap. 4), and a ***universal, modular approach,*** resulting in the ***GSE procedural concept*** (see Chap. 5). The GSE thinking model and GSE procedural concept are used synergistically in the problem-solving cycle. Consequently, the required dialectic between thinking and action in the universal solution of specific problems is standardized in the GSE approach (see Chap. 6).

Thus, the ***Generic Systems Engineering approach*** of Fig. 7.1 can be described as ***Systems Engineering in a new guise.*** The challenges in the research and further development of the GSE approach are still given by digitization as well as the changes in our society. Like other SE approaches (e.g. the international INCOSE Similar), the GSE approach needs to be tested and further developed with regard to various issues. Three of these topics will be briefly reflected upon as an outlook on the further development needs of the GSE approach as well as other approaches:

1. Create benefits for *all* users:

How can GSE be specifically enabled to provide innovative problem solutions also for very large societal challenges? So far, it has been proven that GSE can be used beneficially in the socio-technical area for companies as well as company networks. But can the approach be scaled up even further? And what does it mean for the GSE thinking model if not a socio-technical, but a purely sociological system is to be depicted? What methods exist in disciplines not yet investigated, such as biology, the humanities or economics, that are relevant for SE users? Their classification into the modules, their compatibility with the GSE thinking model or the networking with the methods of engineering sciences within an approach concept are still unclear.

But also within engineering sciences, new challenges arise, if only through new standards and laws as well as the design, implementation and use of IT infrastructures with their operational data.

2. Maintain/increase practicality:

The SE industry needs an SE approach that is not only model-based and allows simulations, multi-disciplinary analyses and immersive visualization. The key word is standardization, so that data and information flows can flow smoothly. Here, questions of ensuring data or information quality are often still unresolved. Also, the availability of data or their time-limited use as well as traceability for manipulation checks etc. are open points. Especially with changing partnerships, the use of data as today's business model and increasing product and producer liability cases in value creation networks, these topics are becoming increasingly urgent. The need of the SE user for trustworthy cooperation in a digital ecosystem is not only a very big technical challenge, but also affects the effectiveness as well as the efficiency of the modeling and approach concepts of SE approaches.

3. Handle uncertainty in the market:
The GSE as well as other SE approaches are confronted with the needs of users to be able to react as anticipatively and effectively as possible to ever increasing market dynamics and uncertainty. To what extent the VUKA principles and agile project management can be integrated into a cross-disciplinary SE—not just the software industry—in a way that is as beneficial and user-friendly as possible is to be considered.

In conclusion, it can be stated that new challenges to the GSE can be found everywhere. After the challenges posed by cyber-physical systems, the handling of cyber-physical production systems, the development and implementation of business models in data sharing communities such as the Mobility Data Space, funded by the German federal government (DRM Datenraum Mobilität GmbH 2022), is now following. The challenges continue to develop in various directions, the systems are merging more and more with each other, in the physical and the cyber world as well as between both worlds. Reason enough to also consider the SE approaches and their communities together and to strive for cooperation and a growing together also with the SE approaches. The compatibility of the research funded by the Federal Ministry of Education and Research on Advanced Systems Engineering (ASE) (AdWiSE) with the GSE seems to be quite possible, but needs to be checked.

References

AdWiSE: Advanced Systems Engineering. Hg. v. AdWiSE. Online verfügbar unter https://www.advanced-systems-engineering.de/, zuletzt geprüft am 22.11.2022.

Bahill, A. Terry; Gissing, Bruce (1998): Re-evaluating systems engineering concepts using systems thinking. In: *IEEE Transactions on Systems, Man, and Cybernetics, Part C (Applications and Reviews)* 28 (4), S. 516–527.

Bielefeld, Ovidiu (2020): Entwicklung einer Methodik für eine modellbasierte und ganzheitliche Fehleranalyse. Dissertation. Wuppertal: Bergische Universität Wuppertal.

DRM Datenraum Mobilität GmbH (2022): Mobility Data Space. Unter Mitarbeit von DRM Datenraum Mobilität GmbH. Hg. v. DRM Datenraum Mobilität GmbH. Online verfügbar unter https://mobility-dataspace.eu/, zuletzt geprüft am 22.11.2022.

Dumitrescu, Roman; Riedel, Oliver; Gausemeier, Jürgen; Albers, Albert; Stark, Rainer (2021): Engineering in Deutschland—Status quo in Wirtschaft und Wissenschaft. Ein Beitrag zum Advanced Systems Engineering. Padaborn. Online verfügbar unter www.advanced-systems-engineering.de, zuletzt geprüft am 06.05.2021.

Gausemeier, J.; Dumitrescu, R.; Steffen, D.; Czaja, A.; Wiederkehr, O.; Tschirner, C. (2013a): Systems Engineering in der industriellen Praxis. Universität Paderborn: Heinz Nixdorf Institut.

Gausemeier, Jürgen; Gaukstern, Tobias; Tschirner, Christian (2013b): Systems Engineering Management Based on a Discipline-Spanning System Model. In: Conference on Systems Engineering Research (CSER 13) (Hg.): Proceedings, 19.–22. März 2013b: Elsevier B.V.

Gausemeier, Jürgen; Lindemann, Udo; Schuh, Günther: Planung der Produkte und Fertigungssysteme für die Märkte von morgen. Ein praktischer Leitfaden für mittelständische Unternehmen des Maschinen- und Anlagenbaus ; [Abschlussbericht des Verbundprojekts Strategische

Produkt- und Prozessplanung, BMBF Rahmenprogramm „Forschung für die Produktion von morgen"]. Frankfurt am Main, Hannover. Online verfügbar unter http://edok01.tib.uni-hannover.de/edoks/e01fb05/50495301Xl.pdf.

Haberfellner, Reinhard; Daenzer, Walter F. (1999): Systems engineering. Methodik und Praxis. 10., durchges. Zürich: Verl. Industrielle Organisation.

Haberfellner, Reinhard; Weck, Olivier de; Fricke, Ernst; Vössner, Siegfried (2012): Systems engineering-grundlagen und anwendung. 12. In: *Auflage. Zürich: Orell Füssli,* 978-3280040683.

Haberfellner, Reinhard; Weck, Olivier L. de; Fricke, Ernst; Vössner, Siegfried (2019): Systems engineering. Fundamentals and applications. 14. überarbeitete Auflage. Cham: Springer International Publishing; Birkhäuser.

Heinrichsmeyer, Marius (2020): Entwicklung eines zielgerichteten Fehlerursachensuch- und Lösungsalgorithmus [FusLa]. Dissertation. Wuppertal: Bergische Universität Wuppertal.

Huber, M.; Schlieper, M.; Schlüter, N.; Winzer, P.; Aust, M. (2014): Vitamin für die Produktentwicklung-Virtuelles Anforderungsmanagement im Innovationsprozess. In: *Qualität und Zuverlässigkeit (QZ)* 59, S. 26–29.

INCOSE (2021): A Complexity Primer for Systems Engineers. Hg. v. INCOSE. USA (INCOSE-TP-2021-007-01). Online verfügbar unter https://www.incose.org/docs/default-source/ProductsPublications/a-complexity-primer-for-systems-engineers.pdf, zuletzt geprüft am 07.11.2022.

INCOSE (2022): Systems Engineering Vision 2035. Hg. v. INCOSE. USA. Online verfügbar unter https://www.incose.org/about-systems-engineering/se-vision-2035, zuletzt geprüft am 07.11.2022.

Lindemann, Udo (Hg.) (2016): Handbuch Produktentwicklung. München: Carl Hanser Verlag.

Mamrot, Michel; Marchlewitz, Stefan; Nicklas, Jan-Peter; Winzer, Petra (2014): Using systems engineering for a requirement-based design support for autonomous robots. In: 2014 IEEE International Conference on Systems, Man, and Cybernetics (SMC). IEEE, S. 3115–3120.

Mistler, Marian (2021): Entwicklung eines Vorgehenskonzeptes zum modellbasierten agilen Anforderungsmanagement (Requirements Engineering und Requirements Management) für Organisationen—REMOt. Dissertation. Wuppertal: Bergische Universität Wuppertal.

Müller, Nico; Schlund, Sebastian; Winzer, Petra (2010): Modellierung komplexer mechatronischer Systeme anhand des Demand Compliant Design. In: E. Schnieder, U. Jumar und C. Diedrich (Hg.): Entwurf komplexer Automatisierungssysteme. Beschreibungsmittel, Methoden, Werkzeuge und Anwendungen ; 11. Fachtagung mit Tutorium, 25. bis 27. Mai 2010 in Magdeburg, Denkfabrik im Wissenschaftshafen. Magdeburg: Ifak, S. 79–88.

Schlund, Sebastian (2011): Anforderungsaktualisierung in der Produktentwicklung. Entwicklung einer Methodik zur Aktualisierung von Anforderungen durch die Einbindung anforderungsrelevanter Ereignisse. Aachen: Shaker.

Schlund, Sebastian; Ott, Stefan; Winzer, Petra (2008): Method integration to enhance the quality capability of the product development process. In: 10th QMOD Conference. Quality Management and Organiqatinal Development. Our Dreams of Excellence; 18–20 June; 2007 in Helsingborg; Sweden. Citeseer.

Schlüter, Nadine; Winzer, Petra (2015): Systems Engineering für die Entwicklung der Theorie zu Verlässlichkeit von Systemen. In: *Tag des Systems Engineering: Verteiltes Arbeiten mit ganzheitlicher Kontrolle,* S. 409.

Sitte, J.; Winzer, P. (2005): Demand Compliant Design of Robotic System. In: Jason Gu und Peter X. Liu (Hg.): 2005 International Conference on Mechatronics and Automation. July 20 to August 1, 2005, Niagara Falls, Ontario, Canada : conference proceedings. Piscataway, NJ: IEEE, S. 1953–1958.

Sitte, J.; Winzer, P. (2011): Systemmodellierung im Fokus von Generic Systems Engineering. In: Gesellschaft für Systems Engineering e.V. (Hrsg), Tag des Systems Engineering.

Weilkiens, Tim (2019): Systems Engineering mit SysML/UML. Anforderungen, Analyse, Architektur. 3., überarb. und aktualisierte Aufl. Heidelberg: dpunkt.verlag.

Winzer, Petra (Hg.) (2006): Generic-Management und Möglichkeiten der Stakeholderintegration. Aachen: Shaker (Berichte zum Generic-Management, 1/2006). Online verfügbar unter http://www.gbv.de/dms/zbw/508406439.pdf.

Winzer, Petra (2015): Generic System Description and Problem Solving in Systems Engineering. In: *IEEE Systems Journal* 11 (4), S. 2052–2061. https://doi.org/10.1109/JSYST.2015.2428811.